Petroleum Refining Technology Series

Speight

lesigned to address the current processes used by the refining industry and take the reader s that are necessary for crude oil evaluation and refining. Technological advancements and are highlighted in each of the volumes.

ks

lting, and Distillation in Petroleum Refining

alytic Processes

d Hydrocracking Processes in Refining Technology

in Refining Technology

s and Product Improvement

Hydrogen Prod
in Refining

This book details the various approaches to the
need for hydrogen is addressed, and then the di
practical and accessible guide is written for m
graduate students transitioning into the refining

Key Features:

- Describes hydrogen purification methods
  and fully describing process operations.
- Describes the types of feedstock that car
  of each process.
- Details commercial processes, including
  siders next-generation processes and deve

Series Editor: James C

This series of books is
through the various st
processing innovation

Refinery Feedst
James G. Speight

Dewatering, Des
James G. Speight

Thermal and Ca
James G. Speight

Hydrotreating a
James G. Speight

Solvent Processe
James G. Speight

Refinery Produc
James G. Speight

Hydrogen Production Processes in Refining Technology

James G. Speight

CRC Press
Taylor & Francis Group
Boca Raton London New York

CRC Press is an imprint of the
Taylor & Francis Group, an **informa** business

Cover design credit: shutterstock

First edition published 2024
by CRC Press
2385 NW Executive Center Drive, Suite 320, Boca Raton FL 33431

and by CRC Press
4 Park Square, Milton Park, Abingdon, Oxon, OX14 4RN

CRC Press is an imprint of Taylor & Francis Group, LLC

ISBN: 9781032033105 (hbk)
ISBN: 9781032033143 (pbk)
ISBN: 9781003186694 (ebk)

DOI: 10.1201/9781003186694

Typeset in Times
by Deanta Global Publishing Services, Chennai, India

Contents

Preface

This series of eight books is designed to present descriptions of (1) the development of technologies for a variety of feedstocks (including the viscous feedstocks which are often referred to as heavy feedstocks) utilizing advanced pretreatment processing and hydrotreating, (2) an analysis of the catalyst deactivation mechanism for developing optimum technologies for processing feedstocks with low reactivity, (3) the development of advanced technologies applicable to the viscous feedstocks, (4) the development of advanced hydrocracking processes for heavy feedstock upgrading, (5) development of innovative upgrading processes for the viscous feedstocks, and (6) the role of biomass and carbonaceous waste refinery of the future.

Each book is a stand-alone volume that will bring the reader further up to date and adds more data as well as processing options that may be the processes of the evolving 21st century.

As the seventh book in the series, this book will focus on the various processes used to produce hydrogen as well as the various forms for hydrogen purification that are used to enable the products to meet the sales specifications.

Chapter 1 presents an introduction and overview of the use and the need for hydrogen in crude oil refineries in order for the reader to place the use of hydrogen in the correct context of the refinery. In fact, the chapter also presents descriptions to show that hydrogen is key in allowing refineries to comply with the latest product specifications and environmental requirements for fuel production being mandated by markets and governments and helping to reduce the carbon footprint of refinery operations.

Chapter 2 deals with the need for hydrogen to upgrade the viscous feedstocks (referred to as 'bottom of the barrel processing') such as heavy crude oil, extra heavy crude oil, tar sand bitumen, and process residua. Thus, the focus of this chapter is on the various catalyst-based hydrogenation technologies for the conversion of viscous feedstocks to saleable products. Non-catalytic processing options such as the coking technologies and solvent deasphalting technologies are not discussed here.

Chapter 3 presents descriptions of the various refinery processes (including gasification processes) by which hydrogen is produced for use in other parts of the refinery as well as for energy, and it is often produced from process by-products that may not be the prime products of the refinery. Such by-products might be the highly aromatic, heteroatom-containing, and metal-containing bottoms (often referred to as 'deasphalter tar') from a deasphalting unit or from a mild hydrocracking process. However attractive this may seem, there will be a need to incorporate a gas cleaning operation to remove any environmentally objectionable components from the hydrogen gas.

Chapter 4 presents descriptions of the various non-typical feedstocks that can be used (such as coal, refinery coke, heavy crude oil, extra heavy crude oil, tar sand bitumen, and process residua) by gasification to produce a variety of gases, including synthesis gas (a mixture of carbon monoxide and hydrogen). Gasification is a process that converts organic carbonaceous feedstocks into carbon monoxide, carbon dioxide, and hydrogen by reacting the feedstock at high temperatures (>700°C, >1290°F), without combustion (the process is also often referred to as partial oxidation), with a controlled amount of oxygen and/or steam to produce a gaseous product (often referred to as synthesis gas or syngas) which is also used as a fuel.

Chapter 5 presents an overview with some degree of detail of the production of hydrogen from biomass and the various types of waste. The gasification of these feedstocks has been the focus of research in recent years to estimate the efficiency and performance of the gasification process using biomass and carbonaceous waste as a means of producing useful gases. Thus, in the current context, the gasification of biomass and waste can evolve into a widely used thermochemical process for obtaining products with more value and potential applications than the raw material itself.

Chapter 6 focuses on the various purification processes that are used to produce high-purity hydrogen and which are an integral part within the concept of production of a usable product (hydrogen) for use in the refinery and also for sale to the consumer. These processes can be categorized as (1) physical methods and (2) chemical methods. The former methods (i.e. the physical methods for hydrogen purification) include adsorption methods such as pressure swing adsorption (more commonly referred to as PSA), temperature swing adsorption (more commonly referred to as TSA), vacuum adsorption, low-temperature separation methods (such as cryogenic distillation and low-temperature adsorption), and membrane separation methods (such as the use of an inorganic membrane and an organic membrane), while the latter methods (i.e. the chemical methods for hydrogen purification) involve the use of a metal hydride separation and methods that use a variety of catalysts.

Chapter 7 presents an overview of the associated hydrogen network, which is limiting refinery throughput and operating margins. Furthermore, higher purity of the hydrogen within the refinery network is becoming more important to boost hydrotreater capacity, achieve product value improvements, and lengthen catalyst life cycles. In addition, this chapter presents an overview of the methods of transporting hydrogen, including (1) pipeline transportation, (2) transportation by tube trailers, (3) transportation by liquefied hydrogen tankers, and an overview of the transportation infrastructure.

Chapter 8 presents the potential operations where hydrogen would fit into the refinery of the future. Hydrogen is a key enabler, allowing refineries to comply with the latest product specifications and environmental requirements for fuel production being mandated by the market and governments, and helping to reduce the carbon footprint of their plants.

Each chapter will present to the reader the various steps that are applicable to the production of hydrogen and the methods by which this important gas can be produced and purified as well as processing options that may well be the processes of choice in the 21st century refinery.

By understanding the use of hydrgoen for product improvement, this book will satisfy the needs of engineers and scientists at all levels from academia to the refinery and help them understand the refining processes and prepare for the new changes and evolution of the industry.

The target audience includes engineers, scientists, and students who want an update on crude oil processing and the direction of the industry in the next fifty years. Such personnel include (1) professionals in the refining industry, (2) technicians in the refining industry, (3) industry management personnel who need to understand the various processes and the role of these processes in producing the desired feedstocks for further processing and the use of solvents to produce saleable products, and (4) the contribution academic staff and graduate students who are moving into the refining industry. Any non-technical readers, with help from the extensive glossary, will also benefit from the series.

Dr. James G. Speight
Laramie, Wyoming, USA
December 2023

About the Author

Dr. James G. Speight has a BSc and PhD in Chemistry; he also holds a DSc in Geological Sciences and a PhD in Petroleum Engineering. He has more than 50 years of experience in areas associated with (1) the properties, recovery, and refining of conventional petroleum, heavy oil, and tar sand bitumen, (2) the properties and refining of natural gas, and (3) the properties and refining of biomass, biofuels, biogas, and the generation of bioenergy. His work has also focused on environmental effects, environmental remediation, and safety issues associated with the production and use of fuels and biofuels. He is the author (and co-author) of more than 100 books related to petroleum science, petroleum engineering, biomass and biofuels, and environmental sciences.

Although he has always worked in private industry and focused on contract-based work, Dr. Speight has served as a Visiting Professor in the College of Science, University of Mosul (Iraq), Visiting Professor in Chemical Engineering at the Technical University of Denmark, and the University of Trinidad and Tobago as well as adjunct appointments at various universities. He has also served as a thesis examiner for more than 25 theses.

As a result of his work, Dr. Speight has been honored as the recipient of the following awards: (1) Diploma of Honor, United States National Petroleum Engineering Society for *Outstanding Contributions to the Petroleum Industry,* 1995; (2) Gold Medal of the Russian Academy of Sciences for *Outstanding Work in the Area of Petroleum Science,* 1996; (3) Einstein Medal of the Russian Academy of Sciences *in Recognition of Outstanding Contributions and Service in the Field of Geologic Sciences,* 2001; (4) Gold Medal – Scientists without Frontiers, Russian Academy of Sciences *in Recognition of His Continuous Encouragement of Scientists to Work Together across International Borders,* 2005; (5) Gold Medal – Giants of Science and Engineering, Russian Academy of Sciences, *in Recognition of Continued Excellence in Science and Engineering,* 2006; (6) Methanex Distinguished Professor, University of Trinidad and Tobago *in Recognition of Excellence in Research,* 2007; and (7) The American Excellence Award for *Excellence in Client Solutions* from the United States Institute of Trade and Commerce, Washington, DC, 2018.

1 Hydrogen Use in Refineries

1.1 INTRODUCTION

In refineries hydrogen is mainly used for hydrogenation processes, such as hydrodesulfurization, or where the more viscous refinery feedstocks (such as heavy crude oil, extra heavy crude oil, tar sand bitumen, and distillation or process residua) are cracked to produce lower boiling less viscous feedstocks. Also, refineries use hydrogen as a means of removing sulfur from the viscous feedstocks and the ensuing products in hydrodesulfurization processes (Parkash, 2003; Speight, 2005; Gary et al., 2007; Speight, 2011a, 2014a; Hsu and Robinson, 2017; Speight, 2017).

As a result, the demand for hydrogen by refineries has increased with the demand for low-sulfur products both domestically and internationally as sulfur-content regulations have become more stringent. While hydrogen is still produced within a refinery system, other sources of hydrogen are sought, such as hydrogen purchased from merchant suppliers rather than from increased hydrogen production on-site at the refinery.

In addition, the consumption of hydrogen for single refinery sites can differ significantly depending on the employed refinery units. Moreover, hydrogen has proven to be indispensable for refining operations, especially with the increasing use of viscous feedstocks (such as heavy crude oil, extra heavy crude oil, tar sand bitumen, and residue from the distillation processes) by refineries (Table 1.1). The residues are the remaining fraction left after the distillation of crude oil and can be either an atmospheric residuum or a vacuum residuum (see Table 1.2).

In fact, hydrogen is a key refinery component that allows refineries to comply with the latest product specifications and, moreover, the latest environmental requirements for product production (especially the specifications and requirements for fuels) that has been mandated not only by market demand but also by government regulations in order to reduce the carbon footprint of the various types of refinery operations (Speight, 2014a, 2017, 2021, 2023).

In addition, an important property of many refinery feedstocks and refinery products is the hydrogen content and, thus, the use of hydrogen during refining in hydrotreating processes (for example, desulfurization) and in hydroconversion processes (for example, hydrocracking) will be an important aspect of the refinery of the future (Parkash, 2003; Speight, 2005; Gary et al., 2007; Speight, 2011a, 2014a; Hsu and Robinson, 2017; Speight, 2017). Although hydrogen recycle gas may contain up to 40% by volume of other gases (usually hydrocarbons), hydrotreater catalyst life is a strong function of hydrogen partial pressure. The level of purity of the hydrogen purity at the reactor inlet determines the catalyst life by maintaining the desulfurization kinetics at lower operating temperatures and reducing carbon laydown on the catalyst (Ancheyta, 2007).

Thus, a critical issue facing the modern refinery is the changing characteristics of the refinery feedstocks that has changed the process options for the conversion of the feedstocks (especially the viscous feedstocks, including the distillation residua) into salable products (insofar as the properties of the product must meet the necessary specifications for optimum performance. This is especially true for the production of transportation fuels under the umbrella of (1) increasingly more stringent clean fuel regulations, (2) decreasing heavy fuel oil demand, and (3) the increasing need for refineries to accept viscous feedstocks, which are often referred to as 'heavy sour feedstocks' (because of the high sulfur content), as the starting feedstock for these products (Table 1.1, Table 1.2).

Thus, this chapter is an introduction, with an overview, of the use and the need for hydrogen in crude oil refineries. This will assist the reader to place the use of hydrogen in the correct context of the refinery. In fact, hydrogen is key in allowing refineries to comply with the latest product

DOI: 10.1201/9781003186694-1

TABLE 1.1
General Descriptions of Conventional Crude Oil, Heavy Crude Oil, Extra Heavy Crude Oil, Tar Sand Bitumen, and Distillation Residua*

Conventional Crude Oil
Mobile in the reservoir; API gravity: typically >25°
High-permeability reservoir
Primary recovery
Recovery
Heavy Crude Oil
More viscous than conventional crude oil; API gravity: on the order of 10 to 20°
Mobile in the reservoir
High-permeability reservoir
Secondary recovery
Tertiary recovery (enhanced oil recovery – EOR, e.g., steam stimulation)
Extra Heavy Crude Oil
Similar properties to those of tar sand bitumen; API gravity: typically <10°
Mobile in the reservoir
High-permeability reservoir
Secondary recovery
Tertiary recovery (enhanced oil recovery – EOR, e.g., steam stimulation)
Tar Sand Bitumen
Immobile in the deposit; API gravity: typically <10°
High-permeability reservoir
Mining (often preceded by explosive fracturing)
Steam assisted gravity draining (SAGD)
Solvent methods (VAPEX)
Extreme heating methods
Residuum (pl.: residua)

*This list is not intended for use as a means of classification.

TABLE 1.2
Examples of the Properties of Atmospheric Resids and the Corresponding Vacuum Resids

Crude Oil	API Gravity	Sulfur % w/w	Nitrogen % w/w	C7-Asphaltenes % w/w
Arabian Light. >650°F	17.7	3.0	0.2	1.8
Arabian Light, >1,050°F	8.5	4.4	0.5	4.3
Arabian Heavy, >650°F	11.9	4.4	0.3	8.0
Arabian Heavy, >1,050°F	7.3	5.1	0.3	10.0
Alaska, North Slope, >650°F	15.2	1.6	0.4	2.0
Alaska, North Slope, >1,050°F	8.2	2.2	0.6	4.0

specifications and environmental requirements for fuel production being mandated by markets and governments and helping to reduce the carbon footprint of refinery operations.

In fact, hydrogen is key in allowing refineries to comply with the latest product specifications and environmental requirements for fuel production being mandated by the market and by the various government regulations thereby helping to reduce the carbon footprint of refinery operations.

1.2 HYDROGEN IN REFINERIES

The chemical nature of the refinery feedstocks has always played (and will remain) the major role in determining the hydrogen requirements of a refinery. For example, the less dense, less viscous (high density) more paraffinic crude oils will require somewhat less hydrogen for upgrading to, say, a naphtha product than a more viscous (low-density) asphaltic crude oil (Speight, 2000; 2014a, 2017). For example, hydrogen demand with the viscous feedstocks (depending upon the feedstock composition and properties requiring an estimated 400 to 1,200 scf (standard cubic feet) per barrel of viscous feedstock compared to less than 200 scf of hydrogen per barrel of conventional light crude oil (Parkash, 2003; Gary et al., 2007; Speight, 2014a; Hsu and Robinson, 2017; Speight, 2017). It follows that the hydrotreating of the viscous feedstocks (i.e., heavy crude oils) and residua (which, by definition, is a hydrogen-dependent process) needs substantial amounts of hydrogen as part of the processing requirements.

In general, considerable variation exists from one refinery to another in the balance between hydrogen produced and hydrogen consumed in the refining operations, i.e.:

Hydrogen Produced	*Hydrogen Used*
Catalytic reforming	Hydrotreating
Process off-gases	Hydrocracking
	Product improvement
	Upgrading viscous feedstocks

However, what is more pertinent to the present text is the excessive amounts of hydrogen that are required for hydroprocessing operations, whether these be hydrocracking or the somewhat milder hydrotreating processes. For effective hydroprocessing, a substantial hydrogen partial pressure must be maintained in the reactor and, in order to meet this requirement, an excess of hydrogen above that consumed by the process must be fed to the reactor. Part of the hydrogen requirement is met by recycling a stream of hydrogen-rich gas. However, the need remains to generate hydrogen as makeup material to accommodate the process consumption of 500 to 3,000 scf/bbl depending upon whether the heavy feedstock is being subjected to a predominantly hydrotreating (hydrodesulfurization) or to a predominantly hydrocracking process.

In some refineries, the hydrogen needs can be satisfied by hydrogen recovery from catalytic reformer product gases, but other external sources are required. However, for the most part, many refineries now require on-site hydrogen production facilities to supply the gas for their own processes. Most of this non-reformer hydrogen is manufactured either by steam-methane reforming or by oxidation processes. However, other processes, such as steam-methanol interaction or ammonia dissociation, may also be used as sources of hydrogen. Electrolysis of water produces high-purity hydrogen, but the power costs may be prohibitive.

As environmental regulations tightened, the technology matured, and more viscous feedstocks streams were hydrotreated. Thus, in the early refineries, the hydrogen for hydroprocesses was provided as a result of catalytic reforming processes in which dehydrogenation is a major chemical reaction and, as a consequence, hydrogen gas is produced. The light ends from the catalytic reformer contain a high ratio of hydrogen to methane, so the stream is freed from ethane and/or propane to get a high concentration of hydrogen in the stream.

The hydrogen gas is recycled though the reactors to provide the atmosphere necessary for the chemical reactions and prevent the formation of carbonaceous deposits on the catalyst, thereby extending the operating life of the catalyst.

However, as refineries and refinery feedstocks evolved during the last four decades, the demand for hydrogen has increased and reforming processes are no longer capable of providing the quantities of hydrogen necessary for feedstock hydrogenation. Within the refinery, other processes are used as sources of hydrogen. Thus, the recovery of hydrogen from the by-products of the coking units, visbreaker units, and catalytic cracking units is also practiced in some refineries.

In coking units and visbreaker units, heavy feedstocks are converted to crude oil coke, oil, light hydrocarbons (benzene, naphtha, liquefied petroleum gas), and gas. Depending on the process, hydrogen is present in a wide range of concentrations. Since coking processes need gas for heating purposes, adsorption processes are best suited to recover the hydrogen because they feature a very clean hydrogen product and an off-gas suitable as fuel.

Catalytic cracking is the most important process step that is used to produce low-boiling products from gas oil and increasingly from vacuum gas oil and heavy feedstocks. In catalytic cracking the molecular mass of the main fraction of the feed is lowered, while another part is converted to coke that is deposited on the hot catalyst. The catalyst is regenerated in one or two stages by burning the coke off with air that also provides the energy for the endothermic cracking process. In the process, paraffin derivatives and naphthene derivatives are cracked to produce olefin derivatives and alkane derivatives while shorter chain length, mono-aromatic compounds are dealkylated without ring cleavage, di-aromatics and poly-aromatics are dealkylated and converted to coke. Hydrogen is formed in the last type of reaction, whereas the first two reactions produce light hydrocarbons and therefore require hydrogen. Thus, a catalytic cracker can be operated in such a manner that enough hydrogen for subsequent processes is formed.

In reforming processes, naphtha fractions are reformed to improve the quality of the naphtha fraction (which is a blend stock for gasoline production) (Speight, 2000; 2014a, 2017). The most important reactions occurring during this process are the dehydrogenation of naphthene derivatives to aromatic derivatives. This reaction is endothermic and is favored by low pressures and the reaction temperature lies in the range of 300 to 450°C (570 to 840°F). The reaction is performed on platinum catalysts, with other metals, e.g., rhenium, as promoters.

The trend to increase the number of hydrogenation (*hydrocracking* and/or *hydrotreating*) processes in refineries (Dolbear, 1998) coupled with the need to process the heavier oils, which require substantial quantities of hydrogen for upgrading, has resulted in vastly increased demands for this gas.

As reforming processes changed from a fixed-bed system to a cyclic system to a continuous regeneration, process pressures have dropped and hydrogen production per barrel of reformate has tended to increase. However, hydrogen production as a by-product is not always adequate to the needs of the refinery and other processes are necessary. Thus, hydrogen production by steam reforming or by partial oxidation of residua has also been used, particularly where heavy crude oil is available. Steam reforming is the dominant method for hydrogen production and is usually combined with pressure swing adsorption (PSA) to purify the hydrogen to greater than 99% by volume (Bandermann and Harder, 1982) (Figure 1.1).

The gasification of viscous feedstocks and coke to produce hydrogen and/or power may become an attractive option for refiners (Campbell, 1997; Dickenson et al., 1997; Fleshman, 1997; Gross and Wolff, 2000). The premise that the gasification section of a refinery will be the *garbage can* for deasphalter residues, high-sulfur coke, as well as other refinery wastes is worthy of consideration.

Several other processes are available for the production of the additional hydrogen that is necessary for the various heavy feedstock hydroprocessing sequences, and it is the purpose of the present chapter to present a general description of these processes. In general, most of the external hydrogen is

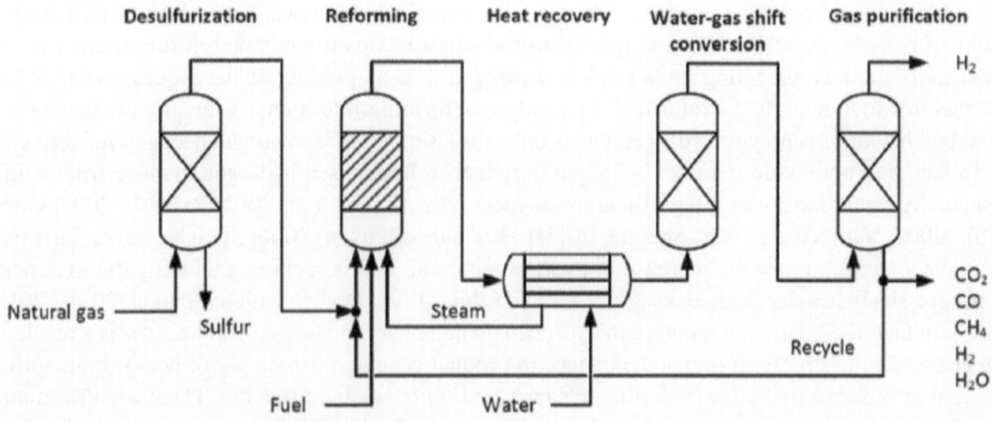

FIGURE 1.1 The Steam Reforming Process.

Source: OSHA Technical Manual, Section IV, Chapter 2: Petroleum Refining Processes. www.osha.gov/dts/osta/otm/otm_iv/otm_iv_2.html.

manufactured by steam-methane reforming or by oxidation processes. Other processes such as ammonia dissociation, steam-methanol interaction, or electrolysis are also available for hydrogen production, but economic factors and feedstock availability assist in the choice between processing alternatives.

Finally, it is worthy of mention at this point that the use pof hydrogen in refinery reactors can diminish the potential for fouling which can be caused by a number of different mechanisms and which include (1) particles in the feedstock, (2) particle formation, (3) corrosion fouling, (4) coking, (5) aggregation and flocculation, (6) phase separation, (7) particle deposition, (8) deposit growth, aging, and hardening, (9) auto-retardation and erosion or removal. In addition to these stages, the rate of fouling and the prediction of fouling factor must also be considered (Speight, 2015).

The presence of contaminants in the feedstock can have a significant effect on the fouling that can occur in a variety of chemical processes. The types of contaminants may include oxygen, nitrogen, ammonia (NH_3), hydrogen sulfide (H_2S), hydrogen cyanide (HCN), olefin derivatives, organic sulfide derivatives, organic chloride derivatives, organometallic compounds, and high molecular weight compounds such as paraffin wax as well as resin constituents and asphaltene constituents. Individual metals, which may exist as metal salts in the feedstock, can catalyze different a variety of reactions. The concentrations of such metals are typically very low, not exceeding few parts per million but small concentrations of certain metals can have a significant effect the catalysis of fouling-related reactions.

The two most common methods for producing hydrogen are (1) the steam-methane reforming process, (2) the gasification of viscous carbonaceous materials including coal, and (3) the pyrolysis of various carbonaceous feedstocks (Speight, 2013a, 2014b). These methods are presented and described in more detail in a later chapter of this book (Chapter 3).

1.3 HYDROGEN PRODUCTION

To maintain the context of this book, it would be remiss not to introduce to the reader the general methods by which hydrogen is currently produced in refineries. More details of the methods used to produce hydrogen are presented elsewhere in this book (Chapter 3) as well as acknowledgment of more novel methods of hydrogen production (Chapter 4, Chapter 5) that could well be part of the refinery of the future (Conte et al., 2001).

By way of introduction, there are two forms of hydrogen production in the refinery which are (1) hydrogen production using the steam methane reforming process, also known as the SMR process, and (2) hydrogen production as a by-product of other chemical processes. Natural gas is used almost exclusively as feedstock for on-site hydrogen production in steam methane reforming units while other units such as reforming units produce hydrogen as a by-product of the process, such as the process in which naphtha is 'reformed' by catalytic reforming into higher value high-octane products, but this latter supply of hydrogen meets only a fraction of their hydrogen needs of the refinery.

In fact, the more widespread of hydrogen in refineries has caused hydrogen to move from being a specialty operation to an integral feature of most refineries (Raissi, 2001; Parkash, 2003; Gary et al., 2007; Vauk et al., 2008; Speight, 2014a; Hsu and Robinson, 2006; Speight, 2017). This has been made necessary by the increase in hydrotreating and hydrocracking, including the treatment of progressively heavier feedstocks (Scherzer and Gruia, 1996; Dolbear, 1998; Speight, 2014a, 2017, 2023). In fact, the continued increase in hydrogen demand over the last several decades is a result of the conversion of crude oil to match changes in product slate and the supply of heavy, high-sulfur oil, and in order to make lower-boiling, cleaner, and more salable products. There are also many reasons other than product quality for using hydrogen in processes adding to the need to add hydrogen at relevant stages of the refining process and, most important according to the availability of hydrogen (Bezler, 2003; Miller and Penner, 2003; Ranke and Schödel, 2003).

With the increasing need for clean fuels, the production of hydrogen for refining purposes requires a major effort by refiners. In fact, the trend to increase the number of hydrogenation (hydrocracking and/or hydrotreating) processes in refineries coupled with the need to process the heavier oils, which require substantial quantities of hydrogen for upgrading because of the increased use of hydrogen in hydrocracking processes, has resulted in vastly increased demands for this gas. The hydrogen demands can be estimated to a very rough approximation using API gravity and the extent of the reaction, particularly the hydrodesulfurization reaction (Speight, 2000, 2014a, 2017). But accurate estimation requires equivalent process parameters and a thorough understanding of the nature of ach process. Thus, as hydrogen production grows, a better understanding of the capabilities and requirements of a hydrogen plant becomes ever more important to overall refinery operations as a means of making the best use of hydrogen supplies in the refinery.

1.3.1 Hydrogen Produced by Refinery Processes

Many refineries produce some by-product hydrogen from the catalytic reforming of naphtha (to produce higher value high-octane products) there is considerable variation that exists from one refinery to another in the balance between hydrogen produced and hydrogen consumed in the refining operations. However, what is more pertinent to the present text is the excessive amounts of hydrogen that are required for hydroprocessing operations, whether these be hydrocracking or the somewhat milder hydrotreating processes. For effective hydroprocessing, a substantial hydrogen partial pressure must be maintained in the reactor and, in order to meet this requirement, an excess of hydrogen above that consumed by the process must be fed to the reactor. Part of the hydrogen requirement is met by recycling a stream of hydrogen-rich gas. However, the need remains to generate hydrogen as makeup material to accommodate the process consumption of 500 to 3,000 scf/bbl depending upon whether the heavy feedstock is being subjected to a predominantly hydrotreating (hydrodesulfurization) or to a predominantly hydrocracking process.

In the past, the predominant method of hydrogen production in a refinery was through the catalytic reforming process, but there may insufficient hydrogen produced in this manner to satisfy the needs of the refinery. Consequently, an external source of hydrogen is necessary to meet the daily hydrogen requirements of any process where the heavier feedstocks are involved which is accompanied by various energy changes and economic changes to the refinery balance sheet (Cruz and de Oliveira Junior, 2008).

The most common, and perhaps the best, feedstocks for steam reforming are low-boiling saturated hydrocarbon derivatives that have a low sulfur content, including natural gas, refinery gas, liquefied petroleum gas (LPG), and low-boiling naphtha. Thus, A transitional solution is required to convert certain hydrocarbon derivatives and these hydrocarbon derivatives must be available through existing infrastructures such as the natural gas pipeline. While the present market demands are (in this context) focused on hydrogen for refinery operations, other applications including the reforming of methanol for portable power applications such as laptop computers, cellular phones, and personnel digital equipment are also discussed. Processing natural gas containing sulfur derivatives requires (1) adsorbents for desulfurization, (2) heterogeneous catalysts for either the autothermal process or the steam reforming process, (3) the water gas shift reaction, and (4) the preferential oxidation of carbon monoxide combustion (Farrauto et al., 2003).

Natural gas typically contains more than 90% v/v methane and ethane with only a small amount (typically amounting to a less than 15% v/v) of propane and higher boiling hydrocarbons (Mokhatab et al., 2006; Speight, 2014a, 2019). Natural gas may (or most likely will) contain traces of carbon dioxide with some nitrogen and other impurities. Purification of natural gas, before reforming, is usually relatively straightforward. Traces of sulfur must be removed to avoid poisoning the reformer catalyst; zinc oxide treatment in combination with hydrogenation is usually adequate.

The so-called 'light refinery gas' which can contain a substantial amount of hydrogen, can be an attractive steam reformer feedstock since it is produced as a by-product. Processing of refinery gas will depend on its composition, particularly the levels of olefin derivatives and of propane and higher molecular weight hydrocarbon derivatives. Olefin derivatives, which can cause problems by forming coke in the reformer, are converted to saturated compounds in the hydrogenation unit. Higher boiling hydrocarbon derivatives in refinery gas can also form coke, either on the primary reformer catalyst or in the preheater.

Refinery gas from different sources varies in suitability as hydrogen plant feed. Catalytic reformer off-gas, for example, is saturated, very low in sulfur, and often has high hydrogen content. The process gases from a coking unit or from a fluid catalytic cracking unit are much less desirable because of the content of unsaturated constituents. In addition to olefin derivatives, these gases contain substantial amounts of sulfur that must be removed before the gas is used as feedstock. These gases are also generally unsuitable for direct hydrogen recovery, since the hydrogen content is usually too low. Hydrotreater off-gas lies in the middle of the range. It is saturated, so it is readily used as hydrogen plant feed. Content of hydrogen and heavier hydrocarbons depends to a large extent on the upstream pressure. Sulfur removal will generally be required.

For the most part, many refineries now require on-site hydrogen production facilities to supply the gas for their own processes. Most of this non-reformer hydrogen is manufactured either by steam-methane reforming or by oxidation processes. However, other processes, such as steam-methanol interaction or ammonia dissociation, may also be used as sources of hydrogen. Electrolysis of water produces high-purity hydrogen, but the power costs may be prohibitive.

The primary sources of hydrogen in the modern refinery are (1) the off gas from catalytic reforming units, (2) an on-site hydrogen production facility, and (3) off-site hydrogen that is purchased from a hydrogen production facility. In the catalytic reforming units, the hydrogen yields are primarily a function of the properties of the feed naphtha, severity, catalyst and operating pressure. On the other hand, hydrogen plants produce hydrogen primarily through the steam reforming and water gas shift reactions:

$$CH_4 + H_2O \rightarrow 3H_2 + CO$$
$$CO + H_2O \rightarrow H_2 + CO_2$$

The three most used processes for hydrogen purification are (1) pressure swing adsorption, often referred to as PSA), (2) the use of semipermeable membranes, and (3) cryogenic separation, also known as 'the cold box separation method (Speight, 2023).

The pressure swing adsorption unit produces high-purity hydrogen. During the separation process, hydrogen is preferentially drawn from the feed gas, so the residue gas is enriched in hydrocarbon derivatives. Heat the membrane feed gas above its dewpoint to avoid condensation during operation. For cryogenic units, the capacity can be increased by adding a fuel gas compressor without modifications to the cold box itself.

However, hydrogen production as a by-product is not always adequate to the needs of the refinery and other processes are necessary (Lipman, 2011; Speight, 2011a, 2014a; Preuster et al., 2017). Thus, hydrogen production by steam reforming or by partial oxidation of residua has also been used, particularly where heavy crude oil is available. Steam reforming is the dominant method for hydrogen production and is usually combined with pressure swing adsorption (PSA) to purify the hydrogen to greater than 99% v/v. However, the process parameters need to be carefully defined in order to optimize capital cost. An unnecessarily stringent specification in the hydrogen purity may cause undesired and unnecessary capital cost – an example is the residual concentration of nitrogen that should not be less than 100 ppm.

The most common, and perhaps the best, feedstocks for steam reforming are low-boiling saturated hydrocarbons that have a low sulfur content, including natural gas, refinery gas, liquefied petroleum gas (LPG), and low-boiling naphtha.

However, one of the key variables in operating the steam-methane reforming unit is maintaining the proper steam to carbon ratio, which can be difficult when natural gas containing various hydrocarbons (Mokhatab et al., 2006; Speight, 2014a) is used as the feedstock and if the feedstock is typically a mixture of refinery fuel gas and natural gas, the composition is not constant. In some cases, the catalyst can be completely destroyed, and the unit will need to be shut down to change the catalyst thereby causing a disruption in the hydrogen supply.

Refinery gas, which may contain a substantial amount of hydrogen, can be an attractive steam reformer feedstock since it is produced as a by-product. Processing of refinery gas will depend on its composition, particularly the levels of olefin derivatives and of propane and higher molecular weight (higher boiling) hydrocarbon derivatives. Olefin derivatives, which can cause problems by forming coke in the reformer, are converted to saturated compounds in the hydrogenation unit. Higher boiling hydrocarbons in refinery gas can also form coke, either on the primary reformer catalyst or in the preheater.

The gasification of crude oil residua, crude oil coke, and other feedstocks such as biomass (Speight, 2008, 2011a, 2011b, 2014a) to produce hydrogen and/or power may become an attractive option for refiners. The premise that the gasification section of a refinery will be the *garbage can* for deasphalter residues, high-sulfur coke, as well as other refinery wastes is worthy of consideration. Other processes such as ammonia dissociation, steam-methanol interaction, or electrolysis are also available for hydrogen production, but economic factors and feedstock availability assist in the choice between processing alternatives.

Hydrogen is produced for use in other parts of the refinery as well as for energy and it is often produced from process by-products that may not be of any use elsewhere. Such by-products might be the highly aromatic, heteroatom (i.e. the non-carbon and non-hydrogen atoms such as nitrogen, oxygen, and sulfur), and metal containing reject from a deasphalting unit or from a mild hydrocracking process. However attractive this may seem, there will be the need to incorporate a gas cleaning operation to remove any environmentally objectionable components from the hydrogen stream.

When the hydrogen content of the refinery gas is greater than 50% v/v, the gas should first be considered for hydrogen recovery, using a membrane (Brüschke, 1995, 2003) or pressure swing adsorption unit. The tail gas or reject gas that will still contain a substantial amount of hydrogen can then be used as steam reformer feedstock. Generally, the feedstock purification process uses three different refinery gas streams to produce hydrogen. First, high-pressure hydrocracker purge gas is purified in a membrane unit that produces hydrogen at medium pressure and is combined with medium pressure off-gas that is first purified in a pressure swing adsorption unit. Finally, low-pressure off-gas is compressed, mixed with reject gases from the membrane and pressure swing adsorption units, and used as steam reformer feed.

Various processes are available to purify the hydrogen stream but since the product streams are available as a wide variety of composition, flows, and pressures, the best method of purification will vary. And there are several factors that must also be taken into consideration in the selection of a purification method. These are: (1) hydrogen recovery, (2) product purity, (3) pressure profile, (4) reliability, and (5) cost, an equally important parameter is not considered here since the emphasis is on the technical aspects of the purification process.

1.3.2 Hydrogen from Biomass and Waste

Although presented in more detail in a later chapter (Chapter 5), it would be a serious omission at this point if mention was not made of the use of biomass as a source of hydrogen (Table 1.3, as well as other potential sources of hydrogen) since biomass could well be a major feedstock in the future refinery (Chadeesingh, 2011; Ramzan et al., 2011; Speight, 2011; Lee et al., 2014a).

In fact, in the near term, biomass is anticipated to become the most likely renewable organic substitute for crude oil and the future source of hydrogen (Ayers et al., 2019; Gonzalez-Garay et al., 2022). Biomass is available from a wide range of sources, such as animal wastes, municipal solid wastes, crop residues, short rotation woody crops, agricultural wastes, sawdust, aquatic plants, short rotation herbaceous species (e.g., switch grass), wastepaper, corn, and many others (Demirbaş, 2006).

In biomass and waste conversion processes, a hydrogen-containing gas is produced in a manner similar to the gas produced by the gasification of coal (Speight, 2013a, 2014b). In fact, gasification and pyrolysis are considered the most promising medium-term technologies for the commercialization of the production of hydrogen from biomass and waste.

However, biomass feedstocks are products with inconsistent composition and the production methods vary according to biomass type, location, and climatic variations.

Addition of steam and/or oxygen in the gasification process results in the production of synthesis (also known as 'syngas') with a hydrogen-to-carbon monoxide ratio on the order of 2/1, the latter used as feedstock to a Fischer-Tropsch reactor to make produce higher molecular weight hydrocarbon derivatives (suitable for the production of gasoline and diesel fuel) or to a water gas shift reactor for hydrogen production (Asadullah et al., 2002).

In the biomass-to-hydrogen or the waste-to-hydrogen process, superheated steam (at a temperature on the order of 900°C, 1,650°F) has been used to reform dry biomass to achieve high hydrogen yields. However, the gasification process provides significant amounts of tar products (complex mixtures of high molecular weight aromatic hydrocarbon derivatives) in the product gas even when the conversion is operated at a temperature on the order of 800 to 1,000°C (1,470 to 1,830°F) range. A secondary reactor, which utilizes calcined dolomite and/or a nickel-containing catalyst, is used to catalytically clean and upgrade the product gas (Asadullah et al., 2002). For hydrogen production, a water gas shift process can be employed to increase the hydrogen concentration followed by a separation process to produce pure hydrogen (Weber et al., 2006).

TABLE 1.3
Simplified Flow Scheme for the Production of Hydrogen from Biomass

Biomass → drying and chipping → gasification	
Gasification → gas cleaning → cleaned gas	
Cleaned gas → reforming	
	Water gas shift
	Separation
Separation → hydrogen	

Also, hydrogen is a product of the metabolism of microorganisms using biomass and organic waste as raw materials (Tasleem and Tahir, 2020; Du et al., 2021). Based on the type of microorganisms and their metabolic mechanisms, the biological hydrogen-production technology includes water splitting production, photo-fermentative production, dark fermentative production, and hydrogen production combined with photo-fermentation and dark fermentation (Cao et al., 2020; Du et al., 2021).

Also, hydrogen can be produced by photoelectrochemical (PRC) means from water using sunlight and a special class of semiconductor materials. In the process, Photocatalyst powders or electrodes can produce photo-generated carriers by absorbing solar energy, so they decompose water into hydrogen and oxygen. Photocatalytic hydrogen production can be subdivided mainly into (1) heterogeneous photocatalytic hydrogen production and (2) photo-electrochemical hydrogen production. (Du et al., 2021) Another water-splitting method uses high temperatures generated by solar concentrators (mirrors that focus and intensify sunlight) or nuclear reactors to drive a series of chemical reactions to split water into hydrogen and oxygen. Also, hydrogen can be produced by pyrolytic methods (Muradov, 2003; Demirbaş and Arin, 2004; Demirbaş, 2006; Ni et al., 2006). The process takes place in the absence of oxygen and air, and therefore the potential for the formation of dioxin derivatives can be mitigated.

By way of explanation, the dioxin derivatives (Figure 1.2) a group of chemical compounds that are persistent organic pollutants in the environment which are mostly by-products of burning or various industrial processes or, in the case of dioxin-like polychloro-biphenyl derivatives, unwanted minor components of intentionally produced mixtures.

Since no water or air is present in the process, carbon oxides (e.g., carbon monoxide or carbon dioxide) are not produced thereby eliminating the need for secondary reactors (such as a water gas shift reactor or a partial oxidation reactor. Consequently, this process offers a significant reduction in emissions. The reaction can be generally described by the following equation:

$$CnHm + \text{heat} \rightarrow nC + 0.5mH2$$

Based on the temperature range, pyrolysis processes are divided into low (up to 500°C), medium (500 to 800°C; 930 to 1,470°F), and high temperatures (in excess of 800°C, 1,470°F). Fast pyrolysis is one of the latest processes for the transformation of organic material into products with higher energy content.

The application of the copyrolysis of a mixture of (for example) coal with organic wastes is also of interest since this process lighten the burden of wastes in waste disposal (waste and pure plastics, rubber, cellulose, paper, textiles, and wood).

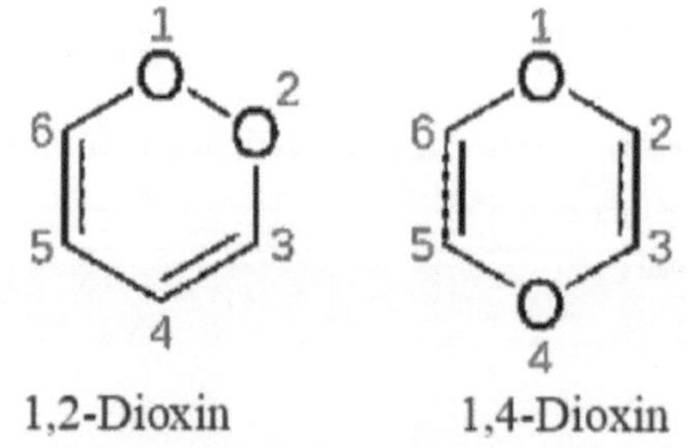

FIGURE 1.2 Structures of the 1,2-Dioxin and 1,4-Dioxin Showing the Numbering of the Ring Systems.

1.4 HYDROPROCESSES

The term 'hydroprocesses' includes a variety of catalytic processes including hydrotreating and hydrocracking for the removal of sulfur, oxygen, nitrogen and metals from feedstocks and from products. These processes are critical for the production of low sulfur fuels to reduce emissions, as removing the sulfur reduces the emissions of sulfur oxides (SOx: sulfur dioxide, SO_2, and sulfur trioxide, SO both pof which can lead to the formation of acidic species in the environment when fuels are combusted.

The use of hydrogen in refinery processes represented a significant advance in refining technology during the 20th century and is now an inclusion in most refineries (Figure 1.3). Hydrogenation processes for the conversion of crude oil fractions and crude oil products may be classified as 'destructive hydrogenation' and 'nondestructive hydrogenation'.

The major goal of hydroconversion processes of various feedstocks is cracking with desulfurization, metal removal, denitrogenation, and asphaltene conversion. Hydrotreating viscous feedstocks is a means of reducing the feedstocks to, for example, produce specification low sulfur residual fuel oil. In many cases, application of hydrotreating technology to heavy feedstocks may also cause some cracking and, by inference, application of hydrocracking to viscous feedstocks will also cause desulfurization and denitrogenation.

By way of recall (Speight, 2023), hydrotreating and hydrocracking are similar processes insofar as both processes are designed to remove specific impurities such as sulfur, nitrogen, oxygen, halides, and metals, which act as catalyst poisons from feedstocks in order to produce a range of saleable contaminant-free products of suitable molecular weight and boiling range (Table 1.4). Hydrotreating also is used to upgrade the quality of fractions by converting olefin derivative and diolefin derivatives to paraffin derivatives and to reduce gum formation in fuels.

In addition to the treated products, the processes also generate a stream of gaseous products which are cooled on exiting the reactor and then the separated hydrogen is recycled back to the

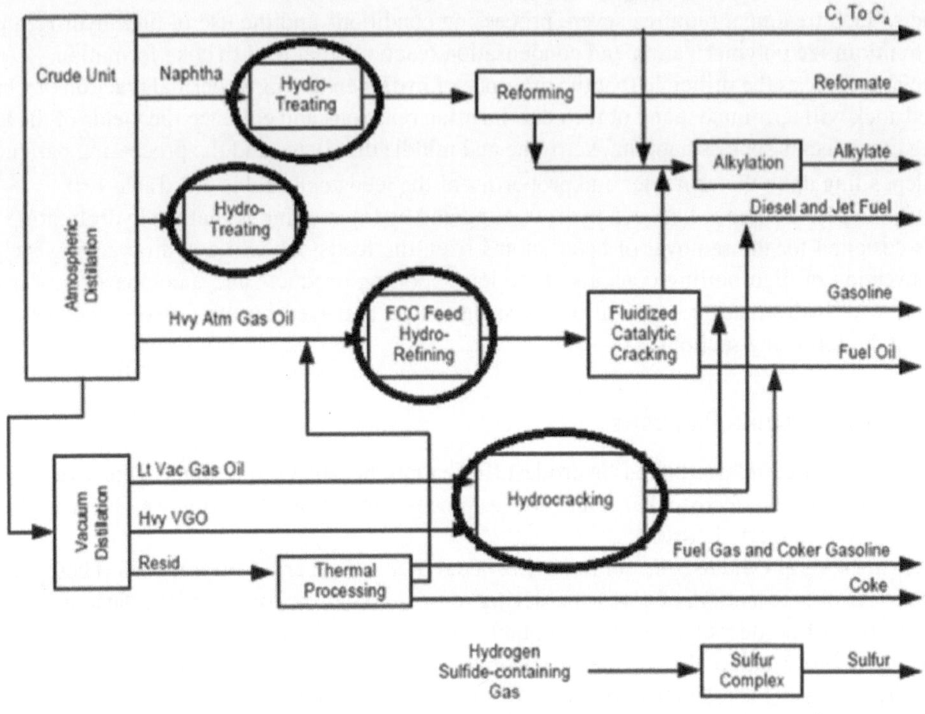

FIGURE 1.3 Examples of the Relative Placement of Hydroprocesses in a Refinery.

TABLE 1.4
Comparison of the Purpose of Hydrotreating and Hydrocracking

Hydrotreating
Remove heteroatoms (N, O, S)
Saturate carbon-carbon bonds
Minimal cracking and conversion
Products suitable for final use or blending
Hydrocracking
High conversion
Some methane, ethane, propane, and butane produced
Produce lower molecular weight paraffin derivatives
Hydrocarbon distillates (full range depending on the feedstock)

reactor. In fact, the off-gas streams are typically rich in hydrogen sulfide and fuel gas constituents. The fuel gas and hydrogen sulfide (H_2S) streams are sent to the sour gas treatment and sulfur recovery units.

Nondestructive hydrogenation (also known as hydrotreating and often referred to as 'simple hydrogenation') is generally used for the purpose of improving product quality without appreciable alteration of the boiling range of the product stream relative to the boiling range of the feedstock. Mild processing conditions are employed so that only the more unstable materials are attacked. Nitrogen, sulfur, and oxygen compounds undergo reaction with hydrogen to remove ammonia, hydrogen sulfide, and water, respectively.

On the other hand, destructive hydrogenation (hydrogenolysis, hydrocracking) is characterized by the conversion of feedstock higher molecular weight constituents to lower-boiling value-added products. Such treatment requires severe processing conditions and the use of high hydrogen pressures to minimize polymerization and condensation reactions that lead to coke formation.

The process uses the principle that the presence of hydrogen during a thermal reaction of a crude oil feedstock will terminate many of the coke-forming reactions and enhance the yields of the lower boiling components such as naphtha, kerosene and middle distillates and the processing parameters vary depending upon the character and properties of the feedstock (Table 1.5, Table 1.6).

Thus, the key difference between hydrotreating and hydrocracking is that the hydrotreating process is designed for the removal of heteroatoms from the feedstock while hydrocracking includes the conversion of high-boiling constituents to lower boiling products and a comparison of hydrocracking with hydrotreating is useful in assessing the parts played by these two processes is presented in the following sections.

1.4.1 Hydrotreating Processes

Hydrotreating (often referred to as 'hydrodesulfurization' because of the major objective of sulfur removal from process feedstocks) refers to the process operations that remove sulfur and other impurities from refinery feedstocks.

There are several valid reasons for removing heteroatoms from crude oil fractions. These include: (1) reduction, or elimination, of corrosion during refining, handling, or use of the various products, (2) production of products having an acceptable odor and specification, (3) increasing the performance and stability of naphtha, and hence gasoline which is produced from naphtha and other blend stocks, (4) decreasing smoke formation in kerosene, and (5) reduction of heteroatom content in fuel oil to a level that improves burning characteristics and is environmentally acceptable.

TABLE 1.5
Summary of Typical Hydrogen Application and Production Process in a Refinery

Hydrogen Applications	Description
Naphtha hydrotreater	Hydrogen is used desulfurize naphtha from atmospheric distillation
	Must hydrotreat the naphtha before sending it to a catalytic reformer unit
Distillate hydrotreater	Desulfurizes distillates after atmospheric or vacuum distillation
	In some units, aromatic derivatives are hydrogenated to cycloparaffin derivatives or alkane derivatives
Hydrodesulfurization	Sulfur compounds are hydrogenated to hydrogen sulfide which can be used as a feedstock for a Claus plant
Hydroisomerization	Normal (straight-chain) paraffins are converted into iso-paraffins to improve the product properties (e.g., octane number)
Hydrocracker	Uses hydrogen to upgrade higher molecular weight (higher boiling fractions into lower molecular weight (lower boiling) more valuable products
Hydrogen Production	
Catalytic reformer	Used to convert the naphtha-boiling range molecules into higher octane reformate; hydrogen is a by-product
Steam-methane reformer	Produces hydrogen for the hydrotreaters or hydrocracker
	Also, steam reforming of higher molecular weight hydrocarbons produces hydrogen from low-boiling hydrocarbons other than methane
Recovery refinery off-gases	Process gas often contains hydrogen in the range up to 50% v/v
Gasification of residua	Recovery from synthesis gas (syngas) produced in gasification units
Gasification of coke	Recovery from synthesis gas (syngas) produced in gasification units
Partial oxidation process	Analogous to gasification process
	Produces synthesis gas from which hydrogen can be isolated

Heteroatom removal, as practiced in various refineries, can take several forms (Speight, 2000) such as (1) the concentration in residua during distillation, (2) concentration in coke as a result of coking processes, or chemical removal (acid treating, caustic treating (such as the 'sweetening' or 'finishing' processes) (Speight, 200, 2014a, 2017). Nevertheless, the heteroatom removal from crude oil feedstocks is almost universally accomplished by the catalytic reaction of hydrogen with the feedstock constituents.

The commercial processes for treating, or finishing, crude oil fractions with hydrogen all operate in essentially the same manner insofar as the feedstock is heated and passed with hydrogen gas through a tower or reactor filled with catalyst pellets. The reactor is maintained at a temperature of 260 to 425°C (500 to 800°F) at pressures from 100 to 1,000 psi, depending on the particular process, the nature of the feedstock and the degree of hydrogenation required.

After leaving the reactor, any excess hydrogen is separated from the liquid product and recycled through the reactor after removal of hydrogen sulfide. The liquid product is passed into a stripping

TABLE 1.6
Examples of the Parameters Used for Hydroprocessing Various Feedstocks

Conditions
Solid Acid Catalyst (silica-alumina with rare earth metals, various other options)
Temperature: 260 to 450°C (500 to 845°F(solid/liquid contact)
Pressure: 1,000 to 6,000 psi hydrogen
Frequent catalysts renewal for heavier feedstocks
Gas oil: catalyst life up to 3 years
Heavy oil/tar sand bitumen: catalyst life less than 1 year
Feedstocks
Distillates
Refractory (aromatic) streams
Coker oils
Cycle Oils
Gas oils
Residua (as a full hydrocracking or hydrotreating option)
In some cases, asphaltic constituents (S, N, and metals) removed by deasphalting
Products
Lower molecular weight paraffins
Some methane, ethane, propane, and butane
Hydrocarbon distillates (full range depending on the feedstock)
Residual tar (recycle)
Contaminants (asphaltic constituents) deposited on the catalyst as coke or metals
Variations
Fixed-bed (applied to viscous feedstocks)
Fluid bed (applied to viscous feedstocks)
Ebullating bed (suitable for viscous feedstocks)

tower where steam removes dissolved hydrogen and hydrogen sulfide and, after cooling, the product is taken to product storage or, in the case of feedstock preparation, pumped to the next processing unit.

More specifically, in a typical catalytic hydrodesulfurization unit, the feedstock is de-aerated, mixed with hydrogen, preheated in a fired heater (315 to 425°F; 600 to 800°F) and then charged (under a pressure of up to 1,000 psi) through a fixed-bed catalytic reactor. In the reactor, the sulfur-containing constituents and the nitrogen-containing constituents in the feedstock are converted, respectively, into hydrogen sulfide and ammonia. The reaction products leave the reactor and, after cooling, are sent to a liquid-gas separator.

The liquid stream is the product from hydrotreating and is normally sent to a stripping column for removal of hydrogen sulfide and other undesirable components. In cases where steam is used for stripping, the product is sent to a vacuum drier for removal of water. The desulfurized products can be used as a blend stock or used as a feedstock for a catalytic reforming unit.

Hydrofining can be applied to lubricating oils, naphtha, and gas oils. The process parameters range from 205 to 425°C (400 to 800°F) and from 50 to 800 psi and depending on the properties of the and the degree of treatment required. Feedstocks that are (1) higher-boiling feedstocks, (2) have a high sulfur content, and (3) require maximum sulfur removal typically require higher temperatures and pressures in the reactor.

After emerging from the reactor, the product is cooled and separated from the excess hydrogen which is then recycled through the reactor. The treated oil is pumped to a stripper tower where hydrogen sulfide, formed by the hydrogenation reaction, is removed by steam, vacuum, or flue gas,

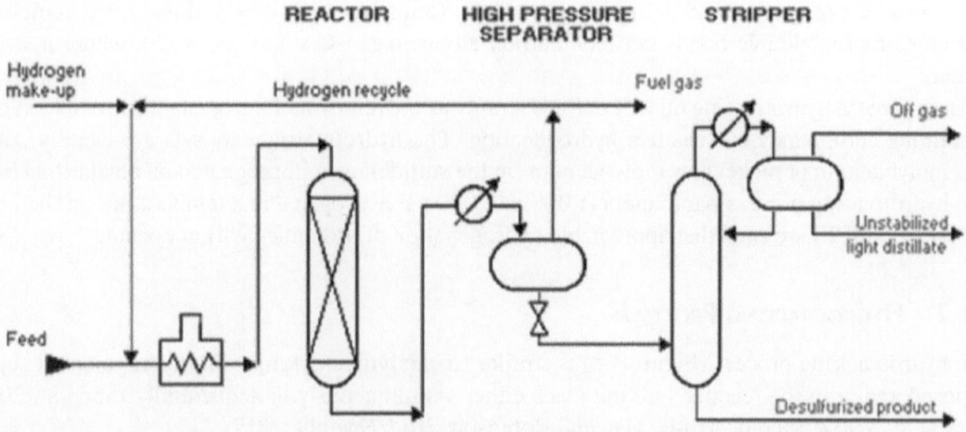

FIGURE 1.4 A Distillate Hydrotreater for Hydrodesulfurization.

Source: OSHA Technical Manual, Section IV, Chapter 2: Petroleum Refining Processes. www.osha.gov/dts/osta/otm/otm_iv/otm_iv_2.html.

and the finished product leaves the bottom of the stripper tower. The catalyst is not usually regenerated; it is replaced after about 1 year's use.

Distillate hydrotreating (Figure 1.4) involves introducing the feedstock to the reactor, together with hydrogen in the presence of catalysts such as tungsten-nickel sulfide, cobalt-molybdenum-alumina, nickel oxide-silica-alumina, and platinum-alumina.

Most processes employ cobalt-molybdenum catalysts which generally contain about 10% w/w molybdenum oxide and less than 1% w/w of cobalt oxide supported on alumina. The temperatures employed are in the range of 260 to 345°C (500 to 655), while the hydrogen pressures are about 500 to 1,000 psi (Ancheyta and Speight, 2007).

Typically, the reaction is a vapor phase reaction but may be a mixed-phase reaction. The advantages are that (1) sulfur is removed from the catalytic cracking feedstock, and corrosion is reduced in the cracking unit, (2) carbon formation during cracking is reduced so that higher conversions result, and (3) the cracking quality of the gas oil fraction is improved.

The reaction parameters for hydrotreating processes differ depending upon the properties of the feedstock and catalysts used in the process. Hydrotreating can be used to improve the burning characteristics of distillates such as kerosene in which the aromatic constituents are converted into naphthene constituents, which are cleaner-burning compounds. Lube-oil hydrotreating uses catalytic treatment of the oil with hydrogen to improve product quality. The objectives in mild lube hydrotreating include saturation of olefin derivatives and improvements in color, odor, and acid nature of the oil. Operating temperatures are typically below 315°C (600) and operating pressures are typically below 800 psi. Severe lube hydrotreating, at temperatures on the order of 315 to 400°C (600 to 750°F) and employ hydrogen pressures up to 3,000 psi, is capable of saturating aromatic rings, along with sulfur and nitrogen removal, to impart specific properties not achieved at mild conditions.

Hydrotreating also can be employed to improve the quality of pyrolysis naphtha (of referred to as 'pyrolysis gasoline' or 'pygas'), which is a by-product from the manufacture of ethylene $CH_2 = CH_2$). Traditionally, as the name might imply, the outlet for the naphtha has been as a blend stock for gasoline, which is a suitable route in view of the high-octane number of this product. The quality of 'pygas', which has a high content of di-olefin derivatives, can be improved by hydrotreating, whereby conversion of di-olefin derivatives into mono-olefin derivatives (or to fully saturated alkane derivatives) provides an acceptable product as a blending stock for gasoline production.

By way of explanation and definition, a di-olefin (often referred to as a 'diene') is a compound that contains two double bonds between carbon atoms, e.g., >C = C-C_n-C = C<, where n = 1 or higher.

In summary, hydrotreating may be defined simply as the removal of nitrogen – sulfur and oxygen-containing compounds by selective hydrogenation. The hydrotreating catalysts are usually cobalt plus molybdenum or nickel plus molybdenum (in the sulfide) form impregnated on an alumina base. The hydrotreating process parameters (1,000 to 2,000 psi hydrogen and a temperature on the order of 370°C, (700°F) are such that appreciable hydrogenation of aromatics will not occur.

1.4.2 Hydrocracking Processes

The hydrocracking process (Figure 1.5) is similar to catalytic cracking, with hydrogenation superimposed and with the reactions taking place either simultaneously or sequentially (Speight, 2005; Gary et al., 2007; Speight, 2014a; Hsu and Robinson, 2017; Speight, 2017).

Hydrocracking was initially used to upgrade low-value distillate feedstocks, such as (1) cycle oils from a catalytic cracking unit, which are usually products with a high content of aromatic derivatives and are not always recycled to extinction for economic reasons, (2) thermal and coker gas oils, and (3) heavy-cracked and straight-run naphtha. These aforementioned feedstocks are often difficult to process either by catalytic cracking or by reforming, since they are often characterized by a high polycyclic aromatic content and/or by a high concentration of the two principal catalyst poisons – sulfur-containing compounds and nitrogen-containing compounds.

Hydrocracking employs high pressure, high temperature, and a catalyst and the process is typically employed when the feedstocks are difficult to process by either catalytic cracking or reforming. These feedstocks are characterized usually by high content of polycyclic aromatic derivatives

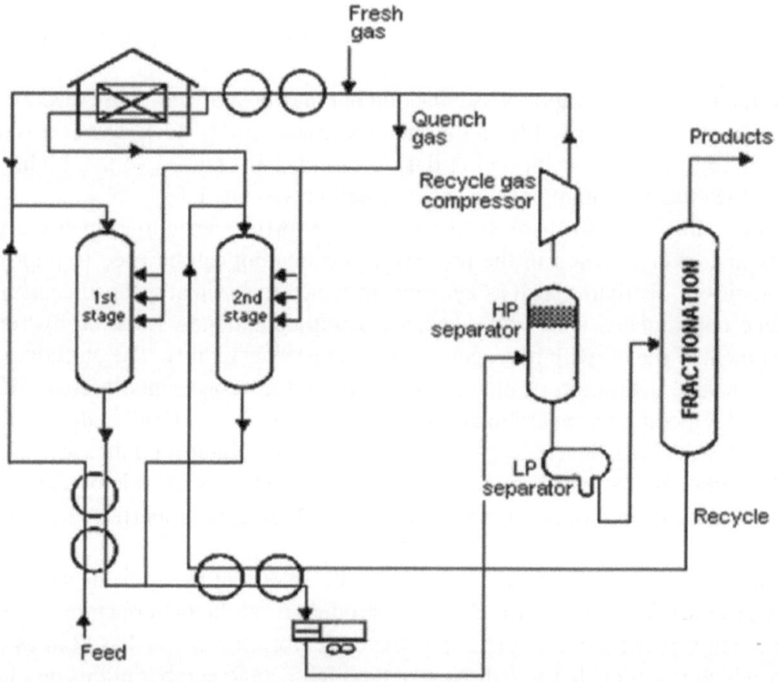

FIGURE 1.5 A Single-Stage or Two-Stage (Optional) Hydrocracking Unit.

Source: OSHA Technical Manual, Section IV, Chapter 2: Petroleum Refining Processes. www.osha.gov/dts/osta/otm/otm_iv/otm_iv_2.html.

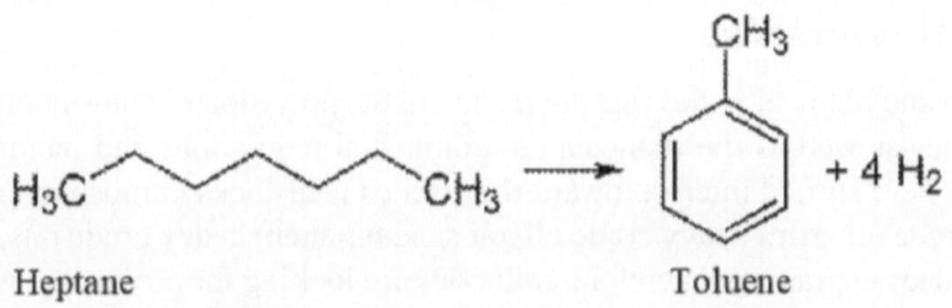

FIGURE 1.6 A Simple Example of the Dehydrocyclization Reaction.

and/or a high concentration of the two principal catalyst poisons, sulfur-containing derivatives, and nitrogen-containing derivatives.

In the process, high molecular weight aromatic feedstocks can be converted into lower molecular weight products under a wide range of very high pressures (1,000 to 2,000 psi) and fairly high temperatures (400 to 815°F; 750 to 1,500°F), in the presence of hydrogen and special catalysts. On the other hand, when the feedstock has a high content of paraffin derivatives (such as waxes), the primary function of hydrogen is to prevent the formation of polycyclic aromatic compounds through dehydrocyclization reactions. A simple example of the dehydrocyclization reaction is the conversion of heptane to toluene with the elimination of hydrogen (Figure 1.6).

Another important role of hydrogen in the hydrocracking process is to reduce the formation of tar formation and also to prevent buildup of coke on the catalyst. Hydrogenation also serves to convert sulfur and nitrogen compounds present in the feedstock to hydrogen sulfide and ammonia.

Typically hydrocracking is a process with options: single stage hydrocracking or two-stage hydrocracking. In the first stage of the process (Figure 1.5), the preheated feedstock is mixed with recycled hydrogen and sent to the first-stage reactor, where catalysts convert sulfur and nitrogen compounds to hydrogen sulfide and ammonia. The hydrogen is recycled to the feedstock. The liquid is charged to a fractionator. Depending on the products desired (such as gasoline components, jet fuel components, and gas oil), the fractionator is operated to produce the desired products. In fact, the constituents that boil in the kerosene boiling range (on the order of 150 to 300°C; 300 to 575°F) can be taken as a separate side-stream product or included in the fractionator bottoms with the gas oil fraction, for example (1) light gas oil 270 to 320°C (520 to 610°F), or (2) heavy gas oil: 320 to 425°C (610 to 800°F)

The fractionator bottoms are mixed with a hydrogen stream and sent to the to the process second stage and, since this bottoms fraction has already been subjected to some hydrogenation, cracking, and reforming in the first stage, the operations of the second stage process parameters are more severe and employ higher temperatures and higher pressures). The second stage product is separated from the hydrogen and sent to a fractionating unit.

Often, especially with the influx of heavy feedstocks into refineries, hydrocracking is a two-stage process combining catalytic cracking and hydrogenation, wherein heavier feedstocks are cracked in the presence of hydrogen to produce more desirable products. Hydrocracking also produces relatively large amounts of iso-butane for alkylation feedstock and the process also performs isomerization for pour-point control and smoke-point control, both of which are important in high-quality jet fuel.

In summary, hydrocracking is an extremely versatile process which can be utilized in many different ways such as conversion of the high-boiling aromatic streams which are produced by catalytic cracking or by coking processes. To take full advantage of hydrocracking the process must be integrated in the refinery with other process units (Figure 1.3).

1.4.3 Slurry Hydrocracking

The supply chain and political issues that are related to the provision of conventional crude oil feedstocks to refineries as well as the stringent environmental regulations and increasing demand for lighter distillates have shifted interest toward the viscous feedstocks (atmospheric residua, vacuum residua, heavy crude oil, extra heavy crude oil, tar sand bitumen) heavy crude oils, and atmospheric and vacuum residues upgrading. Therefore, refineries are looking for promising hydrogen addition-based technologies.

Slurry phase hydrocracking is a technology that can be employed to achieve a high conversion of the viscous feedstocks while minimizing the formation of by-products such as gas and coke. The process (i.e., the slurry phase hydrocracking process) is operated at high temperature (>400°C, >750°F) and high pressure of hydrogen (typically >1,500 psi). More specifically, in the slurry-phase process, the feedstock is premixed with the nano-sized catalyst particles, and the mixture is then introduced into the reactor and the hydrogen gas is injected from the bottom of the reactor. As for all catalysts, the activity of slurry catalyst is influenced by (1) the type and number of the active sites, (2) the dispersion of the catalyst in the feedstock, (3) the type of metal used in the catalyst, and (4) the concentration of the catalyst.

The slurry-phase catalyst technology employs high operating pressures and can achieve near complete conversion (if not, complete conversion, in many instances) of the feedstock (especially if the feedstock is a viscous feedstock), while producing finished saleable products (Motaghi et al., 2010a, 2010b; Speight, 2014a, 2017).

The process offers flexibility with regard to the feedstock and can accept a variety of viscous feedstocks. The typical overall performance of the slurry-phase process is (1) diminished coke formation, (2) metal removal, (3) sulfur removal, and (4) nitrogen removal. In addition, the process has the ability to reach total conversion of the feedstock because of the recycling of unconverted products and the intimate contact of the feedstock constituents with the dispersed catalyst.

Metals that have been screened as potential slurry catalysts and include transition metal-based catalysts derived from vanadium, tungsten, chromium, and iron. Homogeneous catalysts that are based on hydrocracking technology have been developed for upgrading of heavy crude and tar sand bitumen (Kriz and Ternan, 1994; Speight, 2014a, 2017). In this process the hydrocracking catalyst is homogeneously dispersed as a colloid with particles similar in size to that of asphaltene molecule, which results in high conversion of asphaltene constituents (Bhattacharyya and Mezza, 2010; Bhattacharyya et al., 2011).

1.5 COMMERCIAL PROCESSES REQUIRING HYDROGEN

In the modern refinery, the use of hydrogen is demonstrating significant growth and the large reserves of viscous feedstocks (including distillation residua) will increase the hydrogen demand ever further. The main processes include hydrotreating of various refinery streams and hydrocracking of viscous feedstocks.

Hydrotreating technology is a well-used refinery processes that is designed to remove contaminants such as compounds containing sulfur and/or nitrogen and/or metals as well as condensed ring aromatic derivatives. The process feedstocks vary from naphtha fractions to vacuum residua as well as the naturally occurring viscous feedstocks (i.e., heavy crude oil, extra heavy crude oil, tar sand bitumen) and the products in most applications are often environmentally acceptable hydrocarbons products.

The crude oil industry often employs two-stage processes in which the feedstock undergoes both hydrotreating and hydrocracking. In the first, or pretreating, stage the main purpose is conversion of nitrogen compounds in the feed to hydrocarbons and to ammonia by hydrogenation and mild hydrocracking (Speight and Moschopedis, 1979; Reynolds and Beret, 1989). Typical conditions are 340 to 390°C (650 to 740°F), 150 to 2,500 psi (1 to 17 MPa), and a catalyst contact time of 0.5 to 1.5 h; up to 1.5% w/w hydrogen is absorbed, partly by conversion of the nitrogen compounds but chiefly by aromatic compounds that are hydrogenated.

It is most important to reduce the nitrogen content of the product oil to less than 0.001% w/w (10 parts per million). This stage is usually carried out with a bifunctional catalyst containing hydrogenation promoters, for example, nickel and tungsten or molybdenum sulfides, on an acid support, such as silica-alumina. The metal sulfides hydrogenate aromatics and nitrogen compounds and prevent deposition of carbonaceous deposits; the acid support accelerates nitrogen removal as ammonia by breaking carbon-nitrogen bonds. The catalyst is generally used as 1/8 × 1/8-inch (0.32 × 0.32 cm) or 1/16 × 1/8-inch (0.16 × 0.32 cm) pellets, formed by extrusion.

Most of the hydrotreating cracking is accomplished in the first stage. Ammonia and some naphtha are usually removed from the first-stage product, and then the remaining first-stage product, which is low in nitrogen compounds, is passed over the second-stage catalyst. Again, typical conditions are 300 to 370°C (600 to 700°F), 1,500 to 2,500 psi (10 to 17 MPa) hydrogen pressure, and 0.5 to 1.5 hours contact time; 1 to 1.5% w/w hydrogen may be absorbed. Conversion to naphtha or kerosene is seldom complete in one contact with the catalyst, so the lighter oils are removed by distillation of the products and the heavier, higher-boiling product combined with fresh feed and recycled over the catalyst until it is completely converted.

Hydrodesulfurization processes are based on the addition of hydrogen to the feedstock constituents at an elevated temperature, and an elevated pressure and temperature in the presence of a catalyst. The occurrence of the various chemical reactions depends upon the type of feedstock, the type of catalyst, and the process parameters giving rise to reaction such as (1) desulfurization, (2) denitrogenation, (3) saturation of olefin derivatives, and (4) minor-to-high levels of cracking. Generally, the hydrodesulfurization process can be represented by the following simple equations:

$$[S]_{feedstock} + H_2 \rightarrow \text{hydrocarbon product} + H_2S$$

$$[N]_{feedstock} + H_2 \rightarrow \text{hydrocarbon product} + NH_3$$

$$[{>}C = C{<}] + H_2 \rightarrow$$

$$R^1CH_2CH_2CH_2CH_2R^2 \rightarrow R^1CH = CH_2 + R^2CH = CH_2 + H_2$$

The process becomes hydrodesulfurization upon the conversion of some (or all) of the sulfur-containing constituents to lower molecular low-to-no-sulfur products.

On a chemical basis, the hydrotreatment process may be subdivided into (1) selective hydrogeneration, (2) first reaction stage, (3) second reaction stages, and (4) entry of hydrogen into the reaction sequences but, in many cases, the reactions do not occur in this order.

The following sub-sections present examples of the types of process (by name and listed alphabetically rather than in any order of importance or preference) that commonly require the use of hydrogen in the modern refinery.

1.5.1 Autofining Process

The autofining process is a process that is used for reducing the sulfur content of feedstocks by treating the feedstock (such as naphtha, gas oil, and lubricating oil) in the presence of a catalyst. This process differs from other hydrorefining processes in that an external source of hydrogen is not required and sufficient hydrogen to convert sulfur-containing constituents to hydrocarbon derivatives and hydrogen sulfide is obtained by dehydrogenation of naphthene derivatives that are present in the feedstock.

The operating parameters in the reactor range from a temperature on the order of 340 to 425°C (650 to 800°F) at a hydrogen pressure on the order of 100 to 200 psi depending on the kind of feedstock and the degree of treatment required. The catalyst is regenerated with steam and air at 200-hour to 1,000-hour intervals, depending on whether the feedstock to the units is a low-boiling feedstock or a high-boiling feedstock.

1.5.2 Ferrofining Process

The ferrofining process is a mild hydrogen-treating process that was developed to treat distilled and solvent-refined lubricating oils. The process eliminates the need for acid and clay treatment. The catalyst is a three-component material on an alumina base with low hydrogen consumption and a life expectancy of 2 years or more. Process operations include heating the hydrogen-oil mixture and charging it to a downflow catalyst-filled reactor. Separation of oil and gas is a two-stage operation whereby gas is removed to the fuel system. The oil is then steam stripped to control the flash point and dried in vacuum, and a final filtering step removes catalyst fines.

1.5.3 Gulf HDS Process

The Gulf HDS process is a regenerative fixed-bed process that has been developed to upgrade crude oil residua (and other viscous feedstocks) by catalytic hydrogenation to low-sulfur heavy fuel oils or to high-quality feedstocks for catalytic cracking units.

The primary purpose of the process is the desulfurization of the feedstock constituents (which can be whole crude oil, and uncracked or cracked residua) and, therefore, quality improvement, but if the operating conditions and catalysts are varied, low-boiling products can be produced, and the viscosity of higher boiling material can be lowered. If there is the need for long on-stream cycles, this can be achieved by reducing any potential random hydrocracking reactions to a minimum.

1.5.4 Hydrofining Process

The hydrofining process is used for reducing the sulfur content of sulfur-containing viscous feedstocks by being treating the feedstock in the presence of a catalyst (Niu et al., 2020). The process may be used to upgrade straight run products, cracked products, coker-derived naphtha, and distillate boiling-range streams thereby increasing the supply of solvent naphtha, and other naphtha-type products as well as feedstocks for catalytic reforming units. The sulfur content of kerosene can be reduced with improved color, odor, and wick-char characteristics. Cracked gas oil having a moderately high sulfur content can be converted to diesel fuel and also to high-quality fuel oil by reduction in sulfur content and by the elimination of the constituents that form gum and carbon residues. This process can be applied to naphtha and gas oil (including feedstocks boiling in the lubricating oil range (on the order of between 300 to 600°C, 570 to 1,110°F).

Hydrogen is typically obtained from catalytic reforming units. The operating parameters in the reactor range from 205 to 425°C (400 to 800°F) and from 50 to 800 psi, depending on the kind of feedstock and the degree of treatment required. Higher boiling feedstocks, high sulfur content, and maximum sulfur removal require higher temperatures and pressures. After passing through the reactor, the product is cooled and separated from the excess hydrogen, which is then recycled through the reactor. The product is pumped to a stripper tower, where hydrogen sulfide formed by the hydrogenation reaction is removed by steam, vacuum, or flue gas and the finished product leaves the bottom of the stripper tower.

Lube oil hydrofining is a catalytic technology to prepare lube base stocks for further processing or it may be used as a base-stock finishing step. The process is usually integrated Exol N technology, into an Exolfining configuration, to treat the waxy raffinate from extraction upstream from the lube dewaxing unit. As a finishing step, the lube hydrofining process may be used to treat a dewaxed oil product with similar color and oxidation stability benefits. Lube hydrofining, when used on lube distillate feedstocks can enhance the ensuing solvent extraction performance.

Lube hydrofining technology generally produced a product with (1) improved color, (2) improved oxidation stability, (3) improved antioxidant response, (4) better odor, (5) better surface properties,

(6) optimal sulfur reduction, (7) significant viscosity index (VI) increases at high desulfurization levels, and (8) reduction in toxicity.

Generally, the lube hydrofining process is suitable for feedstocks from all crude sources and the process conditions are moderate. The feedstock to the lube hydrofining reactor is the hot bottoms stream from the raffinate recovery tower.

The reactor effluent is flashed to recover the unreacted hydrogen treat gas at high pressure as well as to separate the hydrogen sulfide and ammonia resulting from the hydrofinishing reactions. The hydrofinished oil is then steam stripped and dried under vacuum. As an additional option, the steam stripper tower may be designed to correct the volatility of the hydrofinished oil product.

The wax hydrofining process is used to produce wax from a lube oil plant that can be employed as a food-grade wax that can be used for wax paper, fruit coatings, paper cups, as a seal for medicines, and for similar applications where high purity is required. In fact, the wax hydrofining process is the final critical step for processing low-oil-content waxes from a dewaxing-deoiling process to produce wax with low odor, improved color, and stability. Also, this technology is applicable to either crystalline wax from lube distillates or microcrystalline wax from deasphalted oils. The wax hydrofining option processes low oil content (<1% w/w) wax from conventional solvent lube plant dewaxing-deoiling units under moderate hydroprocessing conditions.

Feedstock constituents containing sulfur and nitrogen are converted to hydrogen sulfide and ammonia, respectively while multi-ring aromatic constituents are saturated, and trace acids and solvent residues are removed. The reactor effluent is flashed to separate hydrogen, hydrogen sulfide and ammonia, and the wax is steam stripped and dried to improve volatility and color. Both crystalline waxes and microcrystalline waxes can be processed in the same unit, although the types of constituents that occur in microcrystalline waxes typically require more severe operating conditions.

The SCANfining process has been designed to achieve high sulfur removal without excessive olefin saturation and octane loss. Similarly, the 'diesel oil deep desulfurization' process (sometimes referred to as the DODD process) as well as the GO-fining process and residfining process are used for upgrading gas oils and the viscous feedstocks, as feedstocks for catalytic cracking units and also to use as low sulfur fuel oil (Song, 2002).

Finally, as an additional option, most feedstocks to a fluid catalytic cracking unit are high sulfur and, in some cases, have a high ratio of residue blended into the gas oil which results in the production of a high sulfur high-olefin naphtha. This kind of naphtha can be treated in a hydrofiner and the hydrofined naphtha is of good quality containing more saturated hydrocarbons, through which high sulfur content problem of naphtha is resolved. Typical operating conditions are a temperature on the order of 280 to 320°C (535 to 610°F) and a pressure on the order of 290 to 1,160 psi with a liquid hourly space velocity (LHSV) in the range 1.5 to –3.0 h^{-1} and a hydrogen-oil ratio (v/v) of 200 to 400

Catalysts for the process typically contain tungsten oxide and/or molybdenum oxide, nickel oxide, and cobalt oxide which are supported on alumina (Al_2O_3). Catalysts with a lower metal content may exhibit higher activity at low temperature (Xia et al., 2000; Tanaka, 2004; Kumagai et al., 2005).

1.5.5 Isomax Process

The isomax process is a two-stage, fixed-bed catalyst process (each stage has a separate hydrogen recycling system) that operates in a temperature range on the order of 205 to 370°C (400 to 700°F) and a hydrogen pressure on the order of 500 to 1,500 psi using, for example, with middle distillate feedstocks. In the first stage, nitrogen, oxygen, sulfur, and metal contaminants are removed from the feedstock, while in the second stage the feedstock is hydrocracked into the desired hydrocarbon fractions.

The process parameters that are employed depend on the feedstock and product requirements, and hydrogen consumption is of the order of 1,000 to 1,600 ft^3 bbl^{-1} of feedstock processed. Conversion may be balanced to provide products for variable requirements, and recycling can be taken to extinction if necessary. Fractionation of the product stream can be achieved in a number of different ways to yield desired products.

1.5.6 Ultrafining Process

Ultrafining is a regenerative, fixed-bed, catalytic process to desulfurize and hydrogenate refinery stocks from naphtha up to and including lubricating oil. The catalyst is cobalt-molybdenum on alumina and may be regenerated in situ using an air-stream mixture. Regeneration requires 10 to 20 hours and may be repeated 50 to 100 times for a given batch of catalyst; catalyst life is 2 to 5 years depending on feedstock.

1.5.7 Unifining Process

This is a regenerative, fixed-bed, catalytic process to desulfurize and hydrogenate refinery distillates of any boiling range. Contaminating metals, nitrogen compounds, and oxygen compounds are eliminated, along with sulfur. The catalyst is a cobalt-molybdenum-alumina type that may be regenerated in situ with steam and air.

1.5.8 Unionfining Process

The unionfining process is a hydrotreating technology that is employed to produce low-sulfur, color stable diesel fuel as well as to reduce the amount of aromatic constituents in the product and (in the case of a diesel product) to improve the cetane number.

The process employs downflow trickle bed reactors in which the pressures and temperatures are higher than for distillate and gas oil hydroprocessing. The process performs a range of chemical reactions simultaneously including hydrodesulfurization (HDS), hydrodenitrogenation (HDN), and hydrodemetalation (HDM).

In the process, the reactor feed is heated by heat exchange with the reactor effluent and reactor inlet temperature is controlled by charge heater firing. The desulfurized product is recovered by removal of any low-boiling constituents (light ends) by stripping with steam.

1.6 HYDROGEN FOR PRODUCT IMPROVEMENT

A typical refinery produces a wide variety of different products from every barrel of refinery feedstock that is processed (Table 1.7). Generally, refineries operate to make as much of the high-value, low-boiling products (such as naphtha, gasoline, kerosene, diesel fuel, jet fuel, and fuel oil).

However, the high value products are not always produced as a first-pass product by the refinery and require various forms of improvement to meet the specifications for sales. The products that do not meet the specifications for sales typically require (as it relates to the present context of this book) a form of hydrotreating.

For example, the fractions or streams produced in the refinery often contained (and still do contain) small amounts of impurities that must be removed by means of 'product improvement' processes (often referred to as 'treating processes') that are used not only to finish products for the market but also to prepare feedstocks for other processes in which catalysts would be harmed by impurities. A primary method of product improvement is the use of hydrogen, typically referred to as 'hydrotreating' (Parkash, 2003; Gary et al., 2007; Speight, 2014a, 2016; Hsu and Robinson, 2017; Speight, 2017, 2023).

Hydrogen for product improvement can be extracted from fossil fuels, biomass, and waste (particularly bio-waste), from water, or from a mix of both (Chapter 3, Chapter 4, Chapter 5). Natural gas is currently the primary source of hydrogen production. In the early decades of the crude oil refining industry at the time when the more conventional crude oils were readily available, the primary use of hydrogen was to improve the quality of the refinery products so that these products met the specifications necessary for sales to the consumers.

TABLE 1.7
Examples of the Major Products from a Typical Refinery*

Fuel-Type Product	Use
Aviation gasoline	An engine fuel in light aircraft
Butane	A feedstock for ethylene cracking
Diesel fuel	A fuel for heavy-duty trucks, trains, and heavy equipment
Fuel oil	A residential cooking, heating, and lighting fuel
Gas oil	A fuel in industrial plants and commercial/residential heating
Gasoline	A transportation fuel for passenger cars and light trucks
Jet fuel	A fuel for jet aircraft
Kerosene	A residential cooking, heating, and lighting fuel
Light naphtha	Low-boiling naphtha; a feedstock into ethylene crackers
LPG	Liquefied petroleum gas; a blend of propane and butane
Propane	A feedstock for ethylene cracking
Residual fuel oil	A fuel in power generation and for large ocean-going ships; also referred to as (bunker fuel)
Non-Fuel Product	*Use*
Aromatics	Feedstock for the petrochemicals industry
Asphalt	Road paving; manufacture of building materials (e.g., roof shingles)
Base oil	Used to make lubricating oils for use in machinery and engines
Coke	A fuel for power plants and cement plants or used to manufacture electrodes and anodes
Grease	Used as a solid lubricating oil, mostly in industrial uses
Propylene	Feedstock for the petrochemicals industry
Wax	A feedstock to specialty wax production (as slack wax) or finished wax product
White oil	A colorless, odorless, tasteless oil used by the food, cosmetics, and pharmaceuticals industries
White spirit	Naphtha range product used as an industrial or household solvent
Sulfur	A feedstock to the petrochemicals industry

*Listed alphabetically and not by use or by any preference.

Hydrotreating achieves the following objectives: (1) removal of impurities, such as sulfur, nitrogen, and oxygen for the control of a final product specification or (2) the preparation of feed for further processing (naphtha reformer feed and feedstock to a fluid catalytic cracking unit), as well as removal of metals which is typically achieved by use of a separate guard catalytic reactor when the organometallic compounds are hydrogenated and decomposed (demetallized) that results in metal deposition on the catalyst pores (e.g., atmospheric residue desulfurization (ARDS) guard reactor); and saturation of olefin derivatives and their unstable compounds.

Thus, the predominant aims of many chemical treating methods (such as treating with hydrogen) is to remove the impurities, such as sulfur, nitrogen, oxygen, unsaturated compounds and some aromatic derivatives at low temperature to mitigate the formation of coke and other carbonaceous deposits.

The most common impurities are sulfur compounds that are derived from the sulfur compounds that occur in crude oil, such as sulfides (R-S-R') and the foul-smelling mercaptan derivatives (R-SH, also commonly referred to as thiols). Oxygen compounds in the form of carboxylic acids (RCO_2H) and phenols (ArOH, where Ar is an aromatic group) may also be present. Nitrogen-containing compounds derived from those that occur in crude oil are also present. Furthermore, olefin derivatives ($RCH = CHR^1$ where R and R^1 can be the same or different moieties) must also be eliminated from a feedstock or aromatics removed from a solvent, and these olefin derivatives and aromatic derivatives are considered impurities. Similarly, polymerized material, asphaltic material, or resin constituents may be impurities, depending on whether their presence in a finished product is harmful.

Treatment processes for the removal of sulfur-containing and nitrogen-containing compounds are much less severe than the desulfurization and denitrogenation processes. In fact, it is generally recognized that the removal or conversion of sulfur and nitrogen compounds in distillates by treatment processes is usually limited to mercaptan derivatives and the lower molecular weight sulfur compounds. When there is more than a trace amount (>0.1 %) of heteroatoms present, it is often more convenient and economical to resort to such methods as those thermal processes (e.g., hydroprocesses) that bring about a decrease in all types of heteroatomic compounds.

If mercaptan derivatives are also present in the unrefined kerosene, a doctor treatment in addition to lye treatment is required, but poor-quality raw kerosene may require, in addition to these treatments, treatment with sulfuric acid and fuller's earth. The lowest quality raw kerosene requires treatment with strong sulfuric acid, neutralization with lye, and redistillation.

1.7 HYDROGEN FOR VISCOUS FEEDSTOCK REFINING

As the 20th century evolved, an extremely important aspect of the modern refining technology was the ability of the refinery to accept and refine the viscous feedstocks into usable products that adhere to the various environmental regulations, and which do not, in any way, threaten the environment. Over the decades since the 1970s, refineries have evolved to accomplish this goal and, at the time of writing, continue along this evolution part.

By way of explanation, the term 'viscous feedstock' includes the high viscosity high sulfur feedstocks that are now playing a major role in refineries as the more conventional refinery feedstocks not only are being depleted but are also more difficult and expensive to acquire (Speight, 2021). Recognizing that there will be considerable variation, typical analytical data for the viscous feedstocks are:

Carbon: 84–88% w/w
Hydrogen: 9–10% w/w
Nitrogen: 0.5–1.5% w/w
Oxygen: 0.1–0.5% w/w
Sulfur: 2–5% w/w
Metals: 200–1,000 ppm
API gravity: 4–10°

As an additional note, the crude oil coke that is produced in a refinery and used as a fuel or as a feedstock for a gasifier typically has 88–90% w/w carbon and 2–5% w/w hydrogen as well as variable amount of nitrogen, oxygen, sulfur, and metals that are dependent upon the amount of these elements in the feedstock.

1.7.1 Types of Viscous Feedstocks

The definition of crude oil has been varied, unsystematic, diverse, and often archaic (Speight, 2014a, 2017, 2021) because the terminology used to describe crude oil is subject to many conventions of which a major one is chemistry-aligned definitions, company-aligned definitions, engineering-aligned definitions, and definitions aligned with the geological sciences.

In the context of this book, viscous feedstocks are refinery feedstocks such as (1) heavy crude oil, (2) extra heavy crude oil, (3) tar sand bitumen, and (4) distillation residua that have a low volatility, high density, and high viscosity (Chapter 2). Although other feedstocks such as naphtha, kerosene, and gas oil are also used as feedstocks for thermal processes and catalytic processes in the refinery, the focus of this book remains on the viscous feedstocks. Also, viscous feedstocks are richer in polynuclear aromatic.

A viscous feedstock is comprised of complex assortment of different molecular and chemical types with high melting points and high pour points that greatly contribute to the poor fluid properties of the viscous feedstock and, hence, to low mobility (compared to conventional crude oil). The same is true for extra heavy crude oil and tar sand bitumen (Speight, 2013b, 2013c, 2014a, 2019). More generally a viscous feedstock typically has moderate-to-high levels of asphaltene constituents. The asphaltene constituents are not solely responsible for the high specific gravity or low API gravity but do contribute to these properties and may make a contribution to production problems. Briefly the asphaltene constituents are those constituents of oil that are insoluble in n-heptane or n-pentane while the resin constituents are those costudents of oil that are soluble in n-heptane or n-pentane bur are adsorbed from these hydrocarbon solutions on to an adsorbent such as a clay mineral or alumina (Al_2O_3). In addition to the asphaltene constituents, it is also essential to consider the of the resin constituents and the aromatic constituents (especially the aromatic constituents that contain heteroatoms).

While the terminology used for the various refinery feedstocks has been described in detail elsewhere (Speight, 2014a, 2017, 2021), for the sake of completeness and to mitigate the potential for any confusion or misunderstanding, a short description of the various viscous feedstocks is worthy of inclusion at this point.

In the fractionation procedure (Speight, 2014a, 2015a), the liquid hydrocarbon (n-pentane or n-heptane) must be specified since each liquid hydrocarbon gives a different yield of the asphaltene fraction. Both the resin constituents and the asphaltene constituents produce coke in thermal processes (Speight, 2014a, 2017) and, therefore, are capable of interfering with refinery processes by means of the formation of sediment – a phenomenon known as instability or incompatibility.

1.7.1.1 Residua

In contrast to the naturally occurring heavy crude oil, extra heavy crude oil, and tar sand bitumen, there are several processes that produce a residuum (pl. residua, also shortened to resid, pl. resids). The most common residua are those produced in the distillation section of the refinery (viz. the atmospheric residuum and the vacuum residuum) but this should not exclude the residues produced in other refinery processes (i.e., non-distillation processes).

More specifically, a residuum is the non-volatile material produced in a refinery process and the residuum is a manufactured material (rather than a naturally occurring material) insofar as a residuum is produced as a result of crude oil refining (Parkash, 2003; Gary et al., 2007; Speight, 2014a; Hsu and Robinson, 2006; Speight, 2017) (Figure 1.7).

Thus, in the current context, the residue from the distillation section of the refinery can be categorized in the following manner:

Crude Oil	Atmospheric distillation
	Naphtha
	Kerosene
	Middle distillates
	Atmospheric gas oil
	Light atmospheric gas oil
	heavy atmospheric gas oil
	Atmospheric residuum
Atmospheric Residuum	Vacuum distillation
	Vacuum gas oil
	Light vacuum gas oil
	Heavy vacuum gas oil
	Vacuum residuum

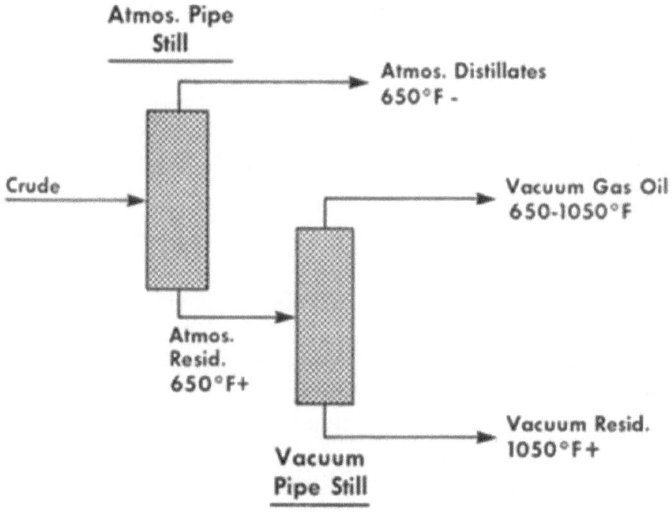

FIGURE 1.7 The Distillation Section of a Refinery.

Source: Speight, J.G. 2017. Handbook of Petroleum Refining. CRC Press, Taylor and Francis Group, Boca Raton, Florida. Page 34.
www.osha.gov/dts/osta/otm/otm_iv/otm_iv_2.html.

For ease of identification and to allay the potential for any confusion, it is often preferable to use the description 'atmospheric' or 'vacuum' as illustrated above when describing the various fractions from the distillation units.

In addition, the properties are variable depending upon the cut point (i.e., either an atmospheric resid or a vacuum resid) (Parkash, 2003; Gary et al., 2007; Speight, 2014a; Hsu and Robinson, 2006; Speight, 2017). Many residua are subject to further refining after production and are worthy of mention her as optional refinery viscous feedstocks either further refined as single feedstocks or as blends with other feedstocks, such as gas oil, as well as conventional whole crude oil, heavy crude oil, extra heavy crude oil, and tar sand bitumen.

Typically, a residuum) is the residue obtained from crude oil after non-destructive distillation has removed all the volatile materials. The temperature of the distillation is usually maintained below 350°C (660°F) since the rate of thermal decomposition of crude oil constituents is minimal below this temperature but the rate of thermal decomposition of crude oil constituents is substantial above 350°C (660°F). If the temperature of the distillation unit rises above 350°C (660°F) as happens in certain units where temperatures up to 395°C (740°F) are known to occur, cracking can be controlled by adjustment of the residence time.

Residua (pl: residuum, often shortened to 'resid') are black, viscous materials that are obtained by distillation of a crude oil under atmospheric pressure (atmospheric residuum) or under reduced pressure (vacuum residuum) and the properties of the residuum do vary with the initial boiling point of the residuum (Parkash, 2003; Gary et al., 2007; Speight, 2014a; Hsu and Robinson, 2017; Speight, 2017). More typically, residua are black, viscous materials that may be near liquid at room temperature (typically, an atmospheric residuum) or solid (typically, a vacuum residuum) depending upon the properties of the crude oil source.

When a residuum is obtained from a feedstock and thermal decomposition has commenced, it is more usual to refer to this product as cracked residuum (sometime referred to as pitch, a name derived from the coal industry). The differences between conventional crude oil and the related residua are due to the relative amounts of various constituents present, which are removed or retained by according to the relative volatility of the constituents.

The vacuum residuum (also commonly referred to as 'vacuum bottoms', typically 950°F $^+$ or 1,050°F $^+$) is the most complex of crude oil and, in many cases, may even resemble heavy crude oil or extra heavy crude oil or tar sand bitumen in composition. Vacuum residua contain the majority of the heteroatoms originally in the crude oil and molecular weight of the constituents range up to several thousand (as near as can be determined but subject to method dependence). The fraction is so complex that the characterization of individual species is virtually impossible, no matter what claims have been made or will be made. Separation of vacuum residua by group type can be difficult because of the multi-substitution of aromatic and naphthenic species as well as by the presence of multiple functionalities in single molecules (Speight, 2014a, 2019, 2021).

Typically, the precipitation of the asphaltene fraction n-pentane or n-heptane precipitation is used as the initial step for the characterization of a viscous feedstock, such as a vacuum residuum (Speight, 2014a, 2015a). The insoluble fraction, the pentane-asphaltene fraction or the heptane-asphaltene fractions, may be as much as 50% by weight of a vacuum residuum. The pentane-soluble or the heptane-soluble portion (often referred to as the maltene fraction) of the residuum is then fractionated by adsorption chromatographically into several solubility or adsorption classes for characterization. However, in spite of claims to the contrary, the method is not a separation by chemical type. However, the separation of the asphaltene fraction does, however, provide a simple way to remove some of the highest molecular weight and most polar components but the asphaltene fraction is so complex that compositional detail based on average parameters is of questionable value.

For the 565°C $^+$ (1,050°F $^+$) fractions of crude oil, the levels of nitrogen and oxygen may begin to approach the concentration of sulfur. These elements consistently concentrate in the most polar fractions to the extent that every molecule contains more than one heteroatom. At this point, structural identification is somewhat fruitless and characterization techniques are used to confirm the presence of the functionalities found in lower boiling fractions such as, for example, acid derivatives, phenol derivatives, non-basic (carbazole-type) nitrogen derivatives, and basic (quinoline-type) nitrogen derivatives (Speight, 2014a).

The nickel and vanadium constituents of the feedstocks that are concentrated into the vacuum residuum (or any viscous feedstock) occur in two forms which are (1) porphyrin derivatives and (2) non-porphyrin derivatives (Figure 1.8) (Reynolds, 1998).

Because the metalloporphyrins can provide insights into crude oil maturation processes, they have been studied extensively and several families of related structures have been identified. On the other hand, the non-porphyrin metals remain not clearly identified although some studies suggest that some of the metals in these compounds still exist in a tetra-pyrrole environment (porphyrin-type environment; Figure 1.8) (Pearson and Green, 1993).

It is more than likely that, in a specific residuum molecule, the heteroatoms are arranged in different functional types thereby contributing to an extremely complex fraction. Thus, considering the potential (and the reality) for the number of different combinations that are possible, the chances of determining every structure in a residuum are minimal and, therefore, because of the

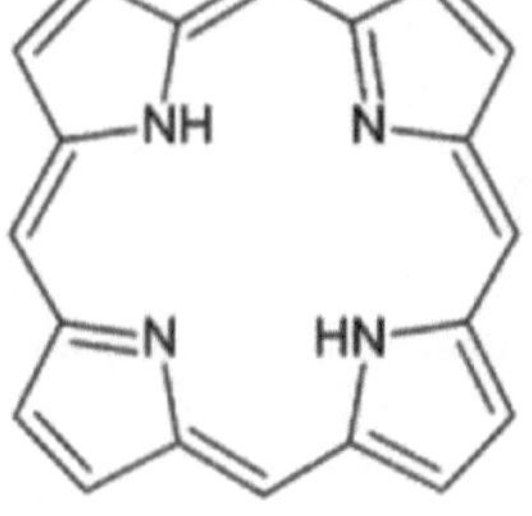

FIGURE 1.8 The Skeleton of the Porphyrin Molecule*. *In the metallo-porphyrin derivatives, the metal (typical nickel, vanadium, and, on occasion, iron) is held in place by bonds with the four nitrogen atoms.

varied reactivity of the residuum (or viscous feedstock) constituents, the structure and behavior of a residuum or any viscous feedstocks cannot, in any way (in spite of claims to the contrary) be represented by a so-called average structure (Speight, 2014a, 2021).

In summary, the chemical composition and physical behavior of a viscous feedstock is complex. Physical methods of fractionation usually indicate high proportions of asphaltene constituents and resin constituents, even in amounts up to 50% (or higher) of the residuum (Parkash, 2003; Gary et al., 2007; Speight, 2014a; Hsu and Robinson, 2006; Speight, 2017). In addition, the presence of ash-forming metallic constituents, including such organometallic compounds as those of vanadium and nickel, is also a distinguishing feature of residua and the heavier oils. Furthermore, the deeper the cut into the feedstock, the greater is the concentration of sulfur and metals in the non-volatile fractions and the greater the deterioration in physical properties.

1.7.1.2 Heavy Crude Oil

The term 'heavy crude oil' (or 'heavy crude oil') has also been arbitrarily used to describe both the heavy crude oil that requires thermal stimulation for recovery from the reservoir as well as the extra heavy crude oil and the bitumen in tar sand formation (i.e., a bituminous sand formation) from which the bitumen is recovered by, for example, a mining operation (Speight, 2014a, 2019).

When crude oil occurs in a reservoir that allows the crude material to be recovered by pumping operations as a free-flowing dark to light colored liquid, it is often referred to as conventional crude oil. Heavy crude oil is a type of crude oil that is different from the conventional crude oil insofar as they are much more difficult to recover from the subsurface reservoir and typically requires the use of thermal methods to reduce the viscosity (and increase the flowability) of the oil (Speight, 2014a, 2019, 2021).

Heavy crude oils are more difficult to recover from the subsurface reservoir than conventional crude oil. The definition of heavy crude oil has been (and continues to be) typically based on the API gravity or viscosity but the form of definition is arbitrary although there have been attempts to rationalize the definition based upon viscosity and API gravity (Speight, 2014a).

For many years, crude oil and heavy crude oil were very defined very generally in terms of physical properties. For example, heavy crude oils were considered to be those crude oils that had gravity somewhat less than 20° API with the gravity of the heavy crude oils falling into the API gravity range 10 to 20°. For example, Cold Lake heavy crude oil has an API gravity equal to 12° and extra heavy crude oils, such as tar sand bitumen, usually have an API gravity in the range 5 to 10° (Athabasca bitumen = 8° API). The similar property applied to distillation residua (or even residue from a non-distillation process) would vary depending upon the temperature at which distillation was terminated (or the process parameters) but, in a very general sense, vacuum residua would fall into a range on the order of 2 to 8° API (Speight, 2000, 2014a).

In fact, the generic term 'heavy crude oil' is often applied to a crude oil that has an API gravity less than 20° API or less than 20° API or, in some cases, less than 25° API. Obviously, the use of the density-API descriptor is subject to variation depending on the company or the country or the limits of accuracy of the test method (by which the API was determined) and there is always the question that refers to the differences between a heavy crude oil having an API gravity of 19.5 and a heavy crude oil having an API gravity of 20.5 (Speight, 2014a, 2015a, 2019). In addition, and to add further confusion to the so-called definition, heavy crude oil is often (a heavy crude oil having an API gravity of 19.5) described as having a sulfur content higher than 2% w/w (Speight, 2000, 2014a, 2017) and, in contrast to conventional crude oil, heavy crude oil is darker in color and may even be black.

Heavy crude oil is typically recognized as a type of crude oil but is different to the other types of crude oil insofar as heavy crude oil differs conventional crude oil (conventional crude oil (often referred to as light oil, i.e. low-boiling (low-density) insofar as the heavy crude oil is more difficult to recover from the subsurface reservoir. The heavy crude oil has a higher viscosity (and lower API gravity) than conventional crude oil and recovery of heavy crude oil usually requires thermal stimulation of the reservoir leading to application of various thermal methods (such as coking) in the refinery for suitable conversion to low-boiling distillates.

There has been the continuing tendency to define heavy crude oil using the API gravity or viscosity, and the definition is quite arbitrary although there have been attempts to rationalize the definition based upon viscosity, API gravity, and density (Speight, 2014a, 2015a, 2019, 2021). For example, heavy crude oil is considered to be the type of crude oils that had gravity somewhat less than 20° API with many heavy crude oils falling into the API gravity range 10 to 15°. For example, Cold Lake heavy crude oil (Alberta, Canada) has an API gravity equal on the order of 12° and tar sand bitumen (Athabasca, Fort McMurray, Alberta, Canada) usually have an API gravity on the order of 5 to 10° (Athabasca bitumen = 8° API). Using this scale, crude oil residua vary depending upon the temperature at which distillation was terminated but the atmospheric residuum is typically in the range 10 to 15° API while the vacuum residuum is in the range 2 to 8° API (Parkash, 2003; Gary et al., 2007; Speight, 2014a; Hsu and Robinson, 2017; Speight, 2017, 2021).

In the simplest terms, heavy crude oil is a type of crude oil which is very viscous and does not flow easily. The common characteristic properties (relative to conventional crude oil) are high specific gravity, low hydrogen to carbon ratios, high carbon residues, and high contents of asphaltenes, heavy metals (i.e., metallic chemical elements that have a relatively high density and are toxic or poisonous at low concentrations; examples of heavy metals include (alphabetically): arsenic, As, cadmium, Cd, chromium, mercury, Hg, Cr, lead, Pb, and thallium, Tl), sulfur, and nitrogen.

Thus, the term heavy crude oil has also been arbitrarily used to describe the viscous crude oil that requires a thermal stimulation method to recover the oil from the reservoir. It must also be remembered that the term 'heavy crude oil' may also include opportunity crude oil, high acid crude oil, and foamy oil and, thus, there is the need for a refinery to be ready to accommodate opportunity crude oils and/or high acid crude oils and/or foamy oils which, for many purposes, may have been shipped to a refinery under the banner of 'heavy crude oil'.

1.7.1.3 Extra Heavy Crude Oil

The term 'extra heavy crude oil' (or 'extra heavy crude oil') is used to describe materials that occur in the solid or near-solid state in the deposit or reservoir and are generally incapable of free flow under ambient conditions. Whether or not such a material exists in the near-solid or solid state in the reservoir can be determined from the pour point and the reservoir temperature (Speight, 2021).

The general difference is that extra heavy crude oil, which may have properties similar to tar sand bitumen in the laboratory under ambient conditions of temperature and pressure (for example, the API gravity is on the order of 8, the sulfur content is on the order of 4.8% w/w, and the C5 asphaltene content is on the order or 17% w/w as well as other properties that are similar to the properties of tars and bitumen) but unlike tar sand bitumen in the deposit the extra heavy crude oil exhibits mobility in the reservoir or deposit (Delbianco and Montanari, 2009; Speight, 2014a, 2019, 2021). Thus, extra heavy crude oil can flow at the reservoir temperature and can be produced without additional viscosity-reduction techniques through variants of conventional processes such as long horizontal wells, or multilateral arrangement of the production wells.

This is the case, for instance, in the Orinoco basin (Venezuela) or in offshore reservoirs of the coast of Brazil but, once outside of the influence of the high reservoir temperature, these oils are too viscous at surface to be transported through conventional pipelines and require heated pipelines for transportation. Alternatively, the oil must be partially upgraded or fully upgraded or diluted with a low-boiling (low-density) hydrocarbon (such as aromatic naphtha) to create a mix that is suitable for transportation (Speight, 2014a, 2017).

Thus, extra heavy crude oil is a material that occurs as in fluid at the high temperature of the reservoir-deposit and, thus, has mobility under temperature conditions in the reservoir-deposit insofar as the temperature of the reservoir-deposit is higher than the pour point of the oil.

On the other hand, much of the extra heavy crude oil located in the Orinoco basin of Venezuela requires recovery methods that are less extreme because of the mobility of the material in the reservoir. Whether the mobility of extra heavy crude oil is due to a high reservoir temperature (that is, the reservoir temperature is higher than the pour point of the extra heavy crude oil) or subject to other factors due to localized conditions in the reservoir. This may also be reflected in the choice of suitable conversion process for extra heavy crude oil or for tar sand bitumen in the refinery.

In the context of this book, the methods outlined in this book for refining viscous feedstocks focus on heavy crude oil with an API gravity of less than 20 with a variable sulfur content (which includes atmospheric resids and vacuum resids) (Parkash, 2003; Gary et al., 2007; Speight, 2014a; Hsu and Robinson, 2006; Speight, 2017). However, it must be recognized that some of the heavy crude oil sufficiently liquid to be recovered by pumping operations and are already being recovered by this method. Refining depends on the characteristics (properties) since these heavy crude oils fall into a range of high viscosity and the viscosity is subject to temperature effects (Speight, 2001, 2014a, 2015a), which is the reason for the application of thermal conditions or dilution with a suitable solvent (such as aromatic naphtha) to enable heavy crude oil to flow in pipeline or within the refinery system.

1.7.1.4 Tar Sand Bitumen

In addition to heavy crude oil and the so-called extra heavy crude oil, there remains an even more viscous material that offers some relief to the potential shortfalls in supply (Meyer and De Witt, 1990; Speight, 2014a, 2016). This is the bituminous material that occurs found in tar sand (oil sand) deposits (referred to as oil sand deposits in Canada.

Tar sand deposits, also variously called oil sand deposits or bituminous sand deposits, vary from a loose sandstone formation to a consolidated sandstone formation or a porous carbonate rock formation that is impregnated with bitumen in the form of a viscous asphaltic oil that has an extremely high viscosity under reservoir conditions.

The term 'tar sand bitumen' (also, on occasion, incorrectly referred to as native asphalt) includes a wide variety of near-black to black materials that occur in the form of near solid to solid in the deposit either (1) with no mineral impurity or (2) with a mineral matter content that may exceed 50% w/w. Bitumen is frequently found filling pores and crevices of sandstone, limestone, or argillaceous sediments, in which case the organic and associated mineral matrix is known as rock asphalt. Even though the term tar sand is incorrectly since tar is a product of the thermal processing of coal (Speight, 2013a), the term has remined in use. Also, it is incorrect to refer to native bituminous materials as tar or pitch even though the word 'tar' is descriptive of the black, non-volatile bituminous material, it is best to avoid its use with respect to natural materials and to restrict the meaning to the volatile or near-volatile products produced in the destructive distillation of such organic substances as coal (Speight, 2013a). In the simplest sense, pitch is the distillation residue of the various types of tar.

Alternatively, bitumen recovery from tar sand deposits can be achieved by conversion of the bitumen to a product in situ (often referred to as in situ upgrading) followed by product recovery from the deposit (Speight, 2013b, 2013c, 2014a, 2016, 2019). The tar sand bitumen is a high-boiling material with little, if any, material boiling below 350°C (660°F) and the boiling range approximates the boiling range of an atmospheric residuum (Speight, 2014a, 2015a, 2019).

There have been many attempts to define tar sand deposits and the bitumen contained therein. In order to define conventional crude oil, heavy crude oil, and bitumen, the use of a single physical parameter such as viscosity is not sufficient. Other properties such as API gravity, elemental analysis, composition, and, most of all, the properties of the bulk deposit must also be included in any definition of these materials. In fact, the most appropriate definition of tar sands is found in the writings of the United States government Speight, 2014a, 2016):

> Tar sands are the several rock types that contain an extremely viscous hydrocarbon which is not recoverable in its natural state by conventional oil well production methods including currently used enhanced recovery techniques.

This definition refers to the character of the bitumen through the method of recovery. Mining methods match the requirements of this definition (since mining is not one of the specified recovery methods) and the bitumen can be recovered by alteration of its natural state such as thermal conversion to a product that is then recovered. In this sense, changing the natural state (the chemical composition) as occurs during several thermal processes (such as some in-situ combustion processes) also matches the requirements of the definition.

By inference and by omission, conventional crude oil and heavy crude oil are also included in this definition. Crude oil is the material that can be recovered by conventional oil well production methods whereas heavy crude oil is the material that can be recovered by enhanced recovery methods. Tar sand currently recovered by a mining process followed by separation of the bitumen by the hot water process. The bitumen is then used to produce hydrocarbon derivatives by a conversion process.

The only commercial operations for the recovery of bitumen from tar sand and its subsequent conversion to liquid fuels exist in the Canadian Province of Alberta where Suncor (initially as Great Canadian Oil Sands before the name change) went on-stream in 1967 and Syncrude (a consortium of several companies) went on-steam in 1977. Thus, throughout this text, frequent reference is made to tar sand bitumen, but because commercial operations have been in place for approximately 50 years it is not surprising that more is known about the Alberta (Canada) tar sand reserves than any other reserves in the world. Therefore, when discussion is made of tar sand deposits, reference is made to the relevant deposit, but when the information is not available, the Alberta material is used for the purposes of the discussion.

Refining heavy crude oil, extra heavy crude oil, tar sand bitumen, and residua depends to a large degree on the composition of the feedstock and there are wide differences in properties between, say, tar sand bitumen and conventional crude oil (Parkash, 2003; Gary et al., 2007; Speight, 2014a; Hsu and Robinson, 2006; Speight, 2017). Generally, the bitumen found in tar sand deposits is an extremely viscous material that is immobile in the deposit and cannot be refined by the application of conventional recovery methods (Speight, 2014a, 2017, 2021). On the other hand, extra heavy crude oil, which is often likened to tar sand bitumen because of similarities in the properties of the two, has a degree of mobility under reservoir or deposit conditions but typically suffers from the same drawbacks as the immobile bitumen in refinery operations.

The term 'oil sand' (the deposit does not contain material that fits the gernal description of 'oil') is also used in the same way as the term tar sand, and these terms are used interchangeably throughout the literature.

Bituminous rock and bituminous sand are those formations in which the bituminous material is found as a filling in veins and fissures in fractured rocks or impregnating relatively shallow sand, sandstone, and limestone strata. The deposits contain as much as 20% bituminous material, and if the organic material in the rock matrix is bitumen, it is usual (although chemically incorrect) to refer to the deposit as rock asphalt to distinguish it from bitumen that is relatively mineral free. A standard test method (ASTM D4) is available for determining the bitumen content of various mixtures with inorganic materials, although the use of word bitumen as applied in this test might be questioned and it might be more appropriate to use the term organic residues to include tar and pitch. If the material is the asphaltite-type or asphaltoid-type, the corresponding terms should be used: rock asphaltite or rock asphaltoid.

Bituminous rocks generally have a porous structure, with the bituminous material in the voids. A much more common situation is that in which the organic material is present as an inherent part of the rock composition insofar as it is a diagenetic residue of the organic material detritus that was deposited with the sediment. The organic components of such rocks are usually refractory and are only slightly affected by most organic solvents.

Tar sand deposits occur throughout the world and the largest deposits occur in Alberta, Canada (the Athabasca, Wabasca, Cold Lake, and Peace River areas), and in Venezuela. Smaller deposits occur in the United States, with the larger individual deposits in Utah, California, New Mexico, and

Kentucky. The term tar sand, also known as oil sand (in Canada), or bituminous sand, commonly describes sandstones or friable sand (quartz) impregnated with a viscous organic material known as bitumen (a hydrocarbonaceous material that is soluble in carbon disulfide). Significant amounts of fine material, usually largely or completely clay, are also present. The degree of porosity in tar sand varies from deposit to deposit and is an important characteristic in terms of recovery processes for the bitumen which is the desirable fraction of the tar sand and from which a variety of distillable products can be produced (Speight, 2013b, 2013c, 2014a, 2016, 2017). However, the bitumen is usually not recoverable by conventional crude oil production techniques.

The properties and composition of the tar sand deposits, and the bitumen contained therein significantly influence the selection of recovery and the bitumen conversion processes and vary among the bitumen from different deposits. In the so-called wet sands or water-wet sands of the Athabasca deposit, a layer of water surrounds the sand grain, and the bitumen partially fills the voids between the wet grains. The Utah tar sand deposits typically lack the water layer and the bitumen in is direct contact with the sand grains and, without any intervening water; such tar sands are typically sometimes referred to as oil-wet sands.

Typically, more than 99% w/w of mineral matter is composed of quartz and clay minerals – the latter are detrimental to refining operations. The typical composition of the P.R. Spring (also written as Peor Spring) tar sand Area has a porosity on the order of 8.4% v/v with the solid/liquid fraction typically being on the order of 90.5% w/w sand, 1.5% w/w fines, 7.5 % w/w bitumen, and 0.5% w/w water. The high concentration of heteroatoms (nitrogen, oxygen, sulfur, and metals) contributes to (1) an increase the viscosity of the bitumen, (2) an increase in the bonding of bitumen with the minerals, (3) a reduced yield of recovered bitumen, and (4) difficulties encountered during processing of the bitumen.

Finally, as an addition to this sub-section and for information purposes, synthetic crude oil is produced by upgrading bitumen to reduce its density and viscosity for transport by pipeline (Speight, 2014a, 2017). A disadvantage to this concept is that supplies of synthetic crude oils are limited by the availability of upgraders at the source of the extraction of the bitumen from the tar sand and that a mixture on the order of 50% v/v/ and 50% v/v synthetic crude oil is required to obtain the desired density and viscosity.

1.7.2 Refining Viscous Feedstocks

Over the past three decades, crude oils available to refineries have generally decreased in API gravity (Speight, 1986; Eccles, 1993; Speight, 2005, 2011a, 2014a). There is, nevertheless, a major focus in refineries on the ways in which heavy feedstocks (such as heavy crude oil and tar sand bitumen) can be converted into low-boiling high-value products (Khan and Patmore, 1997; Speight, 2005; Gary et al., 2007; Rana et al., 2007; Rispoli et al., 2009; Stratiev and Petkov, 2009; Stratiev et al., 2009: et al. Motaghi, 2010a, 2010b; Speight, 2011a, 2014a; Hsu and Robinson, 2017; Speight, 2017). Simultaneously, the changing crude oil properties are reflected in changes such as an increase in asphaltene constituents, an increase in sulfur, metal and nitrogen contents. Pretreatment processes for removing such constituents or at least negating their effect in thermal processes would also play an important role.

However, the limitations of processing these heavy feedstocks depend to a large extent on the amount of higher molecular weight constituents (i.e., asphaltene constituents and resin constituents) that contain most of the heteroatom-containing compounds, which are responsible for high yields of thermal and catalytic coke (Bartholdy and Andersen, 2000; Speight, 2014a). Be that as it may, the essential step required of a modern refinery is the upgrading of heavy feedstocks, particularly atmospheric and vacuum residua.

Upgrading feedstocks such as heavy crude oils and residua began with the introduction of hydrodesulfurization processes (Speight, 2000; Ancheyta and Speight, 2007). In the early days, the goal was desulfurization but, in later years, the processes were adapted to a 10 to 30% partial conversion operation, as intended to achieve desulfurization and obtain low-boiling fractions simultaneously, by increasing severity in operating conditions. However, as refineries have evolved and feedstocks have changed, refining the viscous feedstocks has become a major issue in modern refinery practice and several process configurations have evolved to accommodate the viscous feedstocks (Khan and Patmore, 1997; Speight, 2011a, 2014a, 2017, 201)

The concept of pretreating residua prior to hydrocracking to improve the quality of the products is also practiced (Speight, 2011a, 2014a, 2017). If the feedstock arises from a cracking operation (such as a cracked residuum feedstock), pretreatment of the feedstock by a hydroprocess will be accompanied by some degree of saturation resulting in increased hydrogen consumption.

Finally, there is not one single heavy feedstock upgrading solution that will fit all refineries. Market conditions, existing refinery configuration and available crude prices all can have a significant effect on the final configuration. Furthermore, a proper evaluation however is not a simple undertaking for an existing refinery. The evaluation starts with an accurate understanding of the market for the various products along with corresponding product values at various levels of supply. The next step is to select a set of crude oils that adequately cover the range of crude oils that may be expected to be processed. It is also important to consider new unit capital costs as well as incremental capital costs for revamp opportunities along with the incremental utility, support and infrastructure costs. The costs, although estimated at the start, can be better assessed once the options have been defined leading to the development of the optimal configuration for refining the incoming feedstocks.

1.8 HYDROGEN MANAGEMENT AND DISTRIBUTION

In general, considerable variation exists from one refinery to another in the balance between hydrogen produced and hydrogen consumed in the refining operations (Chapter 7). However, what is more pertinent to the present chapter is the amounts of hydrogen that are required for hydroprocessing operations, whether these be hydrocracking or the somewhat milder hydrotreating processes.

For effective hydroprocessing, a substantial hydrogen partial pressure must be maintained in the reactor and, in order to meet this requirement, an excess of hydrogen above that actually consumed by the process must be fed to the reactor. Part of the hydrogen requirement is met by recycling a stream of hydrogen-rich gas. However, the need still remains to generate hydrogen as makeup material to accommodate the process consumption of 500 to 3,000 scf/bbl depending upon whether the heavy feedstock is being subjected to a predominantly hydrotreating (hydrodesulfurization) or to a predominantly hydrocracking process.

Many existing refinery hydrogen plants use a conventional process, which produces a medium-purity (94–97% v/v) hydrogen product by removing the carbon dioxide in an absorption system and methanation of any remaining carbon oxides. Since the 1980s, most hydrogen plants are built with pressure swing adsorption (PSA) technology to recover and purify the hydrogen to purities above 99.9%. Since many refinery hydrogen plants utilize refinery off gas feeds containing hydrogen, the actual maximum hydrogen capacity that can be synthesized via steam reforming is not certain since the hydrogen content of the off gas can change due to operational changes in the hydrotreaters.

Hydrogen management has become a priority in current refinery operations and when planning to produce lower sulfur naphtha and kerosene (Zagoria et al., 2003; Davis and Patel, 2004; Méndez et al., 2008; Luckwal and Mandal, 2009). Along with increased hydrogen consumption for deeper hydrotreating, additional hydrogen is needed for processing heavier and higher sulfur crude slates. In many refineries, hydroprocessing capacity and the associated hydrogen network is limiting refinery throughput and operating margins.

Improved hydrogen utilization and expanded or new sources for refinery hydrogen and hydrogen purity optimization are now required to meet the needs of the future transportation fuel market and the drive toward higher refinery profitability. Many refineries developing hydrogen management programs fit into the two general categories of either a catalytic reformer supplied network or an on-purpose hydrogen supply.

Multiple hydrotreating units compete for hydrogen – either by selectively reducing throughput, managing intermediate tankage logistics, or running the catalytic reformer sub-optimally just to satisfy downstream hydrogen requirements. Part of the operating year still runs in hydrogen surplus, and the network may be operated with relatively low hydrogen utilization (consumption/production) at 70 to 80%. Catalytic reformer off gas hydrogen supply may swing from 75 to 85% hydrogen purity. Hydrogen purity upgrade can be achieved through some hydrotreaters by absorbing heavy hydrocarbons. But without supplemental hydrogen purification, critical control of hydrogen partial pressure in hydroprocessing reactors is difficult, which can affect catalyst life, charge rates, and/or the yield of naphtha.

More complex refineries, especially those with hydrocracking units, also have on-purpose hydrogen production, typically with a steam methane reformer steam methane reformer that utilizes refinery off gas and supplemental natural gas as feedstock. The steam methane reformer plant provides the swing hydrogen requirements at higher purities (92% to more than 99% hydrogen) and serves a hydrogen network configured with several purity and pressure levels. Multiple purities and existing purification units allow for more optimized hydroprocessing operation by controlling hydrogen partial pressure for maximum benefit. Typical hydrogen utilization is 85% to 95%.

Over the past four decades, the refining industry has been challenged by changing feedstocks and product slate. In the near future, hydroprocessing options (especially for heavy feedstocks) will become increasingly flexible with improved technologies and improved catalysts. The increasing focus to reduce sulfur content in fuels will assure that the role of *desulfurization* in the refinery increases in importance (Babich and Moulijn, 2003). Currently, the process of choice is the hydrotreater, in which hydrogen is added to the fuel to remove the sulfur from the fuel.

As a result of the increased attention for fuel desulfurization (and the need for other low-to-no sulfur products) various new process-concepts are being developed with various claims of efficiency and effectiveness. The major developments in desulfurization three main routes will be (1) advanced hydrotreating (new catalysts, catalytic distillation, processing at mild conditions), (2) reactive adsorption (type of adsorbent used, process design), and (3) oxidative desulfurization (catalyst, process design).

In addition to an increase in the use of 'guard beds' the industry will see an increase in automated demetallization of fixed-bed systems as well as more units that operate as ebullating-bed hydrocrackers (Speight, 2011a, 2014a, 2017).

It is becoming apparent that in the refinery of the future, hydrotreating technology and hydrocracking technology will be the processes of choice for upgrading viscous feedstocks. For cleaner transportation fuel production, the main task is the desulfurization of the naphtha and kerosene fractions.

In fact, hydrocracking will continue to be an indispensable processing technology to modern crude oil refining and petrochemical industry due to its flexibility to feedstocks and product scheme, and high-quality products. Particularly, high quality naphtha, kerosene, and the base oils for the manufacture of lubricating oil can be produced through this technology. The hydrocracker provides a better balance of naphtha distillates which serves to improve the octane quality of the final gasoline product and can supplement the fluid catalytic cracker to upgrade heavy feedstocks. In the hydrocracker, light fuel oil is converted into lighter products under a high hydrogen pressure and over a hot catalyst bed – the main products are naphtha, jet fuel, and diesel fuel.

In summary, the modern refinery faces the challenge of meeting an increasing demand for low-sulfur products as specifications continue to tighten around the world and markets decline for high-sulfur fuel oil. Innovative ideas and solutions to reduce refinery costs must always be considered, including: (1) optimization of the hydrogen management network, (2) multiple feedstock options for hydrogen production, (3) optimization of plant capacity, and last but certainly not least (4) use of hydrogen recovery technologies to maximize hydrogen availability and minimize capital investment.

REFERENCES

Ancheyta, J. 2007. Reactors for Hydroprocessing. In *Hydroprocessing of Heavy Oils and Residua*. J. Ancheyta and J.G. Speight (Editors). CRC Press, Taylor & Francis Group, Boca Raton, FL.

Ancheyta, J., and Speight, J.G. 2007. *Hydroprocessing of Heavy Oils and Residua*. CRC Press, Taylor & Francis Group, Boca Raton, FL.

Asadullah, M., Ito, S.I., Kunimori, K., Yamada, M., and Tomishige, K. 2002. Energy Efficient Production of Hydrogen and Syngas from Biomass: Development of Low-Temperature Catalytic Process for Cellulose Gasification. *Environmental Science and Technology*, 36(20): 4476–4481.

ASTM D4. 2022. *Standard Test Method for Bitumen Content. Annual Book of Standards*. ASTM International, West Conshohocken, PA.

Ayers, K., Danilovic, N., Ouimet, R., Carmo, M., Pivovar, B., and Bornstein, M. 2019. Perspectives on Low-Temperature Electrolysis and Potential for Renewable Hydrogen at Scale. *Annual Review of Chemical and Biomolecular Engineering*, 10: 219–239.

Babich, I.V., and Moulijn, J.A. 2003. Science and Technology of Novel Processes for Deep Desulfurization of Oil Refinery Streams: A Review. *Fuel*, 82(6): 607–631.

Bartholdy, J., and Andersen, S.I. 2000. Changes in Asphaltene Stability During Hydrotreating. *Energy and Fuels*, 14(1): 52–55.

Bandermann, F., and Harder, K.B. 1982. Production of Hydrogen via Thermal Decomposition of Hydrogen Sulfide and. Separation of Hydrogen and Hydrogen Sulfide by Pressure Swing Adsorption. *International Journal of Hydrogen Energy*, 7(6): 471–475.

Bezler, J. 2003. Optimized Hydrogen Production – A Key Process Becoming Increasingly Important in Refineries. Proceedings. DGMK Conference on Innovation in the Manufacture and Use of Hydrogen. Dresden. October 15–17. Page 65.

Bhattacharyya, A., and Mezza, B.J. 2010. Catalyst Composition with Nanometer Crystallites for Slurry Hydrocracking. United States Patent 7,820,135. October 26.

Bhattacharyya, A., Bricker, M.L., Mezza, B.J., and Bauer, L.J. 2011. Process for Using Iron Oxide and Alumina Catalyst with Large Particle Diameter for Slurry Hydrocracking. United States Patent. 8,062,505. November 22.

Cao, L., Yu, I.K., Xiong, X., Tsang, D.C., Zhang, S., Clark, J.H., Hu, C., Ng, Y.H., Shang, J., and Ok, Y.S. 2020. Biorenewable Hydrogen Production Through Biomass Gasification: A Review and Future Prospects. *Environmental Research*, 186: 109547. https://www.sciencedirect.com/science/article/abs/pii/S0013935120304400#:~:text=Gasification%20is%20the%20thermochemical%20conversion,hydrogen%20and%20second%2Dgeneration%20biofuels.

Chadeesingh, R. 2011. The Fischer-Tropsch Process. In *The Biofuels Handbook*. J.G. Speight (Editor). Royal Society of Chemistry, London. Part 3, Chapter 5.

Conte, M., Iacobazzi, A., Ronchetti, M., And Vellone, R. 2001. Hydrogen Economy for a Sustainable Development: State-of-The-Art and Technological Perspectives. *Journal of Power Sources*, 100(1): 171–187.

Cruz, F.E., and De Oliveira Junior, S. 2008. Petroleum Refinery Hydrogen Production Unit: Exergy and Production Cost Evaluation. *International Journal of Thermodynamics*, 11(4): 187–193.

Davis, R.A., and Patel, N.M. 2004. Refinery Hydrogen Management. *Petroleum Technology Quarterly*, Spring: 28–35.

Demirbaş, A. 2006. Hydrogen from Various Biomass Species via Pyrolysis and Steam Gasification Processes. *Energy Sources*, 28(3): 245–252.

Demirbaş, A., and Arin, G. 2004. Hydrogen from Biomass via Pyrolysis: Relationships Between Yield of Hydrogen and Temperature. *Energy Sources*, 26(11): 1061–1069.

Dolbear, G.E. 1998. *Petroleum Chemistry and Refining*. J.G. Speight (Editor). Taylor & Francis, Washington, DC. Chapter 7.

Du, Z., Liu, C., Zhai, J., Guo, X., Xiong, Y., Su, W., and He, G. 2021. A Review of Hydrogen Purification Technologies for Fuel Cell Vehicles. *Catalysts*, 11(3): 393. https://www.mdpi.com/2073-4344/11/3/393.

Eccles, R.M. 1993. Residue Hydroprocessing Using Ebullated-Bed Reactors. *Fuel Processing Technology*, 35(1–2): 21–38.

Farrauto, R., Hwang, S., Shore, L., Ruettinger, W., Lampert, J., Giroux, T., Liu, Y., and Ilinich, O. 2003. New Material Needs for Hydrocarbon Fuel Processing: Generating Hydrogen for the PEM Fuel Cell. *Annual Review of Materials Research*, 33(1): 1–27.

Gary, J.H., Handwerk, G.E., and Kaiser, M.J. 2007. *Petroleum Refining: Technology and Economics*, 5th Edition. CRC Press, Taylor & Francis Group, Boca Raton, FL.

Gonzalez-Garay, A., Bui, M., Freire Ordóñez, D.F., High, M., Oxley, A., Moustafa, N., Sáenz Cavazos, P.A., Patrizio, P., Sunny, N., MacDowell, N., and Shah, N. 2022. Hydrogen Production and Its Applications to Mobility. *Annual Review of Chemical and Biomolecular Engineering*, 13: 501–528.

Hsu, C.S., and Robinson, P.R. (Editors) 2017. *Handbook of Petroleum Technology*. Springer AG, Cham.

Huycke, R., and Zagoria, A. 2003. Refinery Hydrogen Management – The Big Picture. Preprints. DGMK-Conference. Innovation in the Manufacture and Use of Hydrogen. Deutsche Wissenschaftliche Gesellschaft für Erdöl, Erdgas und Kohle eV. German Society for Petroleum and Coal Science and Technology, Dresden, October 15–17.

Khan, M.R., and Patmore, D.J. 1997. Heavy Oil Upgrading Processes. In *Petroleum Chemistry and Refining*. J.G. Speight (Editor). Taylor & Francis, Washington, DC. Chapter 6.

Kriz, J.F., and Ternan, M. 1994. Hydrocracking of Heavy Asphaltenic Oil in the Presence of an Additive to Prevent Coke Formation. United States Patent 5,296,130. March 22.

Kumagai, H., Koyama, H., Nakamura, K., Igarashi, N., Mori, M., and Tsukada, T. 2005. Hydrofining Catalyst and Hydrofining Process. United States Patent. 6,858,132. February 22.

Lee, S., Speight, J.G., and Loyalka, S. 2014a. *Handbook of Alternative Fuel Technologies*, 2nd Edition. CRC Press, Taylor & Francis Group, Boca Raton, FL.

Lipman, T. 2011. An Overview of Hydrogen Production and Storage Systems with Renewable Hydrogen Case Studies. A Clean Energy States Alliance Report. Conducted under US DOE Grant DE-FC3608GO18111. Office of Energy Efficiency and Renewable Energy Fuel Cell Technologies Program, United States Department of Energy, Washington, DC.

Long, R., Picioccio, K., and Zagoria, A. 2011. Optimizing Hydrogen Production and Use. *Petroleum Technology Quarterly*, Autumn: 1–12.

Luckwal, K., and Mandal, K.K. 2009. Improve Hydrogen Management of Your Refinery. *Hydrocarbon Processing*, 88(2): 55–61.

Méndez, C.A., Gómez, E., Sarabia, D., Cerdá, J., De Prada, C., Sola, J.M., and Unzueta, E. 2008. Optimal Management of Hydrogen Supply and Consumption Networks of Refinery Operations. Proceedings. 18th European Symposium on Computer Aided Process Engineering – ESCAPE 18. B. Braunschweig and X. Joulia (Editors). Elsevier, Amsterdam, The Netherlands.

Meyer, R.F., and De Witt, W. Jr. 1990. Definition and World Resources of Natural Bitumens. Bulletin No. 1944. US Geological Survey, Reston, VA.

Meyer, R.F., and Attanasi, E.D. 2003. Heavy Oil and Natural Bitumen: Strategic Petroleum Resources. Fact Sheet No. FS0070-03, 2p. Publisher. United States Geological Survey, Reston, VA.

Miller, G.Q., and Penner, D.W. 2003. Meeting Future Needs for Hydrogen – Possibilities and Challenges. Proceedings. DGMK Conference on Innovation in the Manufacture and Use of Hydrogen. Dresden. October 15–17. Page 7.

Mokhatab, S., Poe, W.A., and Speight, J.G. *Handbook of Natural Gas Transmission and Processing*. Elsevier, Amsterdam.

Motaghi, M., Shree, K., and Krishnamurthy, S. 2010a. Consider New Methods for Bottom of the Barrel Processing – Part 1. *Hydrocarbon Processing*, 89(2): 35–40.

Motaghi, M., Shree, K., and Krishnamurthy, S. 2010b. Consider New Methods for Bottom of the Barrel Processing – Part 2. *Hydrocarbon Processing*, 89(2): 55–88.

Muradov, N.Z. 2003. Emission-Free Fuel Reformers for Mobile and Portable Fuel Cell Applications. *Journal of Power Sources*, 118(1–2): 320–324.

Ni, M., Leung, D.Y.C., Leung, M.K.H., and Sumathy, K. 2006. An Overview of Hydrogen Production from Biomass. *Fuel Processing Technology*, 87(5): 461–472.

Niu, M., Ji, P., Fan, Z., Yan, Y., Li, D., and Li, W. 2020. Hydrofining Process of Coal Tar Based on Four Kinds of Catalyst Grading. *Energy and Fuels*, 34(5): 6510–6517.

Parkash, S. 2003. *Refining Processes Handbook*. Gulf Professional Publishing, Elsevier, Amsterdam.

Preuster, P., Alekseev, A., and Wasserscheid, P. 2017. Hydrogen Storage Technologies for Future Energy Systems. *Annual Review of Chemical and Biomolecular Engineering*, 8: 445–471.

Raissi, A.T. 2001. Technoeconomic Analysis of Area II Hydrogen Production. Part 1. Proceedings. US DOE Hydrogen Program Review Meeting. Baltimore, MD.

Ramzan, N., Ashraf, A., Naveed, S., and Al Malik, A. 2011. Simulation of Hybrid Biomass Gasification Using Aspen Plus: A Comparative Performance Analysis for Food, Municipal Solid and Poultry Waste. *Biomass and Bioenergy*, 35(9): 3962–3969.

Rana, M.S., Sámano, V., Ancheyta, J., and Diaz, J.A.I. 2007. A Review of Recent Advances on Process Technologies for Upgrading of Heavy Oils and Residua. *Fuel*, 86(9): 1216–1231.

Ranke, H., and Schödel, N. 2003. Hydrogen Production Technology – Status and New Developments. Proceedings. DGMK Conference on Innovation in the Manufacture and Use of Hydrogen. Dresden. October 15–17. Page 19.

Reynolds, J.G., and Beret, S. 1989. Effect of Prehydrogenation on Hydroconversion of Maya Residuum. *Fuel Science and Technology International*, 7(2): 165–186.

Rispoli, G., Sanfilippo, D., and Amoroso, A. 2009. Advanced Hydrocracking Technology Upgrades Extra Heavy Oil. *Hydrocarbon Processing*, 88(12): 39–46.

Song, C. 2002. New Approaches to Deep Desulfurization for Ultra-Clean Gasoline and Diesel Fuels: An Overview. Preprints. Division of Fuel Chemistry. *Journal of the American Chemical Society*, 47(2): 438–444.

Speight, J.G., and Moschopedis, S.E. 1979. The Production of Low-Sulfur Liquids and Coke from Athabasca Bitumen. *Fuel Processing Technology*, 2(4): 295.

Speight, J.G. 1986. Upgrading Heavy Feedstocks. *Annual Review of Energy*, 11(1): 253.

Speight, J.G. 2000. *The Desulfurization of Heavy Oils and Residua*, 2nd Edition. Marcel Dekker Inc., New York.

Speight, J.G. 2001. *Handbook of Petroleum Analysis*. John Wiley & Sons Inc., Hoboken, NJ.

Speight, J.G. 2005. Natural Bitumen (Tar Sands) and Heavy Oil. In *Coal, Oil Shale, Natural Bitumen, Heavy Oil and Peat, from Encyclopedia of Life Support Systems (EOLSS), Developed Under the Auspices of the UNESCO*. EOLSS Publishers, Oxford. http://www.eolss.net.

Speight, J.G. 2008. *Synthetic Fuels Handbook: Properties, Processes, and Performance*. McGraw-Hill, New York.

Speight, J.G. 2011a. *The Refinery of the Future*. Gulf Professional Publishing, Elsevier, Oxford.

Speight, J.G. (Editor) 2011b. *The Biofuels Handbook*. Royal Society of Chemistry, London.

Speight, J.G. 2013a. *The Chemistry and Technology of Coal*, 3rd Edition. CRC Press, Taylor & Francis Group, Boca Raton, FL.

Speight, J.G. 2013b. *Heavy Oil Production Processes*. Gulf Professional Publishing Company, Elsevier, Oxford.

Speight, J.G. 2013c. *Heavy and Extra Heavy Oil Upgrading Technologies*. Gulf Professional Publishing Company, Elsevier, Oxford.

Speight, J.G. 2014a. *The Chemistry and Technology of Petroleum*, 5th Edition. CRC Press, Taylor & Francis Group, Boca Raton, FL.

Speight, J.G. 2014b. *Gasification of Unconventional Feedstocks*. Gulf Professional Publishing Company, Elsevier, Oxford.

Speight, J.G. 2015a. *Handbook of Petroleum Product Analysis*, 2nd Edition. John Wiley & Sons Inc., Hoboken, NJ.

Speight, J.G. 2015b. *Fouling in Refineries*. Gulf Professional Publishing Company, Elsevier, Oxford.

Speight, J.G. 2016. *Asphalt Materials Science and Technology*. Butterworth-Heinemann, Elsevier, Kidlington, Oxford and Waltham, MA.

Speight, J.G. 2017. *Handbook of Petroleum Refining*. CRC Press, Taylor & Francis Group, Boca Raton, FL.

Speight, J.G. 2019. *Heavy Oil Recovery and Upgrading*. Gulf Publishing Company, Elsevier, Cambridge, MA.

Speight, J.G. 2021. *Petroleum Refining Technology Series Volume 1: Refinery Feedstocks*. CRC Press, Taylor & Francis Group, Boca Raton, FL.

Speight, J.G. 2023. *Petroleum Refining Technology Series Volume 4: Hydrotreating and Hydrocracking in Petroleum Refining*. CRC Press, Taylor & Francis Group, Boca Raton, FL.

Speight, J.G. 2024. *Petroleum Refining Technology Series Volume 6: Product Improvement Processes*. CRC Press, Taylor & Francis Group, Boca Raton, FL.

Stratiev, D., and Petkov, K. 2009. Residue Upgrading: Challenges and Perspectives. *Hydrocarbon Processing*, 88(9): 93–96.

Stratiev, D., Tzingov, T., Shishkova, I., and Dermatova, P. 2009. Hydrotreating Units Chemical Hydrogen Consumption Analysis: A Tool for Improving Refinery Hydrogen Management. Proceedings. 44th International Petroleum Conference. Bratislava, Slovak Republic. September 21–22.

Tanaka, H. 2004. Process for Producing Hydrofining Catalyst. United States Patent 6,689,712. February 10.

Tasleem, S., and Tahir, M. 2020. Current Trends in Strategies to Improve Photocatalytic Performance of Perovskites Materials for Solar to Hydrogen Production. *Renewable and Sustainable Energy Reviews*, 132: 110073. https://www.sciencedirect.com/science/article/abs/pii/S1364032120303646.

Weber, G., Fu, Q., and Wu, H. 2006. Energy Efficiency of an Integrated Process Based on Gasification for Hydrogen Production from Biomass. *Developments in Chemical Engineering and Mineral Processing*, 14(1–2): 33–49.

Xia, G., Zhu, M., Min, E., Shi, Y., Li, M., Nie, H., Tao, Z., Huang, H., Zhang, R., Li, J., Wang, Z., and Ran, G. 2000. Catalyst Containing Molybdenum and/or Tungsten for Hydrotreating Light Oil Distillates and Preparation Method Thereof. United States Patent 6,037,306. March 14.

Zagoria, A., Huychke, R., and Boulter, P.H. 2003. Refinery Hydrogen Management – The Big Picture. Proceedings. DGMK Conference on Innovation in the Manufacture and Use of Hydrogen. Dresden. October 15–17. Page 95.

2 Hydrogen for Upgrading Viscous Feedstocks

2.1 INTRODUCTION

The demand for crude oil and crude oil products has shown a sharp growth in recent decades (Parkash, 2003; Gary et al., 2007; Speight, 2014a; Hsu and Robinson, 2017; Speight, 2017) and the demand for specialty products, transportation fuels, and fuel oil is forecast to continue to show a steady growth in the future. The simplest means to cover the demand growth in low-boiling products is (1) to increase the imports of the conventional (light, high density) crude oils and (2) to increase the yields of low-boiling crude oil products, but these steps may be limited as the 21st century evolves. In order to accomplish the demand for crude oil products, there is, in addition to using hydrogen for product improvement, also the need for hydrogen for upgrading the viscous feedstocks which (by definition in this text) are feedstocks such as heavy crude oil, extra heavy crude oil, tar sand bitumen, and distillation residue or, for that matter, the viscous residua that are produced during a variety of processing operations (Parkash, 2003; Gary et al., 2007; Speight, 2014a; Hsu and Robinson, 2017; Speight, 2017).

Over the past five decades, the feedstocks available to refineries have shown an overall decrease in API gravity, and as a result, there has been an increased focus on the use of hydrogen to convert the lower-API gravity feedstocks to high-demand, low-boiling, high-value products (Khan and Patmore, 1997; Speight, 2014a, 2017, 2020). Simultaneously, the changing properties of the feedstocks currently accepted by refineries are reflected in several differences, of which the most notable are (1) an increase in asphaltene constituents, (2) an increase in sulfur content, (3) an increase in nitrogen content, and (4) an increase in the metals content which result in diminished yields of distillate products. Pretreatment processes for removing such constituents or at least negating their effect in thermal processes (particularly in catalytic processes) also play an important role.

The newer refinery feedstocks are more commonly referred to, but not always correctly, as 'heavy oils' (or 'heavy crude oils') are referred to in this text as 'viscous feedstocks' (Table 2.1) and which, when subject to refinery processing has been (and still is) referred to as 'resid upgrading' or 'bottom of the barrel processing'. This latter term (i.e., 'bottom of the barrel processing') arose because the phrase referred to processing a non-volatile residuum (typically, but not always, from the distillation section of the refinery) and offered a means by which refiners can lower the feedstock cost by using a viscous feedstock as part of a crude oil blend or to produce more distillate products (Chapter 1).

Upgrading the viscous feedstocks consists of applying a wide range of technologies for separating, upgrading, and converting the feedstocks either directly or using the viscous feedstock as part of a blend with a less viscous feedstock to produce the desired products. However, the production of saleable refinery products is not immediate, and it is at this point that the viscous feedstock (or a blend thereof) requires hydrogen to ensure that the products are those desired (dictated) by the marketplace.

More specifically, the viscous feedstocks are refinery feedstocks that have a lower atomic ratio of hydrogen to carbon (typically on the order of <1.2 compared to >1.4 for the more conventional crude oils) and an API gravity of the order of <20° API gravity and a high viscosity (which indicates a resistance to easy flow), as well as the presence of relatively high amounts of contaminants such as sulfur, nitrogen, oxygen, and metals. Moreover, the viscous feedstocks, while being less expensive to purchase, are more difficult to refine because of the more specialized equipment that is required

DOI: 10.1201/9781003186694-2

TABLE 2.1
Simplified Distinction between Conventional Crude Oil, Heavy Oil, Extra Heavy Crude Oil, and Tar Sand Bitumen

Conventional Crude Oil
Properties: API gravity: >20°; sulfur: <2.0% w/w; metals: <200 ppm
Mobile in the reservoir
High-permeability reservoir
Primary recovery; secondary recovery
Heavy Crude Oil
Properties: API gravity: 10–20°; sulfur: 2.0–4.0% w/w; metals: 200–500 ppm
More viscous than conventional crude oil
Mobile in the reservoir
High-permeability reservoir
Secondary recovery; tertiary recovery (EOR, e.g., steam stimulation)
Extra Heavy Crude Oil
Properties: API gravity: <10°; sulfur: 2.0–5.0% w/w; metals: 400–1,200 ppm
Similar properties to the properties of tar sand bitumen
Mobile in the reservoir/deposit
Reservoir/deposit temperature higher than the pour point of oil
High-permeability reservoir
Secondary recovery; tertiary recovery (EOR, e.g., steam stimulation)
Tar Sand Bitumen
Properties: API gravity: <10°; sulfur: 2.0–5.0% w/w; metals: 400–1,200 ppm
Immobile in the deposit; the deposit lower than the pour point of the bitumen
High-permeability reservoir
Mining (often preceded by explosive fracturing)
Steamassisted gravity drainage (SAGD); solvent methods (VAPEX)
Recovery: Extreme heating methods; innovative methods*
Residua
Properties: API gravity: 5–15°; sulfur: 2.0–5.0% w/w; metals: 100–1,000 ppm
Do not occur naturally; manufactured products produced by refinery distillation
Non-volatile under vacuum distillation conditions
Typically immobile under ambient conditions

*Innovative methods exclude tertiary recovery methods and methods such as SAGD and VAPEX, but does include variants or hybrids thereof.

(such as catalytic cracking and the advanced coking processes) and produce a different mix of refined products than those derived from conventional crude oil (Parkash, 2003; Gary et al., 2007; Speight, 2014a; Hsu and Robinson, 2017; Speight, 2017).

However, in the current market, the price differential between light sweet crude oil and the viscous (high sulfur) feedstocks creates a strong incentive for refiners to lower the feedstock cost by using the viscous feedstock in a blended feedstock. Thus, more commonly, the upgrading process for the viscous feedstocks is really interpreted as management of the atomic hydrogen to carbon ratio. There are two major pathways to manage or change the hydrogen to carbon ratio (i.e., atomic H/C). The first upgrading pathway, variously referred to as 'the carbon rejection pathway' by rejecting carbon from the feedstocks by means of, for example, a thermal coking process produces a carbon-rich by-product, nominally referred to as 'coke' (Parkash, 2003; Gary et al., 2007; Speight, 2014a; Hsu and Robinson, 2017; Speight, 2017).

Difficult-to-refine feedstocks, such as viscous feedstocks are characterized by low API gravity (high density) and high viscosity, high initial boiling point, high carbon residue, high nitrogen content, high sulfur content, and high metals content (Speight, 2007, 2013). In addition to these properties, the viscous feedstocks also have an increased molecular weight and reduced hydrogen content with a relatively low content of volatile saturated and aromatic constituents and a relatively high content of asphaltene and resin constituents that is accompanied by a high heteroatom (nitrogen, oxygen, sulfur, and metals) content. These types of feedstocks (i.e., the viscous feedstocks) are not typically subject to distillation unless contained in the refinery feedstock as a blend with less viscous crude oils.

The limitations of processing the viscous feedstocks depend to a large extent on the tendency for coke formation and the deposition of metals and coke on the catalyst due to the higher molecular weight (low volatility) and heteroatom content. However, the essential step required of refineries is the upgrading of viscous feedstocks. In fact, the increasing supply of heavy crude oils is a matter of serious concern for the crude oil industry. The most efficient and economical solution to this problem will depend to a large extent on individual country and company situations.

Upgrading of viscous feedstock began with the introduction of desulfurization processes (Speight, 2000; Parkash, 2003; Gary et al., 2007; Ancheyta and Speight, 2014a; Hsu and Robinson, 2017; Speight, 2017). In the early days, the goal was desulfurization but, in later years, the processes were adapted to a 10 to 30% partial conversion operation, which was also intended to achieve desulfurization and obtain low-boiling fractions simultaneously, by increasing severity in operating conditions. Refinery evolution has seen the introduction of a variety of cracking processes based on thermal cracking, catalytic cracking, and hydroconversion. Thus, refining viscous feedstocks has become a major issue in modern refinery practice and several process configurations have evolved to accommodate the viscous feedstocks (RAROP, 1991; Khan and Patmore, 1997; Hsu and Robinson, 2006; Gary et al., 2007; Speight, 2007). In fact, process options that are available for upgrading viscous feedstocks can be broadly divided into (1) carbon rejection processes and (2) hydrogen addition processes.

Carbon rejection redistributes hydrogen among the various components, resulting in fractions with increased H/C atomic ratios and fractions with lower H/C atomic ratios. On the other hand, hydrogen addition processes involve reaction heavy crude oils with an external source of hydrogen and result in an overall increase in H/C ratio (Stanislaus and Cooper, 1994; Speight, 2020, 2014a).

The criteria to select one of the routes as an upgrading option depend on several factors which must be analyzed in detail when it comes to considering a project of this nature. For example, the processes termed 'hydrogen addition processes' produce a high yield of products with a commercial value larger than that of the 'carbon rejection processes' but require a larger investment and more natural gas availability to produce the amounts of hydrogen and steam required for these processes.

New processes for the conversion of heavy crude oil and other viscous feedstocks will probably be used in concert with visbreaking, with some degree of hydroprocessing as a primary conversion step. Other processes may replace or augment the deasphalting units in many refineries. Depending on the properties, an option for heavy crude oil, like the early option for tar sand bitumen, is to subject the feedstock to either delayed coking or fluid coking or visbreaking as the 'primary upgrading' step that also includes prior distillation or 'topping' (Speight, 2014a, 2009, 2020). After primary upgrading, the product streams are hydrotreated and combined to form a synthetic crude oil that is shipped to a conventional refinery for further processing into liquid fuels.

However, there is not one single upgrading process for the viscous feedstocks solution that will satisfy the needs of all refineries. Viscous feedstock properties, existing refinery configuration, and desired product slate all can have a significant effect on the final configuration. Furthermore, a proper evaluation, however, is not a simple undertaking for an existing refinery. The evaluation starts with an accurate understanding of the nature of the feedstock along with corresponding conversion chemistry need to be assessed. Once the options have been defined, the development of the optimal configuration for refining the incoming feedstocks can be designed.

There is also recognition that partial or full upgrading as part of the recovery process could be a very beneficial process for leaving the unwanted elements in the reservoir and increasing API gravity (Speight, 2014a, 2017, 2019). While thiis s important in the context of this text insofar as upgrading during recovery will reduce surface refining costs, it is described in more detail elsewhere and will not be repeated here. However, the salient facts of the concept are worthy of note.

The fluid catalytic cracking (RFCC) applied to viscous feedstocks such as residua has been one of the most attractive ways to economically upgrade viscous feedstocks into saleable products such as solvents, transportation fuels, missile distillates, fuel oil, and the various gas oil fractions (which are often used as feedstocks for catalytic cracking processes). There are alternative processing schemes that are applied according to the properties of the feedstock (Phillips and Liu, 2002; Kressmann et al., 2000; Kressmann et al., 2004; Ross et al., 2005; Gary et al., 2007; Speight, 2014a; Stratiev and Petkov, 2009; Bridjanian and Khadem Samimi, 2011; Speight, 2020; Hsu and Robinson, 2017; Speight, 2017).

The limitations of processing these viscous feedstocks depend to a large extent on the tendency for coke formation and the deposition of metals and coke on the catalyst due to the higher molecular weight (low volatility) and heteroatom content. However, the essential step required of refineries is the upgrading of viscous feedstocks. In fact, the increasing supply of heavy crude oils is a matter of serious concern for the crude oil industry.

Upgrading of viscous feedstock began with the introduction of desulfurization processes (Speight, 2000; Hsu and Robinson, 2006; Speight, 2007; Ancheyta and Speight, 2007). In the early days, the goal was desulfurization but, in later years, the processes were adapted to a 10 to 30% partial conversion operation, intended to achieve desulfurization and obtain low-boiling fractions simultaneously, by increasing severity in operating conditions. Refinery evolution has seen the introduction of a variety of cracking processes based on thermal cracking, catalytic cracking, and hydroconversion which are different from one another in the cracking method, cracked product patterns and product properties, and will be employed in refineries according to their respective features. Thus, refining viscous feedstocks has become a major issue in modern refinery practice and several process configurations have evolved to accommodate the viscous feedstocks (Table 2.2) (RAROP, 1991; Khan and Patmore, 1997; Gary et al., 2007; Speight, 2013c, 2014a; Hsu and Robinson, 2017; Speight, 2017).

The criteria used to select one of the routes as an upgrading option for the viscous feedstocks depend on several factors which must be analyzed in detail when it comes to considering a project of this nature. For example, the technologies known as 'hydrogen addition' produce a high yield of

TABLE 2.2
Selected Examples of Recent Process Configurations for Refining Viscous Feedstocks

Thermal Processes	**Comment**
ASCOT process	Delayed coking with deep solvent deasphalting
CHERRY-P process	Feedstock-coal slurry
ET-II process	Feedstock mixed with high-boiling recycle oil
Catalytic Cracking Processes	
ART process	Efficient feedstock-catalyst contact
HOC process	Feedstock: desulfurized residuum
HOT process	Hydrogen produced in the unit by steam-hydrogen reaction
Solvent Processes	
DEMEX process	Less selective solvent than propane
MDS process	Solvent deasphalting and/or desulfurization of feedstock
ROSE process	Deasphaltening of feedstock

products with a commercial value larger than that of the 'carbon rejection' technologies but require the ready availability of natural gas to produce the amounts of hydrogen and steam required for these processes.

New processes for the conversion of the viscous feedstocks will probably be used in concert with visbreaking, with some degree of hydroprocessing as a primary conversion step. Other processes may replace or augment the deasphalting units in many refineries. Depending on the properties, an option for the viscous feedstocks, like the early option for tar sand bitumen, is to subject the feedstock to either delayed coking or fluid coking as the primary upgrading step, with some prior distillation or topping of the feedstock in which volatile constituents are removed (Speight, 2014a, 2017). After primary upgrading, the product streams are hydrotreated and combined to form a 'synthetic crude oil' that is shipped to a conventional refinery for further processing into liquid fuels.

Briefly, and using the processing of tar sand bitumen as an example, a synthetic crude oil (SCO) is a light sweet crude produced by upgrading of tar sand bitumen (Speight, 2013c, 2014a, 2017). The term 'synthetic' distinguishes upgraded bitumen from conventional crude insofar as the synthetic crude oil may contain constituents that are not naturally occurring and are produced by processing.

However, there is not one single upgrading solution for the viscous feedstocks that will fit all refineries. Viscous feedstock properties, existing refinery configuration, and desired product slate all can have a significant effect on the final configuration. Furthermore, a proper evaluation, however, is not a simple undertaking for an existing refinery. The evaluation starts with an accurate understanding of the nature of the feedstock along with corresponding conversion chemistry to be assessed. Once the options have been defined, the development of the optimal configuration for refining the incoming feedstocks can be designed.

There is also recognition that in situ upgrading could be a very beneficial process for leaving the unwanted elements in the reservoir and increasing API gravity. While thiis s important in the context of this text insofar as upgrading during recovery will reduce surface refining costs, it is described in more detail elsewhere and will not be repeated here. However, the salient facts of the concept are worthy of note.

There are two ways that are currently practiced in bringing heavy crude oil to market. The first method is to upgrade the material in the oil field and leave much of the material behind as coke, and then pipeline the upgraded material out as synthetic crude. In this method, the crude is fractionated, and the residue is sent to a coking unit. The products of the coking operation, and in some cases some of the residue, are hydrotreated, which are recombined with the fractionated light materials to form synthetic crude oil that is then transported to market in a pipeline.

The second method is to effect partial upgrading in situ as part of the recovery process. Such an option, to produce an acceptable pipeline material, would be an ideal solution but has a number of limitations (Motaghi et al., 2010a, 2010b, 2010c; Speight, 2013c). For example, the amount of heavy crude oil production (or extra heavy crude oil production) could be limited by the recovery process, and the upgraded products must be compatible with the original or partially changed heavy crude oil. If the products and the original (partially changed) viscous feedstock have limited compatibility, which would limit the amount of dilution and again could limit the effectiveness of the refining process (Speight, 2013c, 2014a, 2017).

However, the increased mobilization of heavy crude oil in the reservoir by partial upgrading is not a new idea and still has many hurdles to overcome before it can be considered close to commercial. The product will be less viscous than the heavy crude oil in place, but some property changes such as high olefin content from cracking are not necessarily positive.

Moreover, the hydrogen requirement for product improvement, which is the hydrotreatment of crude oil products, to ensure that they meet utility and performance specifications, and for the treatment of the viscous feedstocks, is also increasing. Product improvement and refining the viscous feedstocks can not only involve hydrotreatment but also cause changes in the molecular shape of the feedstock constituents or cause changes in molecular size of the constituents. Hence, there is an additional need for hydrogen not only in the operations of the current refineries but also in the refinery of the future (Speight, 2014a, 2017, 2020).

In summary, there are three main approaches for heating the reservoir: (1) steam distillation, (2) mild thermal cracking – visbreaking, and (3) partial combustion (Speight, 2014a, 2017).

Nevertheless, there is (or will be) an obvious future need for partial upgrading during or immediately after recovery. On the other hand, hydrogen addition must be used during upgrading in order to stabilize the upgraded heavy crude oil – which could mean that the cost of partial upgrading is not much reduced as compared to full upgrading. Therefore, the only choice currently is no upgrading or full upgrading. Other goals could be to achieve breakthroughs in upgrading technologies – such as low-energy coking processes that would use far less energy or such as gasification at 800°C (1470°F) which is far less than current commercial temperatures. The technology where changes do occur involves combustion of the feedstock in situ prior to or during recovery operations. The concept of any combustion technology requires that the oil be partially combusted and that thermal decomposition occurs to other parts of the oil. This is sufficient to cause irreversible chemical and physical changes to the oil to the extent that the product is markedly different from the oil in place. Recognition of this phenomenon is essential before combustion technologies are applied to oil recovery.

In short, in situ recovery processes (although less efficient in terms of bitumen recovery relative to mining operations) may have the added benefit of *leaving* some of the more obnoxious constituents (from the processing objective) in the ground. Processes that offer the potential for partial upgrading during recovery are varied but usually follow a surface process. Not that this should be construed as an easy task, there are many disadvantages that arise from attempting in situ upgrading.

Finally, there is not a single in situ recovery process that will be applicable to all reservoirs, and no single recovery process will be able to access all the viscous material in the reservoir (or deposit). To achieve maximum recovery, it will be necessary to apply a combination of different processes. For example, a steam-based recovery process followed by in-situ combustion, followed by in-situ upgrading, followed by bioconversion of the residual hydrocarbons. This type of sequential recovery will require careful planning to ensure that the optimum sequence and timing are applied.

Then to achieve partial upgrading during recovery requires a further sequential operation before a transportable product is produced.

Thus, a major decision at the time of recovery of heavy crude oil, extra heavy crude oil, and tar sand bitumen, is to acknowledge the practical (or impractical) aspects of upgrading during recovery, partial upgrading at the surface, or full upgrading in a conversion refinery. For the purposes of this text only, upgrading in a conversion refinery has been assumed as the means of upgrading one or all of the three viscous feedstocks. In addition, under the action of hydrogen, the constituents of a viscous feedstock will be transformed from high molecular weight constituents to low molecular weight products.

The important numbers for heavy crude oils, extra heavy crude oils, and tar sand bitumen are atomic hydrogen-to-carbon (H/C) ratios on the order of 1.3 to 1.5, while conventional crude oils have an atomic H/C Ratio of the order of 1.6 to 2.0, and the finished product is typically (or is required to be through product specifications) on the order of >2.0. The increase in the atomic H/C ratio in the refinery (of the product over that of the feedstock) is accomplished through the combination of processing steps involving hydrogen addition, which is accomplished through hydrogen addition.

Thus, as well as its use for product improvement, hydrogen is essential for the upgrading of viscous feedstocks which are hydrogen deficient (or carbon rich) and require significant modification to their molecular structure and hydrogen-to-carbon atomic ratio to be useful as various products. Hence, the need for the consideration of the use of hydrogen as a necessary means of processing the viscous feedstocks.

Thus, the focus of this chapter is on hydrogen-based technologies (thereby expanding the need for hydrogen in the modern refinery) for the conversion of distillation residua, heavy crude oil, extra heavy crude oil, and tar sand bitumen.

2.2 VISCOUS FEEDSTOCKS

For the purposes of this text, the term 'viscous feedstocks' is an 'umbrella' term which includes (1) distillation residua, (2) the viscous non-volatile portion from other (non-distillation) processes, (3) heavy crude oil, (4) extra heavy crude oil, and (5) tar sand bitumen (Speight, 2014a, 2017, 2021) In the distillation process, the temperature of the distillation is usually maintained below 350°C (660°F) since the rate of thermal decomposition of crude oil constituents is minimal below this temperature, but the rate of thermal decomposition of crude oil constituents is substantial above 350°C (660°F).

Residua are black, viscous materials that may be liquid at room temperature or almost solid, depending upon the nature of the crude oil and the process by which the residuum was produced.

The term *tar sand* is commonly used to describe a sandstone reservoir that is impregnated with bitumen, a naturally occurring material that is solid or nearly solid and is substantially immobile under reservoir conditions. In fact, tar sand is defined (US Congress, 1976, FE-76-4; Speight, 2005a, 2005b) in the United States as:

> the several rock types that contain an extremely viscous hydrocarbon which is not recoverable in its natural state by conventional oil well production methods including currently used enhanced recovery techniques. The hydrocarbon-bearing rocks are variously known as bitumen-rocks oil, impregnated rocks, tar sands, and rock asphalt.

By inference, conventional crude oil and heavy crude oil are recoverable by well production methods (i.e., primary recovery and secondary recovery methods) (Speight, 2013b, 2016) and by currently used enhanced oil recovery methods (EOR methods) (Speight, 2013b, 2014a, 2016).

The term 'extra heavy crude oil' is a recently evolved term (related to viscosity) of no true scientific meaning. While this type of feedstock may resemble tar sand bitumen and does not flow easily, extra heavy crude oil is recognized as having mobility in the reservoir compared to tar sand bitumen, which is typically incapable of mobility (free flow) under reservoir conditions. For example, the tar sand bitumen located in Alberta, Canada is immobile in the deposit and requires extreme methods of recovery.

On the other hand, much of the extra heavy crude oil located in the Orinoco belt of Venezuela requires recovery methods that are less extreme because of the mobility of the material in the reservoir (Schenk et al., 2009). Whether the mobility of extra heavy crude oil is due to a high reservoir temperature (that is higher than the pour point of the extra heavy crude oil) or due to other factors is subject to localized conditions in the reservoir.

The term 'black oil' is one that has arisen within the past two decades and has seen more common use recently. This term is of unknown scientific and engineering origin and is not used in any way in this report.

2.2.1 Origin and Occurrence

On a worldwide basis, the proved reserve of conventional crude oil has been estimated to be on the order of 1.7 billion barrels (1.7×10^9 bbls, 1.7 thousand million) (BP, 2020). In addition to this substantial resource base, the crude oils available to the modern refinery are quite different in composition and properties from those available 40 to 50 years ago and require different refinery scenarios for conversion to liquid fuels (Speight, 2014a). The current crude oils are somewhat heavier insofar as they have higher proportions of non-volatile constituents, i.e., a higher proportion of the non-volatile residue. In fact, by the older and more arbitrary non-scientific definitions of the mid-20th century, many of the crude oils currently in use would have been classified as heavy crude oils. Changes in feedstock character, such as this tendency to contain higher proportions of non-volatile residua, require adjustments to refinery operations to handle these heavier crude oils to reduce the amount of coke formed during processing and to balance the overall product slate (Speight, 2011, 2014a).

Heavy crude oil occurs in all continents both onshore and offshore, except in Antarctica, and approximate estimates are difficult to assess because of the varied definitions of heavy crude oil and the data (more often than not) includes data for the reserves of tar sand bitumen. Furthermore, it is only recently that the deposits of tar sand have been recognized as a viable energy source, and there have been changes in the way the world views oil reserves, with tar sand reserves coming to be included in the tally of global crude oil and heavy crude oil reserves (Levant, 2010; BP, 2020).

In spite of the high estimations of the reserves of bitumen, the two conditions of vital concern for the economic development of tar-sand deposits are the concentration of the resource, or the percent bitumen saturation in the deposit, and the accessibility of the deposit which is usually measured by the thickness of the overburden. The mining methods are applicable to shallow deposits, characterized by an overburden ratio (i.e., overburden depth to thickness of tar-sand deposit) that is deposit-specific. For example, indications are that for the Athabasca deposit, no more than 15 to 20% v/v of the in-place deposit is available within current concepts of the economics and technology of open-pit mining and advanced non-mining recovery techniques (Speight, 2014a, 2016).

2.2.2 Properties

The properties of viscous feedstocks are variable and typically focused on (1) the API gravity, (2) the content of sulfur-containing constituents, (3) the content of sulfur-containing constituents, and (4) the content of metal-containing constituents (Speight, 2012, 2014a, 2017, 2021). Although variations in these properties do occur, there are general distinctions that can be made. However, the property data is subject to the accuracy of the analytical method and should only be used as a guide and not as a means of classifying the material.

A variety of problems is associated with the utilization of lower quality viscous feedstocks and increased processing. These include both problems in the refinery and problems with product quality. However, understanding the fundamental molecular chemistry of viscous feedstocks is not always adequate to predict processing behavior (Niccum and Northup, 2006; Speight, 2014a, 2017). Very simply, the processing flexibility, especially fluid catalytic cracking, for viscous feedstocks crudes is limited due largely to their metals content and propensity for coke formation and deposition of the metals and coke on the catalyst.

The need for converting heavier feedstocks to valuable lighter products and for producing low-sulfur fuel oil will continue to encourage refiners to evaluate proven conversion process options for viscous feedstocks.

Viscous feedstocks are complex colloidal systems that include (1) a saturates fraction, (2) an aromatics fraction, (3) a resins fraction, and (4) an asphaltene fraction (Speight, 2014a, 2015a, 2017). While the saturates fraction and the aromatics fraction are composed of constituents that form the continuous phase of the system, the resin constituents and the asphaltene constituents form the dispersed phase. During the hydrocracking process, the asphaltene constituents (and the reacted asphaltene constituents) can become incompatible, thereby losing their stability and being separated from the liquid phase to form coke (Speight, 2014a, 2015b, 2017)

2.3 HISTORICAL PERSPECTIVES

The most valuable products from a refinery include low-boiling distillates (such as naphtha from which gasoline is produced) and middle distillates (such as kerosene from which jet fuel and diesel fuel are produced). These products consist predominantly of paraffin hydrocarbon derivatives and contain relatively short paraffin chains and have high atomic hydrogen-to-carbon ratios (H/C ratios).

In this context, the overall goal of the refining industry is to manage the hydrogen content of the refinery products and controlling the H/C ratio of the products requires either lowering the carbon content of the products (i.e., through the use of carbon rejection processes, such as coking processes), or increasing the hydrogen content (through hydrogen addition processes) by maximizing the production of hydrogen in the refinery.

The processes, coking, solvent extraction (e.g., deasphalting), visbreaking, and catalytic cracking reject carbon in the coke (carbonaceous) product so that lower boiling products (with high H/C ratios) can be obtained in these processes. Carbon in the coke, or in the heavier product, is considered to have been rejected (and potentially lost) since the carbonaceous by-products have a much lower value in comparison to the value of the lower boiling products. In contrast, hydrogen addition, as in the processes of hydrogenation and hydrocracking, enables the conversion of all the carbon present in a viscous feedstock to high-value products without rejecting, or sacrificing, any hydrogen that is already present in the feedstock.

Thus, the two common methods for heavy oil upgrading are (1) hydrogenation processes and (2) decarbonization processes. Hydrogenation processes are catalytic processes in which hydrogen is used in combination with various catalysts. On the other hand, the decarbonization processes are thermal processes such as delayed coking, in which coke is a major product, typically at the expense of the distillate products.

Conversion and hydrotreatment units, such as the H-Oil process and the LC Fining process, as well as the various fluid catalytic cracking processes, are options that refiners can use (and have used) to meet future market demands and product specifications (Parkash, 2003; Gary et al., 2007; Speight, 2020, 2014a; Hsu and Robinson, 2017; Speight, 2017). The key to these processes is the use of hydrogen, and the processes options offer solutions to address major challenges such as the variable demand for heavy fuel oil, the need for low sulfur heavy fuel oil, reduced sulfur emissions, reduced sulfur content in gasoline, and the variable diesel-to-gasoline ratio.

There is a promising future for upgrading processes that produce naphtha, kerosene, middle distillate fractions, and gas oil fractions (Parkash, 2003; Gary et al., 2007; Dehghani et al., 2009; Speight, 2020, 2014a; Hsu and Robinson, 2017; Speight, 2017). However, when the viscous feedstock is blended with another refinery feedstock or products, there is a stability limit for the high conversion region of the ebullated bed hydrocracking process (Duddy et al., 2004). Once this limit is reached, the following occur: (1) solids deposition, (2) excessive coke formation, and (3) reduction in distillate yield, which requires unit shutdown for solids removal and cleaning. Careful monitoring of the reaction conditions to maintain the solubility of the constituents will reduce the propensity for deposit formation.

Several kinds of residue upgrading processes are now on-stream and include the well-documented (1) fixed-bed processes (2) moving bed processes, and (3) ebullated bed processes for sulfur removal (Kressmann et al., 1998; Kressmann et al., 2000; Kressmann et al., 2004; Rana et al., 2007; Gillis et al., 2009, 2010; Speight, 2020, 2014a) In addition, slurry-bed process technology is advancing and has proved to be acceptable for viscous feedstock processing. For example, the Uniflex high-conversion slurry hydrocracking technology can achieve non-distillable conversion levels in excess of 90% w/w with distillate yield in excess of 50% v/v and holds yield and economic advantages over conventional residue upgrading technologies, and other applications of the technology (Gillis et al., 2010))

In addition to the tried-and-true fixed-bed, moving bed, and ebullated bed types, which may be adequate for some viscous feedstocks, the evolving focus (that has already commenced) will be on the means by which the catalyst and the feedstock are contacted. Thus, several technologies of interest (which follow alphabetically rather than by preference) are as follows: (1) the CASH process, i.e. the Chevron activated slurry hydroprocessing process, (2) the Microcat RC process, (3) the Intevep HDH (Hydrocracking distillation hydrotreating) process, (4) the Uniflex process, and (5) the Veba-Combi cracking process (Parkash, 2003; Gary et al., 2007; Speight, 2014a; Hsu and Robinson, 2017; Speight, 2017). Also, the use of scavenger additives that have already seen success in the older CANMET process (Pruden, 1978; Pruden et al., 1993) and the Uniflex process (Gillis et al., 2010) will also see a surge in use. As a simple example, a metal oxide (such as calcium oxide) has the ability to react with sulfur-containing feedstock to produce a hydrocarbon (and calcium sulfide). In this context, the Uniflex process (formerly the CANMET process) will be of considerable interest (Gillis et al., 2009, 2010) as a means of decreasing the yield of coke, and integration of such processes with existing coking capacity has been found to show benefits that are not available from the single-process options (Gillis et al., 2009, 2010).

Catalyst shape is another important factor that needs to be continually be developed. Whether the optimal catalyst shape is trilobe, quadralobe, or a unique shape such as the butterfly extrudate, which allows for the diffusion distance of the catalyst. This also helps maintain a low pressure drop of the catalyst bed (Albemarle, 2007), the diffusion distance of the catalyst must be reduced, and a low pressure drop of the catalyst bed must be maintained. By optimizing the combined catalyst system and operating conditions, many sour feedstocks with high contents of metals, sulfur, nitrogen, and high coke-forming propensity (CCR) can be hydrotreated for producing good quality fluid catalytic cracking feedstocks.

2.4 FEEDSTOCK EFFECTS

Trends in quality of the feedstocks to the various refineries and the problems in processing and utilization of the viscous feedstocks are dictated by the changing quality of the feedstocks to refineries, which has declined in the period 1978 to the present. The trend to lower quality refinery feedstocks is likely to continue. The shift in the quality of the refinery feedstocks was not only caused by a shortage of conventional crude oil but also resulted from economic factors which encouraged (even dictated) the utilization of cheaper, more viscous refinery feedstocks. While domestic throughput decreased significantly during this period, the fraction of the non-volatile material being upgraded increased – driven by (1) more viscous refinery feedstocks, (2) a sustained demand for distillate products, (3) a decreased demand for residual fuels, and (4) an increased conversion capacity by industry. This has also been accompanied by an increase in the hydrogen demand with the viscous feedstocks (depending upon the feedstock composition and properties), requiring an estimated 400 to 1200 scf (standard cubic feet) per barrel of viscous feedstock compared to less than 200 scf of hydrogen per barrel of conventional light crude oil (Parkash, 2003; Gary et al., 2007; Speight, 2014a; Hsu and Robinson, 2017; Speight, 2017).

These estimates are not hard and fast amounts of hydrogen but are based on a general review of hydroprocesses and refining viscous feedstocks and are a base from which to work and make more 'accurate' estimates of the hydrogen requirements. The amount of hydrogen required per barrel of feedstock can vary from feedstock to feedstock, even within those feedstocks that have a similar API gravity. An assemblage of general rules is as follows: (1) the lower the API gravity, the more viscous the feedstock, (2) the higher the molecular weight of the constituents, the more viscous the feedstock, (3) the higher the amount of hydrogen required for processing, (4) the more viscous the feedstock, the higher the amount of hydrogen required for processing, (5) the higher the sulfur content, the higher the amount of hydrogen required for processing compared to the hydrogen requirements for a lower sulfur feedstock (Parkash, 2003; Gary et al., 2007; Speight, 2014a; Hsu and Robinson, 2017; Speight, 2017). The analytical laboratory associated with a refinery will have a check-list to show the composition and the hydrogen requirements for each type of feedstock.

More appropriately, it is only by comprehensively considering all the necessary factors, including factors such as the (1) feedstock properties, (2) catalyst performance, (3) product specifications, (4) chemical kinetics, (5) operating conditions, and (6) running period, that optimal results and more definitive values for the hydrogen demand per barrel of refinery feedstock can be achieved. Therefore, further improvement of the various upgrading processes for viscous feedstocks and the development of catalysts, which can tolerate a high content of impurities and metals, are two major challenges for the refineries. Catalyst activity, selectivity, particle size and shape, pore size and distribution, as well as the type of the reactor, have to be optimized according to the properties of the heavy crude oils and the desired purification and conversion levels.

One critical issue that has arisen within the past the three-to-five decades relates to the use of blends of any viscous feedstock with the more typical lighter refinery feedstock that are employed for any unit in a refinery. Compatibility of the constituents of the blends is not always assured and there is excessive laydown of coke (and metals) on the surface (and within the pore system) of the catalyst (Speight, 2014a, 2015b, 2017).

It is now recognized that the choice of one of several pretreatment options for the feedstocks offers process benefits. Such options are: (1) deasphalting to prevent excessive coking on catalyst surfaces, (2) demetallization, i.e. removal of nickel, vanadium, and iron to prevent catalyst deactivation, (3) use of a short residence time as a means of preparing the feedstock, (4) hydrotreating or mild hydrocracking to prevent excessive coking in the fluid catalytic cracking unit, (5) blending with an aromatic gas oil-type to prevent phase separation, and (6) staged partial conversion (Birch and Ulivieri, 2000; Speight, 2000; Gary et al., 2007; Dziabala et al., 2011; Speight, 2020, 2014a; Hsu and Robinson, 2017; Speight, 2017).

There are two ways that are currently practiced in bringing heavy crude oil to market. The first method is to upgrade the material at or near the recovery site and leave behind much of the non-volatile material as coke, and then pipeline the upgraded material out as synthetic crude. In this method, the crude is fractionated, and the residue is coked. In some cases, the products of coking operations may be hydrotreated for further processing.

The second method is to effect partial upgrading in situ as part of the recovery process. Such an option, to produce an acceptable pipeline material, would be an ideal solution but has several limitations (Motaghi et al., 2010a, 2010b, 2010c; Speight, 2013b). For example, the amount of heavy crude oil production could be limited by the recovery process, and the upgraded products must be compatible with the original or partially changed heavy crude oil (Speight, 2016). If the products and the original (partially changed) heavy crude oil have limited compatibility, which would limit the amount of dilution and again could limit the effectiveness of the refining process (Speight, 2020, 2014a).

Refineries remove sulfur from refinery streams to meet sulfur standards for refined products (such as gasoline and diesel fuel) and limit refinery emissions of sulfur oxides (SOx) and nitrogen oxides (NOx). Sulfur and nitrogen removal is accomplished by means of hydrotreating processes, which require substantial amounts of hydrogen, especially for the heavier feedstock that contains higher amounts of sulfur and nitrogen.

Thus, the increasing influx of viscous feedstocks into refineries will increase the need for hydrogen along with the need for effective catalysts that are relatively resistant to poisoning by the viscous feedstock constituents. This will be a continuously occurring issue for the next decade. In this respect, installation of resid (and other feedstock) gasifiers on refinery sites can be expected to be commonplace. This will serve to not only increase the available hydrogen but also to remove low-value coke and waste. It can be anticipated that biomass feedstock will also serve as feedstock to the gasifier to increase hydrogen production. Alternatively, the installation of Flexicoking units will also serve to reduce the amount of resids and coke produced as well as producing hydrogen (Bridjanian and Khadem Samimi, 2011; Speight, 2020; Speight, 2014b).

2.5 UPGRADINGTECHNOLOGY

The difficult-to-refine viscous feedstocks (i.e. heavy crude oil, extra heavy crude oil, tar sand bitumen, and residua) are characterized by a low API gravity (conversely, a high density) and high viscosity, high initial boiling point, high carbon residue, high nitrogen content, high sulfur content, and high metals content (Speight, 2013b, 2014a, 2017). In addition to these properties, the viscous feedstocks also have an increased molecular weight and reduced hydrogen content with a relatively low content of volatile saturated and aromatic constituents and a relatively high content of asphaltene and resin constituents that is accompanied by a high heteroatom (nitrogen, oxygen, sulfur, and metals) content (Khan and Patmore, 1997; Ancheyta and Speight, 2007; Speight, 2014a, 2017). The

Vacuum residuum
- Coking
  - Delayed coking
  - Fluid coking
  - Flexicoking
- Residuum fluid catalytic cracking (Resid FCC)
- Solvent deasphalting
- Resid hydrotreating
  - Ebullated bed process
  - Slurry hydrocracking process

FIGURE 2.1 Schematic Example of the Technologies for Upgrading of Viscous Feedstocks.

criteria to select an upgrading option depend on several factors which must be analyzed in detail when it comes to considering a project of this nature.

There are technologies that have been used in the past (Figure 2.1) that were, at the time, satisfactory for the purpose of upgrading.

Some hydrocrackers are single-stage units with just one reactor that is usually filled with hydrocracking catalyst. However, in order to advance viscous feedstock conversion and to ensure optimum performance and reliability, single-stage and two-stage processes are typically employed in a refinery.

In a two-stage hydrocracking process (Figure 2.2), a mixture of the feedstock and hydrogen is heated and injected into a reactor containing a hydrotreating catalyst, which accelerates the reactions that remove sulfur and nitrogen from the feedstocks as well as saturating the aromatic ring systems.

The entire product from this reactor is then injected into a second reactor containing a hydrocracking catalyst, which promotes the cracking reactions and saturates the cracked products with hydrogen. Any unconverted feedstock is then separated and recycled to the hydrocracking reactor with the option of using a second hydrocracking reactor for further conversion. Volatile products can also be drawn off at the separation steps to maximize distillate production, or they can be processed further (through recycling or second-step hydrocracking) to maximize the yields of distillate products.

More modern technologies, such as hydrotreating and hydrocracking (for which a plentiful supply of hydrogen is needed), along with various thermal technologies, catalytic cracking technologies, and solvent-based technologies, are now available (Table 2.2) are now available to maximize the yield of liquid products (or a higher conversion rate) ranging from solvents and fuels to petrochemical feedstocks. All of which require further processing and additional assets, such as sulfur plants and a hydrogen purification system (Chapter 6) and a hydrogen management system (Chapter 7). The commercial application of these technologies is expanding and is projected to expand even more as the 21st Century evolves.

For example, 'hydrogen addition technologies' produce a high yield of products with a commercial value larger than that of the 'carbon rejection technologies' but require a larger investment and more natural gas availability to produce the amounts of hydrogen and steam required for these processes. Moreover, the deposition of coke on the catalysts from the carbon rejection reaction during thermally driven catalytic upgrading in situ deactivates the catalysts within a short time interval. In addition, the presence of heavy metals (such as vanadium and nickel), sulfur, nitrogen, and other contaminants severely reduces catalytic activity rapidly (Castañeda et al., 2011). Therefore, the minimization of unwanted heteroatom species (such as sulfur, nitrogen, oxygen, and metals) and coke deposition will sustain the catalyst activity and improve the lifetime.

In this respect, the presence of hydrogen not only elevates the extent of the removal of heteroatom impurities but also increases the ability of the process to accelerate the breakdown of higher

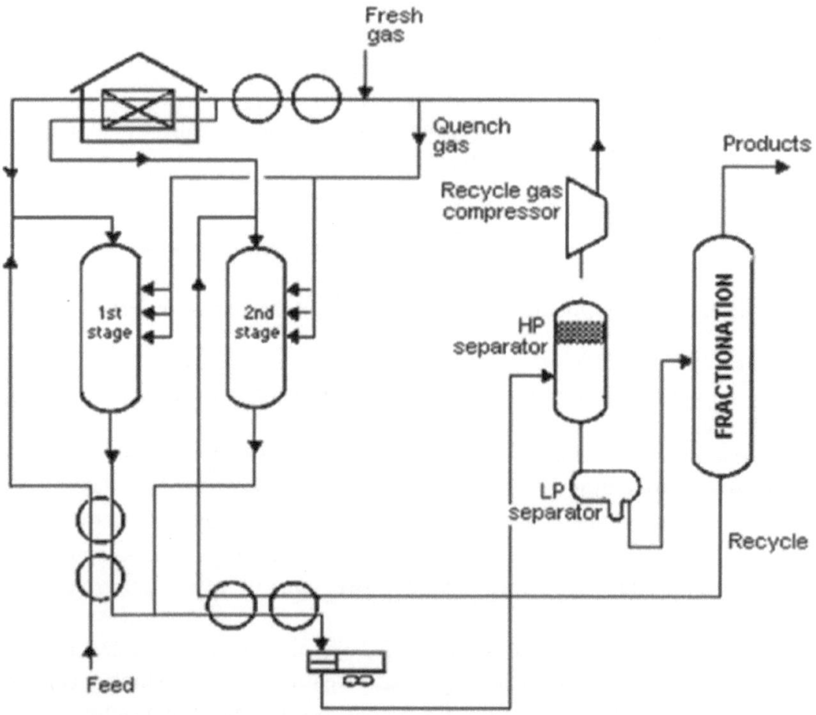

FIGURE 2.2 Two-stage Hydrocracking Process.

molecular weight constituents to lower-boiling products. In this regard, the favored route to avoid instability in the upgraded oil, in situ upgrading in the presence of hydrogen or hydrogen donors would include catalysts (Ovalles et al., 2015; Zhao et al., 2022). For such feedstocks, a platform of different technologies will help refiners address the challenges. The processes that follow are listed in alphabetical order with no other preference in mind.

Residue hydrotreating was originally developed to reduce the sulfur content of atmospheric residues to produce specification low sulfur residual fuel oil. Changes in crude oil quality and product demand, however, have shifted the commercial importance of this technology to include pretreating conversion unit feedstocks to minimize catalyst replacement costs, and coker feedstocks to reduce the yield and increase the quality of the by-product coke fraction. Although residue hydrotreaters are capable of processing feedstocks having a wide range of contaminants, the feedstock's organometallic and asphaltene components typically determine its processability. Economics generally tends to limit residue hydrotreating applications to feedstocks with limitations (dictated by the process catalyst) on the content of nickel plus vanadium.

In many cases, the application of hydrotreating technology to viscous feedstocks may also cause some cracking, and by inference, application of hydrocracking to viscous feedstocks will also cause desulfurization and denitrogenation. Rather than promote unnecessary duplication of the process description, two hydrotreating processes (the RDS and VRDS process options are described together as sub-categories of one process) are described here since these other processes are more amenable to hydrocracking than to hydrotreating.

2.6 CATALYSTS AND PROCESS CHEMISTRY

Hydrocracking catalysts have (1) a cracking function and (2) a hydrogenation-dehydrogenation function. The former (i.e. the cracking function) is provided by an acidic support, whereas the latter (i.e.

the hydrogenation-dehydrogenation function) is provided by active metals. The acidic support can be (1) an amorphous inorganic oxide derivative, such as a silica–alumina support, (2) a crystalline zeolite, such as the modified Y zeolite plus a binder such as alumina, or (3) a mixture of crystalline zeolite and amorphous oxides.

The metals providing the hydrogenation–dehydrogenation function can be noble metals (such as palladium or platinum) or non-noble metal sulfide derivatives from Group VI.A (molybdenum, tungsten) and Group VIII.A (cobalt, nickel) (Figure 2.3). These metals catalyze the hydrogenation of the feedstock as a pretreatment for the cracking and heteroatom removal reactions as well as reducing the potential for coke formation.

If mild hydrocracking is the goal of the process, the catalysts have a composition similar to that of hydrotreating catalysts and consist of Group VI and VIII non-noble metals supported on γ-alumina (such as cobalt, nickel, molybdenum, and tungsten in sulfided form).

More pertinent to the hydroprocessing context, the composition of a hydrocracking catalyst depends on the final product requirements and mode of operation (single-stage process or two-stage process).

Furthermore, the increasing importance of hydrodesulfurization (HDS) and hydrodenitrogenation (HDN) in refinery hydroprocesses in order to produce clean-burning fuels has led to a surge of research on the chemistry and engineering of heteroatom removal, with sulfur removal being the most prominent area of research. Most of the earlier works are focused on (1) catalyst characterization by physical methods, (2) low-pressure reaction studies of model compounds having relatively high reactivity, (3) process development, or (4) cobalt-molybdenum (Co-Mo) catalysts, nickel-molybdenum catalysts (Ni-Mo), or nickel-tungsten (Ni-W) catalysts supported on alumina, often doped by fluorine or phosphorus.

Sulfur and nitrogen occurring in residua are converted to hydrogen sulfide and ammonia in the catalytic reactor, and these gases are scrubbed out of the reactor effluent gas stream.

The need to develop catalysts that can carry out deep hydrodesulfurization and deep hydrodenitrogenation has become even more pressing in view of recent environmental regulations limiting the amount of sulfur and nitrogen emissions. The development of a new generation of catalysts to

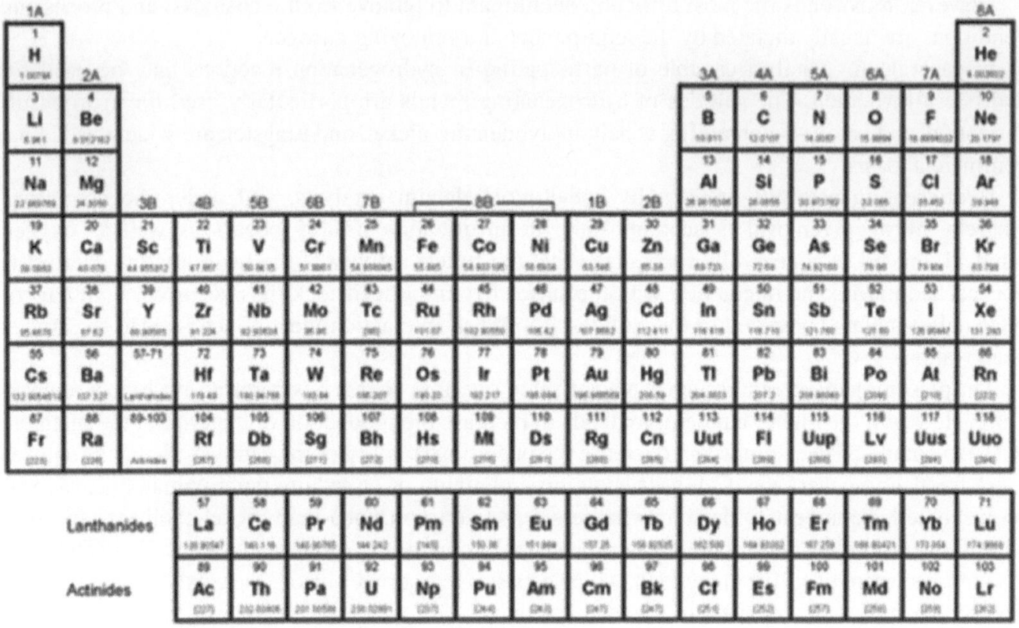

FIGURE 2.3 The Periodic Table of Elements.

achieve this objective of low nitrogen and sulfur levels in the processing of different feedstocks presents an interesting challenge for catalyst development.

Basic nitrogen-containing compounds in a feed diminish the cracking activity of hydrocracking catalysts. However, zeolite catalysts can operate in the presence of substantial concentrations of ammonia, in marked contrast to silica-alumina catalysts, which are strongly poisoned by ammonia. Similarly, sulfur-containing compounds in a feedstock adversely affect the noble metal hydrogenation component of hydrocracking catalysts. These compounds are hydrocracked to hydrogen sulfide, which converts the noble metal to the sulfide form. The extent of this conversion is a function of the hydrogen and hydrogen sulfide partial pressures.

Removal of sulfur from the feedstock results in a gradual increase in catalyst activity, returning almost to the original activity level. As with ammonia, the concentration of hydrogen sulfide can be used to precisely control precisely the activity of the catalyst. Non-noble metal-loaded zeolite catalysts have an inherently different response to sulfur impurities since a minimum level of hydrogen sulfide is required to maintain the nickel-molybdenum and nickel-tungsten in the sulfide state.

Hydrodenitrogenation is more difficult to accomplish than hydrodesulfurization, but the relatively smaller amounts of nitrogen-containing compounds in conventional crude oil made this of lesser concern to refiners. However, the trend to heavier feedstocks in refinery operations, which are richer in nitrogen than the conventional feedstocks, has increased the awareness of refiners to the presence of nitrogen compounds in crude feedstocks. For the most part, however, hydrodesulfurization catalyst technology has been used to accomplish hydrodenitrogenation, although such catalysts are not ideally suited to nitrogen removal. However, in recent years, the limitations of hydrodesulfurization catalysts when applied to hydrodenitrogenation have been recognized, and there are reports of attempts to manufacture catalysts more specific to nitrogen removal (Ho, 1988).

The character of the hydrotreating processes is chemically very simple since they essentially involves removal of sulfur and nitrogen as hydrogen sulfide and ammonia, respectively.

$$R\text{-}S\text{-}R' + H_2 \rightarrow RH + R'H + H_2S$$

$$R\text{-}N(R')\text{-}R'' + 3H_2 \rightarrow RH + R'H + R''H + 2NH_3$$

However, nitrogen is the most difficult contaminant to remove from feedstocks, and processing conditions are usually dictated by the requirements for removing nitrogen.

In general, any catalyst capable of participating in hydrogenation reactions may be used for hydrodesulfurization. The sulfides of hydrogenating metals are particularly used for hydrodesulfurization, and catalysts containing cobalt, molybdenum, nickel, and tungsten are widely used on a commercial basis.

Hydrotreating catalysts are usually cobalt-molybdenum catalysts, and under the conditions whereby nitrogen removal is accomplished, desulfurization usually occurs as well as oxygen removal. Indeed, it is generally recognized that the fullest activity of the hydrotreating catalyst is not reached until some interaction with the sulfur (from the feedstock) has occurred, with part of the catalyst metals converted to the sulfides. Too much interaction may, of course lead to catalyst deactivation.

The poisoning effect of nitrogen can be offset to a certain degree by operating at a higher temperature. However, the higher temperature tends to increase the production of material in the methane (CH_4) to butane ($C4H_{10}$) range and decrease the operating stability of the catalyst, so that it requires more frequent regeneration. Catalysts containing platinum or palladium (approximately 0.5% wet) on a zeolite base appear to be somewhat less sensitive to nitrogen than nickel catalysts, and successful operation has been achieved with feedstocks containing 40 ppm nitrogen. This catalyst is also more tolerant of sulfur in the feed, which acts as a temporary poison, the catalyst recovering its activity when the sulfur content of the feed is reduced.

On such catalysts as nickel or tungsten sulfide on silica-alumina, isomerization does not appear to play any part in the reaction, as uncracked normal paraffins from the feedstock tend to retain their normal structure. Extensive splitting produces large amounts of low-molecular-weight (C_3-C_6) paraffins, and it appears that the primary reaction of paraffins is catalytic cracking followed by hydrogenation to form *iso*-paraffins. With catalysts of higher hydrogenation activity, such as platinum on silica-alumina, direct isomerization occurs. The product distribution is also different, and the ratio of low molecular weight to intermediate molecular weight paraffins in the breakdown product is reduced.

In addition to the chemical nature of the catalyst, the physical structure of the catalyst is also important in determining the hydrogenation and cracking capabilities, particularly for viscous feedstocks (Kobayashi et al., 1987a, 1987b; Fischer and Angevine, 1986). When gas oils and residua are used, the feedstock is present as liquids under the conditions of the reaction. Additional feedstock and hydrogen must diffuse through this liquid before the reaction can take place at the interior surfaces of the catalyst particle.

At high temperatures, reaction rates can be much higher than diffusion rates, and concentration gradients can develop within the catalyst particle. Therefore, the choice of catalyst porosity is an important parameter. When feedstocks are to be hydrocracked to liquefied petroleum gas (LPG) and naphtha, pore diffusion effects are usually absent. High surface area (about 300 m2/g) and low to moderate porosity (from 12 pore diameter with crystalline acidic components to 50 or more with amorphous materials) catalysts are used. With reactions involving high-molecular-weight feedstocks, pore diffusion can exert a large influence, and catalysts with pore diameters greater than 80 are necessary for more efficient conversion.

Catalyst operating temperature can influence reaction selectivity since the activation energy for hydrotreating reactions is much lower than for hydrocracking reactions. Therefore, raising the temperature in a residuum hydrotreater increases the extent of hydrocracking relative to hydrotreating, which also increases the hydrogen consumption.

Aromatic hydrogenation in crude oil refining may be carried out over supported metal or metal sulfide catalysts depending on the sulfur and nitrogen levels in the feedstock. For hydrorefining of feedstocks that contain appreciable concentrations of sulfur and nitrogen, sulfided nickel-molybdenum (Ni-Mo), nickel-tungsten (Ni-W), or cobalt-molybdenum (Co-Mo) on alumina (-Al_2O_3) catalysts are generally used, whereas supported noble metal catalysts have been used for sulfur- and nitrogen-free feedstocks. Catalysts containing noble metals on Y-zeolites have been reported to be more sulfur tolerant than those on other supports (Jacobs, 1986). Within the series of cobalt-promoted or nickel-promoted group VI metal (Mo or W) sulfides supported on -Al_2O_3, the ranking for hydrogenation is:

$$\text{Ni-W} > \text{Ni-Mo} > \text{Co-Mo} > \text{Co-W}$$

Nickel-tungsten (Ni-W) and nickel-molybdenum (Ni-Mo) on Al2O3 catalysts are widely used to reduce sulfur, nitrogen, and aromatics levels in crude oil fractions through hydrotreating.

Molybdenum sulfide (MoS_2), usually supported on alumina, is widely used in crude oil processes for hydrogenation reactions. It is a layered structure that can be made much more active by the addition of cobalt or nickel. When promoted with cobalt sulfide (CoS), making what is called *cobalt-moly* catalysts, it is widely used in hydrodesulfurization (HDS) processes.

The closely related tungsten compound (WS_2) is used in commercial hydrocracking catalysts. Other sulfides (iron sulfide, FeS, chromium sulfide, Cr_2S_3, and vanadium sulfide, V_2S_5) are also effective and used in some catalysts. A valuable alternative to base metal sulfides is palladium sulfide (PdS). Although it is expensive, palladium sulfide forms the basis for several very active catalysts.

Several different metal complexes exist in the asphaltene fraction of the residuum, and an explicit reaction mechanism of decomposition that would be a perfect fit for all of the compounds is not possible. However, in general terms, the reaction can be described as hydrogen (A) dissolved in the feedstock contacting an organometallic compound (B) at the surface of the hydrotreating catalyst and producing a metal sulfide (C) and a hydrocarbon (D).

$$A + B \rightarrow C + D$$

For example, a medium metal-content feedstock will generally have a lower rate of demetallization compared to high-metal content feedstock.

Obviously, the choice of hydrogenation catalyst depends on what the catalyst designer wishes to accomplish. In order to produce naphtha (a bland stock for gasoline manufacture), extensive cracking of the feedstock constituents is needed to convert a large fraction of the feed to the kinds of molecules that will make a good blending stock for gasoline. For this vigorous cracking, a vigorous hydrogenation component is needed to reduce the potential occurrence of unslatted (olefin-type products, >C = C<) and to remove the nitrogen, oxygen, and sulfur in the form of their respective hydrogenated products (NH_3, H_2O, and H_2S).

Viscous feedstocks contain substantial amounts of metals and coke precursors. These impurities deposit on the surface of the catalyst during the reaction and lead to deactivation of the catalyst. Therefore, the means to inhibit metal depositing on the catalysts and prevent bed plugging to prolong the run length is an important challenge in fixed-bed technology. Recently, catalyst grading schemes have been developed to help solve this problem – to meet the different requirements of the products, the grading technology should involve different kinds of catalysts and each type has its specific objectives (Liu et al., 2009) The operability barriers in all viscous feedstock processing units due to sediment-induced fouling (Speight, 2015b) need to be overcome with innovative catalysts that have the ability (or property) to control sediment formation or coke formation.

In addition, nickel-molybdenum catalysts for applications to viscous feedstocks are already being supplied as higher activity quadralobes because diffusion restrictions are even more critical in these heavy applications.

2.6.1 Physical Composite Properties

The most important property for the feedstock hydroconversion catalyst is pore diameter because the feedstock contains higher molecular weight asphaltene constituents as well as other coke-forming precursors and metal-containing constituents which serve to deactivate the catalyst.

Thus, catalyst development continues to focus on the production of catalysts that can tolerate coke formers and enhance demetallization catalysts without serious deactivation. Demetallizing catalysts are often referred to as metal catchers, and there has been recent progress and developments that are making promising inroads for cracking processes (Zhang et al., 2014). This work is accompanied and complemented by the investigation of the use of scavenger-additives that are evolving as valuable pretreating steps to catalytic cracking units and/or hydrocracking units (Stark and Falkler, 2008; Stark et al., 2008).

2.6.2 Support Materials

An efficient residue hydrotreating process requires the development of new catalysts, including support and active metal, either by updating already-known catalysts or by developing new catalyst formulations.

As a result, catalyst performance has also continued to focus on catalyst supports – especially supports with a high surface area – and the use of hydrogen donors (Zhao et al., 2015, 2022). The properties of the supports are now seen as an important aspect of the porosity of the catalyst, which has a direct effect on catalyst efficiency for the conversion of viscous feedstocks. The pore size must be adequate to accommodate and decompose the higher molecular weight constituents (Hudec, 2011).

2.6.3 Metal Functions for Hydrogenation

Due to the unique structure and surface characteristics of dispersed catalysts, their catalytic activity and selectivity are higher than those of traditional catalysts. With the decrease of the catalyst particle size, their surface area increases gradually, along with adsorption capacity and catalytic performance. These unique effects make the dispersed catalysts not only control the reaction velocity and greatly improve the reaction efficiency but also make the original reaction (Zhao et al., 2015, 2022).

In addition, slurry catalysts, typically molybdenum-based, prepared from oil soluble or water-soluble precursors, continue to be the focus of catalyst development programs because of the success with these catalysts (Rispoli et al., 2009; Motaghi et al., 2010a, 2010b).

2.6.4 Acidic Functions for Cracking

The philosophy of the FCC-catalyst is also undergoing development and is seen as a stepwise process guided by the pore systems of the catalyst (Hudec, 2011). The decision is moving toward having weak acid centers in the macropores of the catalyst particles to ensure cracking of the higher molecular weight species of the feedstock so that the lower molecular weight products then enter the mesopores with stronger acidity. The products from the mesopores sites then enter the zeolite micropore sites (with the strongest acid centers) and crack into smaller molecules to produce naphtha from which gasoline is produced.

2.6.5 Contaminant Removal and Catalyst Stability

Viscous feedstock conversion increases the need to reduce the effects of contaminants such as metals and the higher molecular weight constituents of these feedstocks without reducing catalyst efficiency. The development of catalysts such as Y-type has improved the thermal stability of the catalyst to temperatures on the order of 760°C (1,400°F). There is still the need to fully protect the catalyst from such contaminants, and the use of guard beds positioned prior to the reaction (catalyst) vessel may still be a necessity, depending upon the contaminants in the viscous feedstock. A major aspect of successful technology is to move from off-the-shelf catalyst technology into feedstock–related and process-related customized catalysts.

The integrated technology-catalyst customized solutions (including FCC unit revamps) offered by, for example, Shell Global Solutions and Criterion, can offer refiners quick fix routes to maximizing viscous feedstock unit performance for hydrodesulfurization and fluid catalytic cracking to upgrade viscous feedstocks.

2.6.6 Multiple Catalyst Options

The use of a guard bed prior to the reaction (catalyst) vessel is considered by some refiners to be a multistage catalyst operation. However, the multi-stage catalyst platform combines two or more existing functionalities within the same catalyst particle. In this respect, BASF has introduced and continued to develop the *multi-stage reaction catalysts* (MSRC) platform – which combines two or more existing functionalities within the same catalyst particles (www.catalysts.basf.com/refining).

2.7 REACTORS AND PROCESS PARAMETERS

The refinery feedstocks being used for the production of valuable products (such as transportation fuels, solvents, and petrochemicals) have become more and more viscous. Because of this, many topping refineries have shut down because of their inability to process these heavier crude oils. In comparison, the total capacity of those processes intended for upgrading high-boiling distillates (such as residua) and heavy crude oils has increased (Parkash, 2003; Gary et al., 2007; Speight,

2014a; Hsu and Robinson, 2017; Speight, 2017). Refiners must, therefore, continue to remove substantial portions of sulfur from the lower boiling (lower density) products. However, the viscous feedstocks pose particularly difficult problems, and it is now clear that there are other problems involved in the processing of the viscous feedstocks and that these heavier feedstocks, which are gradually emerging as the liquid fuel supply of the future, need special attention. This is especially applicable not only to the types of reactors used in the refinery but also to the reactor parameters.

Briefly, and by way of definition or clarification, refinery reactors are vessels that are designed to perform the various chemical reactions that take place in the presence of a catalyst, between reactants in the feedstock or reactants in multiple feedstock streams.

In the process, the feedstock is preheated and mixed with hot recycle gas containing hydrogen, and the mixture is passed over the catalyst in the reactor section at temperatures between 290 to 445°C (550 to 850°F) and pressures between 150 and 3,000 psi. The reactor effluent is then cooled by heat exchange, and desulfurized liquid hydrocarbon product and recycle gas are separated at essentially the same pressure as used in the reactor. The recycle gas is then scrubbed and/or purged of the hydrogen sulfide and light hydrocarbon gases, mixed with fresh hydrogen makeup, and preheated prior to mixing with the hot hydrocarbon feedstock.

The operating conditions in distillate hydrodesulfurization are dependent on the feedstock as well as the desired degree of desulfurization or quality improvement. Kerosene and light gas oils are generally processed at mild severity and high throughput, whereas light catalytic cycle oils and thermal distillates require slightly more severe conditions. Higher boiling distillates, such as vacuum gas oils and lube oil extracts, require the most severe conditions.

2.7.1 Reactors

Typically, a refinery reactor is an enclosed volume in which a chemical reaction takes place, often in the presence of a catalyst, between reactants in feedstock or reactants in multiple feed streams.

Reactor configurations within crude oil hydroprocessing units may include catalyst beds that are fixed or moving (Speight, 2000; Ancheyta, 2007; 2014a, 2017). Most hydroprocessing reactors are fixed-bed reactors. Hydroprocessing units with fixed-bed reactors must be shut down to remove the spent catalyst when catalyst activity declines below an acceptable level (due to the accumulation of coke, metals, and other contaminants).

Briefly, a trickle-bed reactor (TBR) consists of a fixed-bed of catalyst particles contacted by a co-current downward gas-liquid flow carrying both reactants and products. When the gas and liquid are fed concurrently upward through the catalyst bed, the system is called a flooded bed reactor (FBR) or upflow reactor. On the other hand, plug flow reactors, also known as tubular reactors, consist of a cylindrical pipe with openings on each end for reactants and products to flow through.

2.7.1.1 Downflow Fixed-Bed Reactor

The reactor design commonly used in hydrodesulfurization of distillates is the fixed-bed reactor design in which the feedstock enters at the top of the reactor and the product leaves at the bottom of the reactor. The catalyst remains in a stationary position (fixed-bed) with hydrogen and crude oil feedstock passing in a downflow direction through the catalyst bed. The hydrodesulfurization reaction is exothermic, and the temperature rises from the inlet to the outlet of each catalyst bed. With high hydrogen consumption and subsequent large temperature rise, the reaction mixture can be quenched with cold recycled gas at intermediate points in the reactor system. This is achieved by dividing the catalyst charge into a series of catalyst beds, and the effluent from each catalyst bed is quenched to the inlet temperature of the next catalyst bed.

The extent of desulfurization is controlled by raising the inlet temperature to each catalyst bed to maintain a constant catalyst activity over the course of the process. Fixed-bed reactors are mathematically modeled as plug-flow reactors with very little back mixing in the catalyst beds. The first catalyst bed is poisoned with vanadium and nickel at the inlet to the bed and may be a cheaper catalyst (*guard bed*). As the catalyst is poisoned in the front of the bed, the temperature exotherm moves down the bed, and the activity of the entire catalyst charge declines, thus requiring a raise in the reactor temperature over the course of the process sequence. After catalyst regeneration, the reactors are opened and inspected, and the high metal content catalyst layer at the inlet to the first bed may be discarded and replaced with fresh catalyst. The catalyst loses activity after a series of regenerations, and consequently, after a series of regenerations, it is necessary to replace the complete catalyst charge. In the case of very high metal content feedstocks (such as residua), it is often necessary to replace the entire catalyst charge rather than to regenerate it. This is due to the fact that the metal contaminants cannot be removed by economical means during rapid regeneration, and the metals have been reported to interfere with the combustion of carbon and sulfur, catalyzing the conversion of sulfur dioxide (SO_2) to sulfate (SO_4^{2-}) that has a permanent poisoning effect on the catalyst.

Fixed-bed hydrodesulfurization units are generally used for distillate hydrodesulfurization and may also be used for residuum hydrodesulfurization but require special precautions in processing. The residuum must undergo two-stage electrostatic desalting so that salt deposits do not plug the inlet to the first catalyst bed, and the residuum must be low in vanadium and nickel content to avoid plugging the beds with metal deposits. Hence, the need for a guard bed in residuum hydrodesulfurization reactors.

During the operation of a fixed-bed reactor, contaminants entering with fresh feed are filtered out and fill the voids between catalyst particles in the bed. The buildup of contaminants in the bed can result in the channeling of reactants through the bed and reduce the hydrodesulfurization efficiency. As the flow pattern becomes distorted or restricted, the pressure drop throughout the catalyst bed increases. If the pressure drop becomes high enough, physical damage to the reactor internals can result. When high-pressure drops are observed throughout any portion of the reactor, the unit is shut down and the catalyst bed is skimmed and refilled.

In the n fixed-bed reactor, the reaction is achieved on the surface of the catalyst pellet inside the reactor, and the pellet acts as a catalyst for the reaction. In the packed bed reactor, the reaction is achieved by finely mixing the feedstock stream through physical mixing.

With fixed-bed reactors, a balance must be reached between reaction rate and pressure drop across the catalyst bed. As catalyst particle size is decreased, the desulfurization reaction rate increases, but so does the pressure drop across the catalyst bed. Expanded-bed reactors do not have this limitation, and small 1/32-inch (0.8 mm) extrudate catalysts or fine catalysts may be used without increasing the pressure drop.

2.7.1.2 Upflow Expanded-Bed Reactor

In the upflow expanded bed reactor, a gas-liquid or a gas-liquid-solid feed is contacted in a vessel containing a packed bed of contact particles. At least a portion of the feedstock is passed upwards through the packed bed, and the mass flow rate of the feed is periodically increased sufficiently to cause a portion of the particles within the vessel to form an expanded region within the packed bed.

Expanded-bed reactors are applicable to distillates, but are commercially used for very heavy, high metals, and/or dirty feedstocks having extraneous fine solids material. They operate in such a way that the catalyst is in an expanded state so that the extraneous solids pass through the catalyst bed without plugging. They are isothermal, which conveniently handles the high-temperature exotherms associated with high hydrogen consumption. Since the catalyst is in an expanded state of motion, it is possible to treat the catalyst as a fluid and to withdraw and add catalyst during operation.

Spent catalyst may be withdrawn and replaced with fresh catalyst on a daily basis. Daily catalyst addition and withdrawal eliminate the need for costly shutdowns to change out catalyst and also result in a constant equilibrium catalyst activity and product quality. The catalyst is withdrawn daily and has a vanadium, nickel, and carbon content that is representative on a macro-scale of what is found throughout the entire reactor. On a micro-scale, individual catalyst particles have ages ranging from that of fresh catalyst to as old as the initial catalyst charge to the unit, but the catalyst particles of each age group are so well dispersed in the reactor that the reactor contents appear uniform.

Commercial expanded-bed reactors normally operates with 1/32-inch (0.8 mm) extrudate catalysts that provide a higher rate of desulfurization than the larger catalyst particles used in fixed-bed reactors. With extrudate catalysts of this size, the upward liquid velocity based on fresh feedstock is not sufficient to keep the catalyst particles in an expanded state. Therefore, for each part of the fresh feed, several parts of product oil are taken from the top of the reactor, recycled internally through a large vertical pipe to the bottom of the reactor, and pumped back up through the expanded catalyst bed.

The turbulence of gas and feedstock passing upward through the expanded catalyst bed is sufficient to cause almost complete back-mixing. Consequently, in order to effect near complete sulfur removal (over 75%), it is necessary to operate with two or more reactors in series. The ability to operate at a single temperature throughout the reactor or reactors, and to operate at a selected optimum temperature rather than an increasing temperature from the start to the end of the run, results in a more effective use of the reactor and catalyst contents. When all these factors are put together, i.e., the use of a smaller catalyst particle size, isothermal, fixed temperature throughout the run, back-mixing, daily catalyst addition, and constant product quality, the reactor size required for an expanded bed is often smaller than that required for a fixed-bed to achieve the same product goals. This is generally true when the feeds have high initial boiling points and/or the hydrogen consumption is very high.

2.7.1.3 Ebullating Bed Reactor

In ebullated bed hydroprocessing, the catalyst within the reactor bed is not fixed (Eccles, 1993). The feedstock enters the bottom of the reactor and flows upward through the catalyst, which is maintained in suspension by the pressure of the fluid feedstock. Thus, ebullating bed reactors are capable of converting the most problematic feeds, such as atmospheric residue, vacuum residua, heavy crude oil, and tar sand bitumen (all of which have a high content of asphaltene constituents as well as metal constituents, sulfur constituents, and constituents ready to form sediment) to lower-boiling, more valuable products while simultaneously removing the contaminants.

Some of the catalyst is withdrawn and replaced with fresh catalyst, while the majority of the catalyst is returned to the reactor.

The main advantages of ebullating bed reactors are (1) high conversion of atmospheric residue, up to 90% v/v, (2) better product quality than many other residue conversion processes, especially delayed coking, and (3) long run length. Catalyst life does not limit these units since fresh catalyst is added and spent catalyst is removed continuously, and (subject only to mechanical issues), the units can typically run for a much longer time than fixed-bed units for the same viscous feedstock.

Examples of processes which use ebullated bed reactors are (1) the H-Oil process and (2) the LC-Fining process. The H-Oil Process is a catalytic hydrocracking process which is used to convert and upgrade petroleum residua and heavy oils, and the LC-Fining process is an ebullating bed residue hydrocracking process, which provides high conversion of residues selectively to liquid products and high removal of metals, sulfur, and coke-forming constituents from viscous feedstocks (Parkash, 2003; Gary et al., 2007; Speight, 2014a; Hsu and Robinson, 2017; Speight, 2017).

2.7.1.4 Demetallization Reactor

The demetallization reactor (also called a 'guard reactor' or a 'guard bed reactor') is a reactor that is placed in line before a hydrocracking reactor in order to remove contaminants, particularly metals, prior to hydrocracking. The demetallization reactor typically employs an inexpensive catalyst (or solid adsorbent) to remove metal-containing constituents from the feedstock. The catalyst support typically is a large pore inorganic material and the feedstock is preferentially demetallized with a low degree of desulfurization.

Feedstocks that have relatively high metal contents (on the order of >300 ppm) substantially increase catalyst consumption because the metals poison the catalyst, thereby requiring frequent catalyst replacement. The usual desulfurization catalysts are relatively expensive for these consumption rates, but there are catalysts that are relatively inexpensive and can be used in the first reactor to remove a large percentage of the metals. Subsequent reactors downstream of the first reactor would use normal hydrodesulfurization catalysts. Since the catalyst materials are proprietary, it is not possible to identify them here. However, it is understood that such catalysts contain little or no metal promoters, i.e., nickel, cobalt, molybdenum. Metal removal on the order of 90% has been observed with these materials.

Thus, one method of controlling demetallization is to employ separate smaller guard reactors just ahead of the fixed-bed hydrodesulfurization reactor section. The preheated feed and hydrogen pass through the guard reactors that are filled with an appropriate catalyst for demetallization, often the same as the catalyst used in the hydrodesulfurization section. The advantage of this system is that it enables replacement of the most contaminated catalyst (*guard bed*), where pressure drop is highest, without having to replace the entire inventory or shut down the unit. The feedstock is alternated between guard reactors while the catalyst in the idle guard reactor is being replaced.

When the expanded-bed design is used, the first reactor could employ a low-cost catalyst (5% of the cost of Co/Mo catalyst) to remove the metals, and subsequent reactors can use the more selective hydrodesulfurization catalyst. The demetallization catalyst can be added continuously without taking the reactor out of service, and the spent demetallization catalyst can be loaded with more than 30% vanadium, which makes it a valuable source of vanadium.

2.7.2 Reactor Parameters

The principal variables affecting the required severity in distillate desulfurization are: (1) hydrogen partial pressure, (2) space velocity, (3) reaction temperature, and (4) feedstock properties.

2.7.2.1 Hydrogen Partial Pressure

The important effect of the extent of the hydrogen partial pressure is to mitigate the occurrence of the reactions that produce coke. Also, if the hydrogen pressure is too low for the required effect at any position within the reaction system, premature aging of the remaining portion of catalyst will be encountered.

One of the issue-related problems in processing high-sulfur and high-nitrogen feedstocks is the large quantity of hydrogen sulfide (H_2S) and ammonia (NH_3) that is produced. Substantial removal of both compounds from the recycle gas can be achieved by the injection of water in which, under the high-pressure conditions employed, both hydrogen sulfide and ammonia are very soluble compared with hydrogen and hydrocarbon gases. The solution is processed in a separate unit for the recovery of anhydrous ammonia and hydrogen sulfide.

2.7.2.2 Space Velocity

By definition, the space velocity is the flow rate of the feedstock per volume of the reactor (or volume of the catalyst), which is an indicator of the number of reactor volumes of feed that can be fed in unit time (for instance, a reactor with a space velocity of 1 hr^{-1} is able to process feed equivalent to one time the reactor volume each hour). Thus, the space velocity (as used in the industry) is the relation between volumetric flow rate and reactor volume in a chemical reactor.

2.7.2.3 Temperature

Typically, an increase in temperature increases the rate of reaction and raises the average kinetic energy of the feedstock constituents. Therefore, a greater proportion of constituent molecules will have the minimum energy necessary for an effective collision and reaction.

As an example for refining, temperatures on the order of 415°C (780°F) and above result in excessive coking reactions and higher than normal catalyst aging rates.

2.7.2.4 Catalyst Life

The rates of refinery reactions can be increased dramatically by introducing a catalyst that increases the rate of a chemical reaction without being used up in the reaction and may actually accomplish the conversion by providing an alternate reaction pathway that has a lower activation energy barrier. Therefore, the use of a catalyst lowers the overall activation energy of the reaction.

Thus, the life of the catalyst depends on the charge stock properties and the degree of desulfurization desired. The only permanent poisons to the catalyst are metals in the feedstock that deposit on the catalyst, usually quantitatively, causing permanent deactivation as they accumulate.

Nitrogen compounds are a temporary poison to the catalyst, but there is essentially no effect on catalyst aging except that caused by a higher temperature requirement to achieve the desired desulfurization. Hydrogen sulfide can be a temporary poison in the reactor gas, and recycle gas scrubbing is employed to counteract this condition.

Providing that pressure drop buildup is avoided, cycles of 1 year or more and ultimate catalyst life of 3 years or more can be expected. The catalyst employed can be regenerated by normal steam-air or recycle combustion gas-air procedures. The catalyst is restored to near fresh activity by regeneration during the early part of its ultimate life. However, permanent deactivation of the catalyst occurs slowly during usage and repeated regenerations, so replacement becomes necessary.

2.8 THE FUTURE

There is no one single process that is a solution for viscous feedstock upgrading which is applicable to all refineries and to all viscous feedstocks. Feedstock properties, existing refinery configuration, catalysts, and desired product slate all can have a significant effect on the final configuration. New processes for the conversion of heavy crude oil feedstocks will probably be used in concert with visbreaking, with some degree of hydroprocessing as a primary conversion step. Other processes may replace or augment the deasphalting units in many refineries. Depending on the properties, an option for heavy crude oil, like the early option for tar sand bitumen, is to subject the feedstock to either delayed coking or fluid coking as the *primary upgrading* step, with some prior distillation or topping (Hedrick et al., 2006; Speight, 2020, 2014a, 2016). After primary upgrading, the product streams are hydrotreated and combined to form a *synthetic crude oil* that is shipped to a conventional refinery for further processing into liquid fuels.

2.8.1 Future Directions

Viscous feedstock upgrading will become more and more profitable as feedstocks to refineries become heavier. In fact, the increasing focus of research and development efforts to reduce sulfur content in fuels will ensure that the role of *desulfurization* in the refinery increases in importance (Babich and Moulijn, 2003). Some hydrogen may be lost to reduce the octane number of the fuel, which is undesirable.

Visbreaking or even hydrovisbreaking – which is visbreaking in an atmosphere of hydrogen or in the presence of a hydrogen donor material (Speight, 2014a, 2017) may see a surge in use as a pretreatment process for viscous feedstock upgrading. Management of the process to produce a liquid product that has been freed of the high potential for coke deposition (by taking the process parameters into the region where sediment forms), either in the absence or presence of hydrogen and/or a metal oxide scavenger, will see the process evolve as a valuable pretreating steps to catalyst cracking or hydrocracking units (Stark and Falkler, 2008; Stark et al., 2008).

A variant of the visbreaking process is the *aquaconversion process* (sometimes referred to as the aquathermolysis process), which is a catalytic visbreaking process that operates in the presence of steam (Marzin et al., 1998; Wu et al., 2010; Ovalles et al., 2015; Zhao et al., 2015). The visbreaking technology is limited in conversion level because of the stability of the resulting product because one process requirement is that the product has to be stable. Standard visbreaking allows only an approximate 2 to 6° API upgrading of the heavy crude and only a limited viscosity reduction, which does not ensure its transport without external diluent. The process pushes this maximum conversion level within the stability specification by adding a homogeneous catalyst in the presence of steam.

Recent investigations (Wu et al., 2010; Desouky et al., 2013) have demonstrated that amphiphilic catalysts (such as iron (III) dodecyl benzene sulfonate and nickel (II) dodecyl benzene sulfonate) exhibit good performance in the catalytic aquathermolysis of heavy crude oil, with a reduction in product viscosity (on the order of 79%) over that of the original heavy crude.

More recently, the concept of using water as the upgrading agent has taken one step further and morphed into a process that uses supercritical water for viscous feedstock conversion in the presence of various catalysts, and it is claimed that the process can lead to a higher yield of light oil and a lower yield of coke from tar sand bitumen (Li et al., 2015). The product yield for bitumen upgrading in supercritical water depends not only on the nature of raw bitumen and reaction conditions, but also on the operation mode. Currently, the operation modes for reactions in SCW include batch, semi-batch, or continuous reactors. The disadvantage of the concept is that the origin of the hydrogen required for upgrading is unclear, but it is known that water (or water vapor) will provide hydrogen to reactive free radicals (Birchall et al., 1970).

The use of microwaves to generate heat in a uniform and controlled fashion is well known, and microwave technology with catalysts has been applied to the upgrading of heavy crude oil (Mutyala et Al., 2010; Lam et al., 2012). For example, catalytic hydroconversion of the non-volatile product from coal liquefaction by using microwave irradiation with a Ni catalyst (Wang et al., 2008), microwave-assisted desulfurization of heavy and sour crude oil using iron powder or iron oxide as a catalyst (Leadbeater and Khan, 2008; Fumoto et al., 2012). In addition, advances have been made in the use of nanoparticle catalysts for asphaltene decomposition because of the properties of nanoparticles such as large surface area, which endows upon the particles the potential to adsorb asphaltene constituents from viscous feedstocks (Nassar, 2010; Nassar et al., 2020, 2011; Pereira-Almao, 2012, 2013; Al-Marshed et al., 2015). Also upgrading of Athabasca bitumen with microwave-assisted catalytic pyrolysis has been documented, and it has been shown that microwave heating with nanoparticle catalysts can be a useful process for upgrading viscous feedstocks such as tar sand bitumen (Jeon et al., 2012).

High conversion refineries will continue to move to gasification of viscous feedstocks for the development of alternative fuels and to enhance equipment usage either through viscous feedstock gasification processes or the installation of Flexicoking units (Bridjanian and Khadem Samimi, 2011; Speight, 2020; Speight, 2014b). A major trend in the refining industry market demand for refined products will be synthesizing fuels from simple basic reactants (e.g. synthesis gas) when it becomes uneconomical to produce super clean transportation fuels through conventional refining processes. Fischer-Tropsch plants together with IGCC systems will be integrated with or even into refineries, which will offer the advantage of high-quality products (Stanislaus et al., 2000; Holt, 2001; Speight, 2013a).

Gasification of residua is an old technology and is used in some refineries to generate hydrogen (Speight, 2014a). Research and development efforts are continuing to determine the most suitable feedstock blends for gasification (Speight, 2014b). As a result, the refinery could be equipped to produce synthesis gas from blends of viscous feedstocks and other feedstocks such as biomass (US DOE, 2002) from which liquid fuels would be manufactured using the Fischer-Tropsch synthesis technology (Wallace et al., 1998; Penrose et al., 1999;

Marano, 2003; Chadeesingh, 2011; Speight, 2011; Chhiti and Kemiha, 2013; Speight, 2013a, 2014b).

There is also increasing recognition that in situ upgrading of tar sand bitumen in the presence of a supported catalyst could be a very beneficial process for leaving the unwanted elements in the reservoir and decreasing the viscosity and increasing API gravity (Abu et al., 2015). While this is important in the context of this text insofar as upgrading during recovery will reduce surface refining costs, it is described in more detail elsewhere (Speight, 2016) and will not be repeated here.

Catalyst development for the various catalytic and hydrogen-related processes (Speight, 2014a) will be key in the modification of processes and the development of new ones to make environmentally acceptable fuels (Rostrup-Nielsen, 2004). Innovations have already occurred in catalyst materials, which have allowed refiners to vastly improve environmental performance, product quality and volume, feedstock flexibility, and energy management without fundamentally changing fixed capital stocks.

Hydrocracking heavy crude oil (5.13% w/w sulfur, 0.78% w/w nitrogen, 12.5° API, 25% w/w asphaltenes) in the presence of this novel bimetallic catalyst (W-Zr) resulted in the production of a product with a lower content of asphaltene constituents (6 to 10% w/w), lower content of resin constituents, as well as lower contents of sulfur (2.8 to 2.4% w/w) and nitrogen (0.6 to 0.7% w/w). Also, the API gravity was increased to 15 to 19° (Schacht et al., 2014).

Scavenger additives that have already seen success in the older CANMET process (Pruden, 1978; Pruden et al., 1993) and the Uniflex process (Gillis et al., 2010) will also see a surge in use. As a simple example, a metal oxide (such as calcium oxide) has the ability to react with sulfur-containing feedstock to produce a hydrocarbon (and calcium sulfide).

$$\text{Viscous feedstock[S]} + CaO \rightarrow \text{hydrocarbon product} + CaS + H_2O$$

In concert with the use of scavenger additives, metals trapping technologies will continue to be the focus of research and development, especially in regard to nickel contamination. These technologies have allowed higher levels of viscous feedstocks to be processed in numerous units, while reducing hydrogen yields and lowering coke yields. Such technologies will be continually researched for further development. As feedstocks deteriorate in quality and metals content of the feed is higher, the need for such catalyst design will be a major factor in upgrading.

Another option under study for the processing of the viscous feedstock crudes is biological upgrading through the use of biological catalysts such as fungi and bacteria. These catalysts are selected and gathered through field operations and reproduced in the laboratory. Biotechnology has been widely used in bio-remediation (Speight and Arjoon, 2012), but, in the case of upgrading, further research and extra-lab test performance are still necessary. There have been studies to characterize metabolic routes associated with desulfurization and the elimination of other types of contaminants. Several institutions have identified microorganisms that work between 50 and 65 °C to degrade crude at atmospheric pressure, while conventional processes require elevated temperatures and pressures. In microbial-enhanced oil recovery processes (McFarland et al., 1998; Setti et al., 1999; Li et al., 2005; Speight, 2014a; El-Gendy and Speight, 2015), microbial technology is exploited in oil reservoirs to improve recovery. In the process, injected nutrients, together with indigenous or added microbes, promote in situ microbial growth and/or the generation of products that mobilize additional oil and move it to producing wells through reservoir repressurization, interfacial tension/oil viscosity reduction, and selective plugging of the most permeable zones.

2.8.2 Biodesulfurization

It would be remiss to ignore an up-and-coming technology that could well be the technology of the future and that is the 'biodesulfurization' technology.

Refiners are continually challenged to produce products with ever-decreasing levels of sulfur. At the same time, the supplies of light, low sulfur crude oil that favor distillate production are limited and even decreasing. Generally, the sulfur content of the feedstocks delivered to refineries continues to rise (Speight, 2020, 2011) with the accompanying decrease in API gravity and an increase in the proportion of residua in the crude oil. These factors require the crude oil to be processed more severely to produce environmentally acceptable gasoline, other transportation fuels, and products. Thus, many refineries are now configured for maximum gasoline production that also includes increasingly processing highly aromatic distillate by-products, such as light cycle oil, for the additional feedstock to produce more distillate.

In microbial enhanced oil recovery processes, microbial technology is exploited in oil reservoirs to improve recovery. In the process, injected nutrients, together with indigenous or added microbes, promote in situ microbial growth and/or the generation of products that mobilize additional oil and move it to producing wells through reservoir repressurization, interfacial tension/oil viscosity reduction, and selective plugging of the most permeable zones.

Biocatalyst desulfurization of crude oil distillates is one of a number of possible modes of applying biologically based processing to the needs of the crude oil industry (McFarland et al., 1998; Setti et al., 1999; Abbad-Andaloussi et al., 2003; Mohebali and Ball, 2008).

The desulfurization product was identified as α-hydroxystyrene. This strain appeared to have the ability to remove organic sulfur from a broad range of sulfur species in gasoline. When straight-run gasoline containing various organic sulfur compounds was treated with immobilized cells of *mycobacterium goodie* for 24 hours at 40°C (104°F), the total sulfur content significantly decreased, from 227 to 71 ppm at 40°C. Furthermore, when immobilized cells were incubated at 40°C (104°F) with *mycobacterium goodie*, the sulfur content of the product (gasoline) decreased from 275 to 54 ppm in two consecutive reactions.

A dibenzothiophene-degrading bacterial strain, *Nocardia sp.*, was able to convert dibenzothiophene to 2-hydroxybiphenyl as the end metabolite through a sulfur-specific pathway (Chang et al., 1998). Other organic sulfur compounds, such as thiophene derivatives, thiazole derivatives, sulfides, and disulfides, were also desulfurized by *Nocardia sp.* When a sample in which dibenzothiophene was dissolved in hexadecane and treated with growing cells, the dibenzothiophene was desulfurized in approximately 80 hours.

The soil-isolated strain microcobe identified as *Rhodococcus erythropolis* can efficiently desulfurize benzonaphthothiophene (Yu et al., 2006). (Figure 2.4)

The desulfurization product was α-hydroxy-ß-phenyl-naphthalene. Resting cells were able to desulfurize diesel oil (total organic sulfur, 259 ppm) after hydrodesulfurization, and the sulfur content of diesel oil was reduced by 94.5% after 24 hours at 30°C (86°F). Biodesulfurization of crude oils was also investigated, and after 72 hours at 30°C (86°F), 62.3% of the total sulfur content in Fushun crude oil (initial total sulfur content, 3,210 ppm) and 47.2% of the sulfur in Sudanese crude oil (initial total sulfur, 1,237 ppm) were removed (see also; see also Abbad-Andaloussi et al., 2003).

However, such projects have been slow to get underway although in situ biosurfactant and biopolymer applications continue to garner interest (Van Hamme et al., 2003). In fact, studies have been carried out on biological methods of removing heavy metals such as nickel and vanadium from crude oil distillate fractions, coal-derived liquid shale, bitumen, and synthetic fuels.

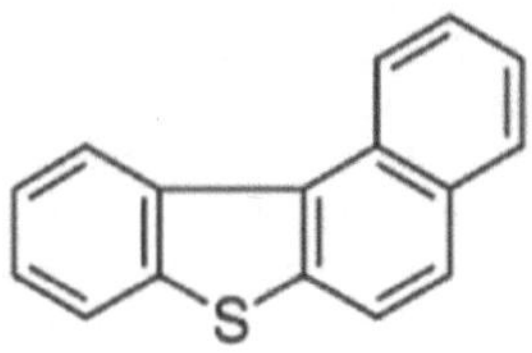

FIGURE 2.4 Benzonaphthothiophene.

For upgrading options, the use of microbes has to show a competitive advantage of enzymes over the tried-and-true chemical methods prevalent in the industry. Currently, the range of reactions using microbes is large but is usually related to the production of bioactive compounds or precursors. But the door is not closed, and the issues of biodesulfurization and bioupgrading remain open for the challenge of bulk crude oil processing. These drawbacks limit the applicability of this technology to specialty chemicals and steer it away from bulk crude oil processing.

Biodesulfurization is, therefore, another technology to remove sulfur from the feedstock. However, several factors may limit the application of this technology. Many ancillary processes novel to crude oil refining would be needed, including a biocatalyst fermentation reactor to regenerate the bacteria. The process is also sensitive to environmental conditions such as sterilization, temperature, and residence time of the biocatalyst. Finally, the process requires the existing hydrotreater to continue in operation to provide a lower sulfur feedstock to the unit and is more costly than conventional hydrotreating. Nevertheless, the limiting factors should not stop the investigations of the concept, and work should be continued with success in mind.

Once the concept has been proven on the scale that a refiner would require, the successful microbial technology will most probably involve a genetically modified bacterial strain for (1) upgrading distillates and other crude oil fractions in refineries, (2) upgrading crude oil upstream, and (3) dealing with environmental problems that face the industry, especially in areas related to spillage of crude oil and products. These developments are part of a wider trend to use bioprocessing to make products and do many of the tasks that are currently accomplished currently by conventional chemical processing. If commercialized for refineries, however, biologically based approaches will be at scales and with economic impacts beyond anything previously seen in industry.

In addition, the successful biodesulfurization process will, most likely, be based on naturally occurring aerobic bacteria that can remove organically bound sulfur in heterocyclic compounds without degrading the fuel value of the hydrocarbon matrix. Because of the susceptibility of bacteria to heat, the process will need to operate at temperatures and pressures close to ambient and also use air to promote sulfur removal from the feedstock.

REFERENCES

Abbad-Andaloussi, S., Warzywoda, M., and Monot, F. 2003. Microbial Desulfurization of Diesel Oils by Selected Bacterial Strains. *Révue Institut Français du Pétrole*, 58(4): 505–513.

Abu, I.I., Moore, R.G., Mehta, S.A., Ursenbach, M.G., Mallory, D.G., Pereira-Almao, P., and Carbognani, L. 2015. Upgrading of Athabasca Bitumen Using Supported Catalyst in Conjunction with In-Situ Combustion. *Journal of Canadian Petroleum Technology*, 54(4): 220–232.

Ancheyta, J., and Speight, J.G. 2007. *Hydroprocessing Heavy Oils and Viscous Feedstocks*. CRC Press, Taylor & Francis Group, Boca Raton, FL.

Babich, I.V., and Moulijn, J.A. 2003. Science and Technology of Novel Processes for Deep Desulfurization of Oil Refinery Streams: A Review. *Fuel*, 82(6): 607–631.

Bellussi, G., Rispoli, G., Landoni, A., Millini, R., Molinari, D., Montanari, E., Moscotti, D., and Pollesel, P. Hydroconversion of Heavy Residues in Slurry Reactors: Developments and Perspectives. *Journal of Catalysis*, 308: 189–200.

Birch, C.H., and Ulivieri, R. 2000. *ULS Gasoline and Diesel Refining Study*. Pervin and Gertz Inc., Houston, TX.

Birchall, J.N., Haszeldine, R.N., and Speight, J.G. 1970. Polyfluorarenes. Part XIII. Further Homolytic Reactions of Polyfluoroiodobenzenes. *Journal of the Chemical Society*, 2187.

BP. 2020. *BP Statistical Review of World Energy*. British Petroleum Company, London June. https://www.energyinst.org/statistical-review.

Bridjanian, H., and Khadem Samimi, A. 2011. Bottom of the Barrel: An Important Challenge of the Petroleum Refining Industry. *Petroleum and Coal*, 53(1): 13–21.

Castañeda, L.C., Muñoz, J.A.D., and Ancheyta, J. 2011. Comparison of Approaches to Determine Hydrogen Consumption during Catalytic Hydrotreating of Oil Fractions. *Fuel*, 90(12): 3593–3601.

Chadeesingh, R. 2011. The Fischer-Tropsch Process. In *The Biofuels Handbook*. J.G. Speight (Editor). The Royal Society of Chemistry, London. Part 3, Chapter 5, Page 476–517.

Chhiti, Y., and Kemiha, M. 2013. Thermal Conversion of Biomass, Pyrolysis and Gasification: A Review. *International Journal of Engineering and Science*, 2(3): 75–85.
Chang, J.H., Rhee, S.K., Chang, Y.K., and Chang, H.N. 1998. Desulfurization of Diesel Oils by a Newly Isolated Dibenzothiophene-Degrading *Nocardia sp.* strain CYKS2. *Biotechnology Progress*, 14(6): 851–855.
Dehghani, A., Sattarin, M., Bridjanian, H., and Mohamadbeigy, K. 2009. Investigation on Effectiveness Parameters in Residue Upgrading Methods. *Petroleum and Coal*, 51(4): 229–236.
Desouky, S., Al-sabagh, A., Betiha, M., Badawi, A., Ghanem, A., and Khalil, S. 2013. Catalytic Aquathermolysis of Egyptian Heavy Crude Oil. *International Journal of Chemical, Molecular, Nuclear, Materials and Metallurgical Engineering*, 7(8): 286–291.
Duddy, J.E., Wisdom, L.I., Kressmann, S., and Gauthier, T. 2004. Understanding and Optimization of Residue Conversion in H-Oil. Proceedings. Third Bottom of the Barrel Technology Conference (BBTC). Antwerp. October 20–21, Page 1003–1022.
Dziabala, B., Thakkar, V.P., and Abdo, S.F. 2011. Combination of Mild Hydrotreating and Hydrocracking for Making Low Sulfur Diesel and High-Octane Naphtha. United States Patent 8,066,867. November 29.
El-Gendy, N.S., and Speight, J.G. 2015. *Handbook of Refinery Desulfurization*. CRC Press, Taylor & Francis Group, Boca Raton, FL.
Fumoto, E., Sato, S., and Takanohashi, T. 2012. Catalytic Cracking of Heavy Oil over a Complex Metal Oxide Catalyst in a Steam Atmosphere. In *Nanocatalysis for Fuels and Chemicals*. Symposium Series Volume 1092,. American Chemical Society, Washington, DC.
Gary, J.H., Handwerk, G.E., and Kaiser, M.J. 2007. *Petroleum Refining: Technology and Economics*, 5th Edition. CRC Press, Taylor & Francis Group, Boca Raton, FL.
Gillis, D., VanWees, M., and Zimmerman, P. 2009. *Upgrading Residues to Maximize Distillate Yields*. UOP LLC, Des Plaines, IL.
Gillis, D., VanWees, M., and Zimmerman, P. 2010. Upgrading Residues to Maximize Distillate Yields with UOP Uniflex Process. *Journal of the Japan Petroleum Institute*, 53(1): 33–41.
Hedrick, B.W., Seibert, K.D., and Crewe, C. 2006. A New Approach to Heavy Oil and Bitumen Upgrading. AM-06-29. UOP 4355B. UOP LLC, Des Plaines, IL.
Holt, N.A.H. 2001. *Integrated Gasification Combined Cycle Power Plants. Encyclopedia of Physical Science and Technology*, 3rd Edition. Academic Press Inc., New York. September.
Hsu, C.S., and Robinson, P.R. (Editors) 2017. *Handbook of Petroleum Technology*. Springer AG, Cham.
Hudec, P. 2011. FCC Catalyst – Key Element in Refinery Technology. Proceedings. 45th International Petroleum Conference. Bratislava, Slovak Republic. June 13.
Jeon, S.G., Kwak, N.S., Rho, N.S., Ko, C.H., Na, J.-G., Yi, K.B., and Park, S.B. 2012. Catalytic Pyrolysis of Athabasca Bitumen in H2 Atmosphere Using Microwave Irradiation. *Chemical Engineering Research and Design*, 90(9): 1292–1296.
Khan, M.R., and Patmore, D.J. 1997. Heavy Oil Upgrading Processes. In *Petroleum Chemistry and Refining*. J.G. Speight (Editor). Taylor & Francis, Washington, DC. Chapter 6.
Kressmann, S., Morel, F., Harlé, V., and Kasztelan, S. 1998. Recent Developments in Fixed-Bed Catalytic Residue Upgrading. *Catalysis Today*, 43(3–4): 203–215.
Kressmann, S., Boyer, C., Colyar, J.J., Schweitzer, J.M., and Viguié, J.C. 2000. Improvements of Ebullated-Bed Technology for Upgrading Heavy Oils. *Oil & Gas Science and Technology. Rev. IFP*, 55(4): 397–406.
Kressmann, S., Guillaume, D., Roy, M., and Plain, C.A. 2004. New Generation of Hydroconversion and Hydrodesulfurization Catalysts. 14th Annual Symposium Catalysis in Petroleum Refining & Petrochemicals King Fahd University of Petroleum & Minerals-KFUPM The Research Institute, Dhahran, Saudi Arabia December 5–6.
Lam, S.S., Russell, A.D., Lee, C.L., and Chase, H.A. 2012. Microwave-Heated Pyrolysis of Waste Automotive Engine Oil: Influence of Operation Parameters on the Yield, Composition, and Fuel Properties of Pyrolysis Oil. *Fuel*, 92(1): 327–339.
Leadbeater, N.E., and Khan, M.R. 2008. Microwave-Promoted Desulfurization of Heavy Oil and a Review of Recent Advances on Process Technologies for Upgrading of Heavy and Sulfur Containing Crude Oil. *Energy and Fuels*, 22(3): 1836–1839.
Levant, E. 2010. *Ethical Oil: The Case for Canada's Oil Sands*. McClelland & Stewart Ltd., Toronto, ON.
Li, F., Xu, P., Feng, J., Meng, L., Zheng, Y., Luo, L., and Ma, C. 2005. Microbial Desulfurization of Gasoline in a Mycobacterium *Goodii X7B* Immobilized-Cell System. *Applied and Environment Microbiology*, 71(1): 276–281.
Li, N., Yan, B., and Xiao, X.-M. 2015. A Review of Laboratory-Scale Research on Upgrading Heavy Oil in Supercritical Water. *Energies*, 8(8): 8962–8989.

Liu, Y., Gao, L., Wen, L., and Zong, B. 2009. Recent Advances in Heavy Oil Hydroprocessing Technologies. *Recent Patents on Chemical Engineeringe*, 2(1): 22–36.

Marzin, R., Pereira, P., Zacarias, L., Rivas, L., McGrath, M., and Thompson, G.J. 1998. Resid Conversion through the Aquaconversion Technology – An Economical and Environmental Solution. SPE Paper No. 1998.086. Society of Petroleum Engineers, Richardson, TX.

McFarland, B.L., Boron, D.J., Deever, W., Meyer, J.A., Johnson, A.R., and Atlas, R.M. 1998. Biocatalytic Sulfur Removal from Fuels: Applicability for Producing Low Sulfur Gasoline. *Critical Reviews in Microbiology*, 24(2): 99–147.

Mohebali, G., and Ball, A.S. 2008. Biocatalytic Desulfurization (BDS) of Petrodiesel Fuels. *Microbiology*, 154(8): 2169–2183.

Motaghi, M., Shree, K., and Krishnamurthy, S. 2010a. Consider New Methods for Bottom of the Barrel Processing – Part 1. *Hydrocarbon Processing*, 89(2): 35–40.

Motaghi, M., Shree, K., and Krishnamurthy, S. 2010b. Consider New Methods for Bottom of the Barrel Processing – Part 2. *Hydrocarbon Processing*, 89(2): 55–88.

Mutyala, S., Fairbridge, C., Jocelyn Paré, J.R., Bélanger, J.M.R., Ng, S., and Hawkins, R. 2010. Microwave Applications to Oil Sands and Petroleum: A Review. *Fuel Processing Technology*, 91(2): 127–135.

Nassar, N.N. 2010. Asphaltene Adsorption on to Alumina Nano-particles: Kinetics and Thermodynamic Studies. *Energy and Fuels*, 24(8): 4116–4122.

Nassar, N.N., Azfar, H., and Pereira-Almao, P. 2011. Application of Nanotechnology for Heavy Oil Upgrading: Catalytic Steam Gasification/Cracking of Asphaltenes. *Energy and Fuels*, 25(4): 1566–1570.

Nassar, N.N., Azfar, H., and Pereira-Almao, P. 2020. Metal Oxide Nano-particles for Asphaltene Adsorption and Oxidation. *Energy and Fuels*, 25: 1017–1023.

Niccum, P.K., and Northup, A.H. 2006. Economic Extraction of FCC Feedstock from Residual Oils. Paper AM-06-18. National Petrochemical & Refiners Association (NPRA) Annual Meeting, Salt Lake City, UT.

Ovalles, C., Rivero, V., and Salazar, A. 2015. Downhole Upgrading of Orinoco Basin Extra-Heavy Crude Oil Using Hydrogen Donors under Steam Injection Conditions. Effect of the Presence of Iron Nanocatalysts. *Catalysts*, 5(1): 286–297.

Parkash, S. 2003. *Refining Processes Handbook*. Gulf Professional Publishing, Elsevier, Amsterdam.

Penrose, C.F., Wallace, P.S., Kasbaum, J.L., Anderson, M.K., and Preston, W.E. 1999. Enhancing Refinery Profitability by Gasification, Hydroprocessing and Power Generation. Proceedings. Gasification Technologies Conference. San Francisco, CA. October. http://www.gasification.org/uploads/eventLibrary/GTC99270.pdf.

Pereira-Almao, P. 2012. In Situ Upgrading of Bitumen and Heavy Oils via Nanocatalysis. *Canadian Journal of Chemical Engineering*, 90(2): 320–329.

Pereira-Almao, P. 2013. Recent Technological Advances in the Application of Nano-Catalytic Technology to the Enhanced Recovery and Upgrading of Bitumen and Heavy Oils. Proceedings. *New Technologies and Alternative Feedstocks in Petrochemistry and Refining*. DGMK Conference. Dresden. October 9–11. http://www.dgmk.de/petrochemistry/abstracts_content21/Pereira.pdf.

Phillips, G., and Liu, F. 2002. Advances in Residue Upgrading Technologies Offer Refiners Cost-Effective Options for Zero Fuel Oil Production. Proceedings. 2002 European Refining Technology Conference. Paris, France. November.

Pruden, B.B. 1978. Hydrocracking Bitumen and Heavy Oils at CANMET. *Canadian Journal of Chemical Engineering*, 56(3): 277–280.

Pruden, B.B., Muir, G., and Skripek, M. 1993. Preprints. *Oil Sands - Our Petroleum Future*. Alberta Research Council, Edmonton, Alberta. Page 277.

Rana, M.S., Sámano, V., Ancheyta, J., and Diaz, J.A.I. 2007. A Review of Recent Advances on Process Technologies for Upgrading of Heavy Oils and Residua. *Fuel*, 86(9): 1216–1231.

Rispoli, G., Sanfilippo, D., and Amoroso, A. 2009. Advanced Hydrocracking Technology Upgrades Extra Heavy Oil. *Hydrocarbon Processing*, 88(12): 39–46.

Ross, J., Roux, R., Gauthier, T., and Anderson, L.R. 2005. Fine-Tune FCC Operations for Changing Fuels Market. *Hydrocarbon Processing*, 84(9): 65–73.

Rostrup-Nielsen, J.R. 2004. Fuels and Energy for the Future: The Role of Catalysis. *Catalysis Reviews*, 46(3&4): 247–270.

Schacht, P., Díaz-García, L., Aguilar, J., Ramírez, S., Pérez-Romo, P., and Espinosa, M. 2014. Upgrading of Heavy Crude Oil with W-Zr Catalyst. *Advances in Chemical Engineering and Science*, 4(2): 250–257.

Schenk, C.J., and Cook, T. Charpentier, R.R., Pollastro, R.M., Charpentier, SA, Klett, T.R., Tennyson, M.E., Kirschbaum, M.A., Brownfield, M.E., and Pitman, J.K. 2009. An Estimate of Recoverable Heavy Oil Resources of the Orinoco Oil Belt, Venezuela. Fact Sheet 2009-3028. United States Geological Survey, U.S. Department of the Interior, Reston, Virginia. October.

Setti, L., Farinelli, P., Di Martino, S., Frassinetti, S., Lanzarini, G., and Pifferia, P.G. 1999. Developments in Destructive and Non-destructive Pathways for Selective Desulfurization in Oil Biorefining Process. *Applied Microbiology and Biotechnology*, 52(1): 111–117.

Speight, J.G. 2000. *The Desulfurization of Heavy Oils and Residua*, 2nd Edition. Marcel Dekker Inc., New York.

Speight, J.G. 2005a. Chemistry and Physics of Natural Bitumen and Heavy Oil. In *Coal, Oil Shale, Natural Bitumen, Heavy Oil and Peat, from Encyclopedia of Life Support Systems (EOLSS), Developed under the Auspices of the UNESCO*. EOLSS Publishers, Oxford. http://www.eolss.net.

Speight, J.G. 2005b. Upgrading and Refining of Natural Bitumen and Heavy Oil. In *Coal, Oil Shale, Natural Bitumen, Heavy Oil and Peat, from Encyclopedia of Life Support Systems (EOLSS), Developed under the Auspices of the UNESCO*. EOLSS Publishers, Oxford. http://www.eolss.net.

Speight, J.G. 2011. *The Biofuels Handbook*. Royal Society of Chemistry, London.

Speight, J.G. 2012. *Crude Oil Assay Database. Knovel*. Elsevier, New York. Online version. http://www.knovel.com/web/portal/browse/display?_EXT_KNOVEL_DISPLAY_bookid=5485&VerticalID=0.

Speight, J.G., and Arjoon, K.K. 2012. *Bioremediation of Petroleum and Petroleum Products*. Scrivener Publishing, Salem, MA.

Speight, J.G. 2013a. *The Chemistry and Technology of Coal*, 3rd Edition. CRC Press, Taylor & Francis Group, Boca Raton, FL.

Speight, J.G. 2013b. *Oil Sand Production Processes*. Gulf Professional Publishing, Elsevier, Oxford.

Speight, J.G. 2013c. *Heavy and Extra Heavy Oil Upgrading Technologies*. Gulf Professional Publishing Company, Elsevier, Oxford.

Speight, J.G. 2014a. *The Chemistry and Technology of Petroleum*, 5th Edition. CRC Press, Taylor and Francis Group, Boca Raton, FL.

Speight, J.G. 2014b. *Gasification of Unconventional Feedstocks*. Gulf Professional Publishing, Elsevier, Oxford.

Speight, J.G. 2015a. *Handbook of Petroleum Product Analysis*, 2nd Edition. John Wiley & Sons Inc., Hoboken, NJ.

Speight, J.G. 2015b. *Fouling in Refineries*. Gulf Professional Publishing, Elsevier, Oxford.

Speight, J.G. 2016. *Introduction to Enhanced Recovery Methods for Heavy Oil and Tar Sand Bitumen*, 2nd Edition. Waltham, MA.

Speight, J.G. 2017. *Handbook of Petroleum Refining*. CRC Press, Taylor & Francis Group, Boca Raton, FL.

Speight, J.G. 2019. *Heavy Oil Recovery and Upgrading*. Gulf Publishing Company, Elsevier, Cambridge, MA.

Speight, J.G. 2020. *Refinery of the Future*, 2nd Edition. Gulf Professional Publishing, Elsevier, Cambridge, MA.

Speight, J.G. 2021. *Petroleum Refining Technology Series Volume 1: Refinery Feedstocks*. CRC Press, Taylor & Francis Group, Boca Raton, FL.

Speight, J.G. 2024. *Petroleum Refining Technology Series Volume 4: Hydrotreating and Hydrocracking*. CRC Press, Taylor & Francis Group, Boca Raton, FL.

Stanislaus, A., Qabazard, H., and Absi-Halabi, M. 2000. Refinery of the Future. Proceedings. 16th World Petroleum Congress, Calgary, Canada. June 11–15.

Stark, J.L., and Falkler, T. 2008. Method for Improving Liquid Yield during Thermal Cracking of Hydrocarbons. United States Patent 7,425,259 September 16.

Stark, J.L., Falkler, T., Weers, J.J., and Zetlmeisl, M.J. 2008. Method for Improving Liquid Yield during Thermal Cracking of Hydrocarbons. United States Patent 7,416,654. August 26.

Stratiev, D., and Petkov, K. 2009. Residue Upgrading: Challenges and Perspectives. *Hydrocarbon Processing*, 88(9): 93–96.

US Congress. 1976. Public Law FEA-76-4. United States Congress. Library of Congress, Washington, DC.

US DOE. 2002. Gasification Technologies. Gasification Markets and Technologies: Present and Future. An Industry Perspective. Report No. DOE/FE-0447. United States Department of Energy, Washington, DC. July 3.

Van Hamme, J.D., Singh, A., and Ward, O.P. 2003. Recent Advances in Petroleum Microbiology. *Microbiology and Molecular Biology Reviews*, 67(4): 503–549.

Wallace, P.S., Anderson, M.K., Rodarte, A.I., and Preston, W.E. 1998. Heavy Oil Upgrading by the Separation and Gasification of Asphaltenes. Proceedings. Gasification Technologies Conference. San Francisco, CA. October. http://www.gasification.org/uploads/eventLibrary/gtc9817p.pdf.

Wang, T.X., Zong, Z.M., Zhang, V.W., Wei, Y.B., Zhao, W., Li, B.M., and Wei, X.Y. 2008. Microwave-Assisted Hydroconversion of Demineralized Coal Liquefaction Residues from Shenfu and Shengli Coals. *Fuel*, 87(4–5): 498–507.

Wu, C., Leu, G., Yao, C., and Jia, X. 2010. In Situ Upgrading Extra Heavy Oil by Catalytic Aquathermolysis Treatment Using a New Catalyst Based Anamphiphilic Molybdenum Chelate. Paper No. SPE-130333-MS. Proceedings. International Oil and Gas Conference and Exhibition. Beijing. June 8–10. Society of Petroleum Engineers, Richarson, TX.

Yu, B., Xu, P., Shi, Q., and Ma, C. 2006. Deep Desulfurization of Diesel Oil and Crude Oils by a Newly Isolated *Rhodococcus Erythropolis Strain. Applied and Environmental Microbiology*, 72(1): 54–58.

Zhang, J., Zhou, Y., Xu, Y., and Tian, H. 2014. Research and Development of Novel Heavy Oil Catalytic Cracking Catalyst RCC-1. *China Petroleum Processing and Petrochemical Technology*, 16(4): 7–11.

Zhao, F., Huang, J., Li, M., Liu, S., Guo, Y., and Zhang, P. 2015. Study on Hydrogen Donors Catalytic Upgrading of Heavy Oil Using Ultradispersed Catalyst. *Journal of Chemical and Pharmaceutical Research*, 7(4): 1370–1377.

Zhao, F., Xu, T., Zhu, G., Wang, K., Xu, X., and Liu, L. 2022. A Review on the Role of Hydrogen Donors in Upgrading Heavy Oil and Bitumen. *Sustainable Energy and Fuels*, 6(8): 1866–1890.https://pubs.rsc.org/en/content/articlelanding/2022/se/d1se01969d/unauth#:~:text=In%20addition%20to%20further%20reducing,hydrogen%20donor%20increases%20with%20time.

3 Hydrogen Production in the Refinery

3.1 INTRODUCTION

The two previous chapters present the role of hydrogen in the general layout of a refinery (Chapter 1) and the need for hydrogen not only for product improvement but also for refining the viscous feedstocks to produce saleable product (i.e., products that meet the necessary specifications of sale).

Thus, hydrogen is an important part of a variety of refining, chemical, and petrochemical processes and, thus, continues to be used (and necessary) in large quantities in refinery feedstocks refining (Parkash, 2003; Gary et al., 2007; Speight, 2011, 2014a, 2016; Hsu and Robinson, 2017; Speight, 2017). In fact, refinery feedstocks refineries are large hydrogen consumers and this consumption is increasing in order to enhance the conversion capacity of the viscous feedstocks (Chapter 2). Some hydrogen is produced within the refinery site from catalytic reforming units and from recovery of hydrogen from off-gas streams.

Throughout the previous chapters, there have been several references and/or acknowledgments of a very important property of refinery feedstocks and refinery products, and that is the production of hydrogen in the refinery and the use of hydrogen during refining in hydrotreating processes, such as desulfurization, and in hydroconversion processes such as hydrotreating and hydrocracking (Table 3.1, Table 3.2, Table 3.3, Figure 3.1) (Parkash, 2003; Gary et al., 2007; Speight, 2011, 2014a, 2016; Hsu and Robinson, 2017; Speight, 2017). In terms of hydrogen availability, there are reports that naturally occurring hydrogen is available and occurs as a free gas, either in layers of the continental crust, deep in the oceanic crust, in volcanic gases, in geysers, or in hydrothermal systems and is present in a wide range of rock formations and geological regions (Wang et al., 2023). However, for the most part, hydrogen is produced as a refinery product and used as such within refinery systems.

Although hydrogen recycle gas may contain up to 40% by volume of other gases (usually hydrocarbon derivatives), hydrotreater catalyst life is a strong function of hydrogen partial pressure. Typical purity increases resulting from hydrogen purification equipment and/or increased hydrogen sulfide removal, as well as tuning hydrogen circulation and purge rates may extend catalyst.

Currently, refineries are under increasing pressure to meet environmental legislation and reduce the emissions of their processes, so it is desirable to produce hydrogen in a cleaner way.

The hydrogen content of the refinery feedstock or the use of hydrogen during refining in hydrotreating processes, such as in the various desulfurization processes and in the hydroconversion processes such as hydrocracking is an important aspect of refining (Speight, 2014a, 2017, 2023). Although the hydrogen recycle gas may contain up to 40% v/v of other gases (usually hydrocarbon but which do not interfere with or change the course of the hydroprocess), hydrotreater catalyst life is a strong function of hydrogen partial pressure. Typical purity increases resulting from hydrogen purification equipment and/or increased hydrogen sulfide removal, as well as tuning hydrogen circulation and purge rates may extend catalyst life up to about 25%.

As hydrogen use has become more widespread in refineries, hydrogen production has moved from the status of a high-tech specialty operation to an integral feature of most refineries by a variety of methods (Table 3.1) (Raissi, 2001; Vauk et al., 2008). This has been made necessary by the increase in hydrotreating and hydrocracking, including the treatment of progressively higher viscosity feedstocks.

DOI: 10.1201/9781003186694-3

TABLE 3.1
General Methods for the Production of Hydrogen

Process	Description
Gas reforming/gasification	Synthesis gas (a mixture of hydrogen, carbon monoxide, and a small amount of carbon dioxide) is created by reacting natural gas with high-temperature steam. The carbon monoxide is reacted with water (the water gas shift reaction) to produce additional hydrogen
	Natural gas reforming using steam accounts for the majority of hydrogen produced in refineries.
Coal/biomass gasification	Synthesis gas can also be created by reacting coal or biomass with high-temperature steam and oxygen in a pressurized gasifier
Electrolysis	An electric current splits water into hydrogen and oxygen. If the electricity is produced by renewable sources, such as solar or wind, the resulting hydrogen will be considered renewable as well, and have numerous emissions benefits. Power-to-hydrogen processes use excess renewable electricity, when available, to make hydrogen through electrolysis
Biomass-derived liquids	Renewable liquid fuels, such as ethanol, are reacted with high-temperature steam to produce hydrogen near the point of end use
Microbial Biomass	Biomass is converted microbially into sugar-rich products that can be fermented to produce hydrogen
Water splitting — 1	In the thermochemical water splitting process, high temperatures generated by solar concentrators or nuclear reactors drive chemical reactions that split water to produce hydrogen
Water Splitting – 2	In the microbial water splitting process, photobiological microbes, such as green algae, consume water in the presence of sunlight and produce hydrogen as a by-product
Water Splitting – 3	In the photoelectrochemical water splitting systems, hydrogen is produced from water using special semiconductors and energy from sunlight

The continued increase in hydrogen demand over the last several decades is a result of the conversion of refinery feedstocks to match changes in product slate and the supply of heavy, high-sulfur oil, and in order to make lower-boiling, cleaner, and more saleable products. There are also many reasons other than product quality for using hydrogen in processes, adding to the need to add hydrogen at relevant stages of the refining process, and most important according to the availability of hydrogen (Bezler, 2003); Miller and Penner, 2003; Ranke and Schödel, 2003).

With the increasing need for clean fuels and other environmentally benign products, the production of hydrogen for refining purposes requires a major effort by refiners. In fact, the trend to increase the number of hydrogenation (*hydrocracking* and/or *hydrotreating*) processes in refineries coupled with the need to process the heavier oils, which require substantial quantities of hydrogen for upgrading because of the increased use of hydrogen in hydrocracking processes, has resulted in vastly increased demands for this gas. The hydrogen demands can be estimated to a very rough approximation using API gravity and the extent of the reaction, particularly the hydrodesulfurization reaction (Speight, 2000, 2014a, 2017). But accurate estimation requires equivalent process parameters and a thorough understanding of the nature of the ach process. Thus, as hydrogen production grows, a better understanding of the capabilities and requirements of a hydrogen plant becomes ever more important to overall refinery operations as a means of making the best use of hydrogen supplies in the refinery.

The chemical nature of the refinery feedstocks has always been a major determinant (and will continue to play) the major role in determining the hydrogen requirements of that refinery. For

TABLE 3.2
Summary of Typical Hydrogen Application and Production Process in a Refinery

Hydrogen Applications

Naphtha hydrotreater:

- Uses hydrogen to desulfurize naphtha from atmospheric distillation; must hydrotreat the naphtha before sending it to a catalytic reformer unit

Distillate hydrotreater:

- Desulfurizes distillates after atmospheric or vacuum distillation; in some units, aromatic constituents are hydrogenated to cycloparaffin derivatives or to alkane derivatives

Hydrodesulfurization:

- Sulfur compounds are hydrogenated to hydrogen sulfide as feed for Claus plants

Hydroisomerization:

- Normal (straight-chain) paraffins are converted into iso-paraffins to improve the product properties (e.g., octane number)

Hydrocracker:

- Uses hydrogen to upgrade heavier fractions into lower boiling, more valuable products

Hydrogen Production

Catalytic reformer:

- Used to convert the naphtha-boiling range molecules into higher octane reformate; hydrogen is a by-product

Steam-methane reformer:

- Produces hydrogen for hydrotreaters or hydrocrackers

Steam reforming of higher molecular weight hydrocarbons:

- Produces hydrogen from low-boiling hydrocarbons other than methane

Recovery from refinery off-gases:

- Process gas often contains hydrogen in the range of up to 50% v/v

Gasification of petroleum residua:

- Recovery from synthesis gas (syngas) produced in gasification units

Gasification of petroleum coke:

- Recovery from synthesis gas (syngas) produced in gasification units

Partial oxidation processes:

- Analogous to gasification process, synthesis gas can be produced from which hydrogen can be isolated

example, the lower boiling more paraffinic refinery feedstock will require somewhat less hydrogen for upgrading to, say, a naphtha product than a heavier more asphaltic refinery feedstocks (Speight, 2014, 2017, 2024). It follows that the hydrodesulfurization of a viscous refinery feedstock (which, by definition, is a hydrogen-dependent process) needs substantial amounts of hydrogen as part of the processing requirements.

In general, considerable variation exists from one refinery to another in the balance between hydrogen produced and hydrogen consumed in the refining operations. However, what is more pertinent to the present text is the excessive amounts of hydrogen that are required for hydroprocessing operations, whether these be hydrocracking or the somewhat milder hydrotreating processes. For effective hydroprocessing, a substantial hydrogen partial pressure must be maintained in the reactor, and in order to meet this requirement, an excess of hydrogen above that actually consumed by the process must be fed to the reactor. Part of the hydrogen requirement is met by recycling a stream of hydrogen-rich gas. However, the need remains to generate hydrogen as makeup material to accommodate the process consumption of 500 to 3000 scf/bbl, depending upon whether the heavy feedstock is being subjected to a predominantly hydrotreating (hydrodesulfurization) or to a predominantly hydrocracking process.

TABLE 3.3
Hydroprocessing Parameters

Conditions
Solid acid catalyst (silica-alumina with rare earth metals, various other options)
Temperature: 260 to 450°C (500 to 845°F solid/liquid contact)
Pressure: 1,000 to 6,000 psi hydrogen
Frequent catalysts renewal for heavier feedstocks
Gas oil: catalyst life up to three years
Heavy oil/tar sand bitumen: catalyst life is less than 1 year
Feedstocks
Distillates
Refractory (aromatic) streams
Coker oils
Cycle oils
Gas oils
Residua (as a full hydrocracking or hydrotreating option)
In some cases, asphaltic constituents (S, N, and metals) are removed by deasphalting
Products
Lower molecular weight paraffins
Some methane, ethane, propane, and butane
Hydrocarbon distillates (full range depending on the feedstock)
Residual tar (recycle)
Contaminants (asphaltic constituents) deposited on the catalyst as coke or metals
Variations
Fixed-bed (suitable for liquid feedstocks)
Ebullating bed (suitable for heavy feedstocks)

In some refineries, the hydrogen needs can be satisfied by hydrogen recovery from catalytic reformer product gases, but other external sources are required because of the inclusion of the viscous feedstocks as refinery feedstocks on a more regular basis Chapter 2 (Speight, 2014, 2017, 2024). However, for the most part, many refineries now require on-site hydrogen production facilities to supply the gas for their own processes. Most of this non-reformer hydrogen is manufactured either by steam-methane reforming or by oxidation processes. However, other processes, such as steam-methanol interaction or ammonia dissociation, may also be used as sources of hydrogen.

As environmental regulations tightened, the technology matured, and heavier streams were hydrotreated.

3.2 PROCESSES REQUIRING HYDROGEN

The demand for hydrogen in the refinery system has increased both in domestic refineries and in international refineries, and as sulfur-content regulations have become more stringent. Much of the growth in hydrogen use at refineries is being met through hydrogen purchased from non-refinery sources rather than from increased production of hydrogen within the refinery site. This increased demand has arisen not only from the changing characteristics of the feedstocks sent to a refinery (Chapter 2), but also because of the increasingly more stringent regulations related to the production of environmentally acceptable products.

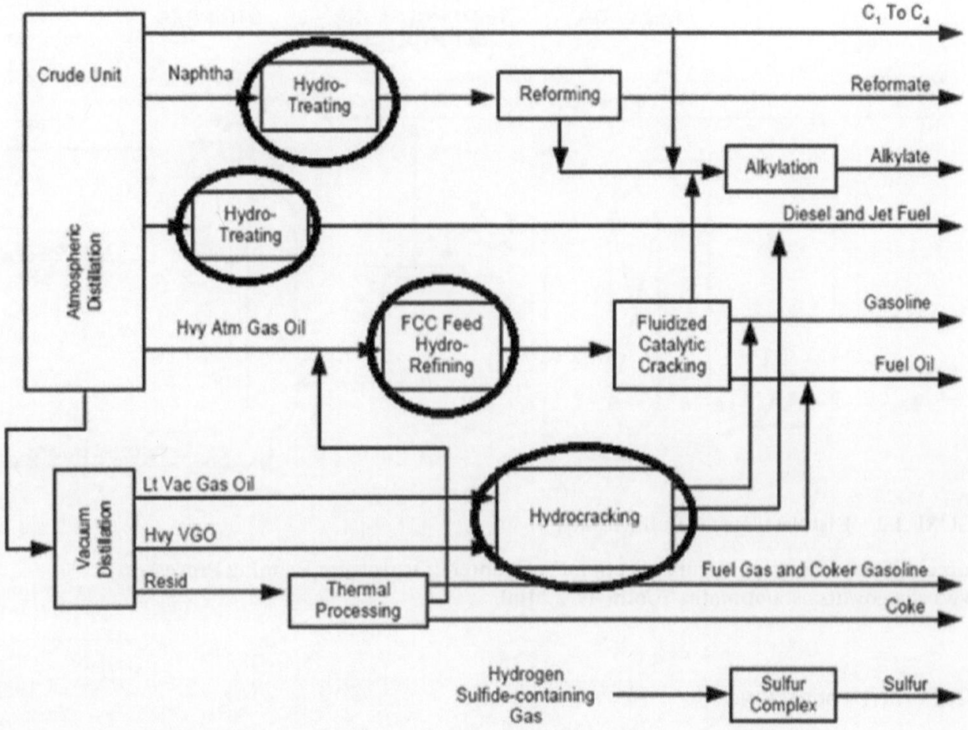

FIGURE 3.1 Examples of the placement of hydroprocesses in a refinery.

Source: OSHA Technical Manual, Section IV, Chapter 2: Petroleum Refining Processes. www.osha.gov/dts/osta/otm/otm_iv/otm_iv_2.html.

3.2.1 Hydrotreating

In a typical catalytic hydrodesulfurization unit, the feedstock is deaerated and mixed with hydrogen, preheated in a fired heater (315 to 425°F; 600 to 800°F) and then charged under pressure (up to 1,000 psi) through a fixed-bed catalytic reactor (Figure 3.2).

In the reactor, the sulfur and nitrogen compounds in the feedstock are converted into hydrogen sulfide and ammonia. The liquid stream is the product from hydrotreating and is normally sent to a stripping column for removal of hydrogen sulfide and other undesirable components. In cases where steam is used for stripping, the product is sent to a vacuum drier for removal of water.

Lube-oil hydrotreating uses catalytic treatment of the oil with hydrogen to improve product quality. The objectives in mild lube hydrotreating include saturation of olefins and improvements in color, odor, and acid nature of the oil. Mild lube hydrotreating also may be used following solvent processing. Operating temperatures are usually below 315°C (600°F) and operating pressures below 800 psi. Severe lube hydrotreating, at temperatures in the range of 315 to 400°C (600 to 750°F) range and hydrogen pressures up to 3000 psi, is capable of saturating aromatic rings, along with sulfur and nitrogen removal, to impart specific properties not achieved at mild conditions.

Hydrotreating can also can be employed to improve the quality of naphtha produced by pyrolysis (often referred to as 'pyrolysis gasoline' or 'pygas), a by-product from the manufacture of ethylene. Traditionally, the outlet for pygas has been as a blend stock for motor gasoline, which is a suitable use in view of its high-octane number.

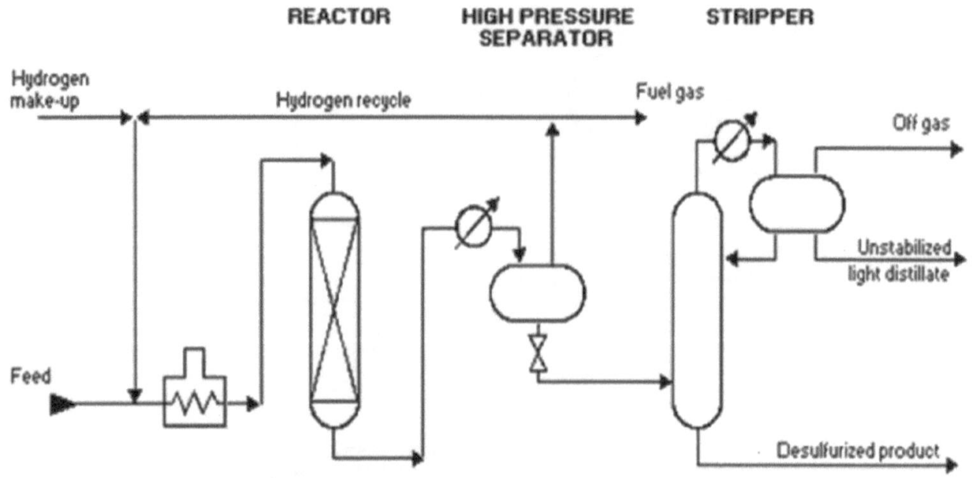

FIGURE 3.2 Distillate Hydrodesulfurization.

Source: OSHA Technical Manual, Section IV, Chapter 2: Petroleum Refining Processes. www.osha.gov/dts/osta/otm/otm_iv/otm_iv_2.html.

3.2.2 Hydrocracking

The hydrocracking process is a two-stage process that also produces a relatively large amount of iso-butane for alkylation feedstock. The process also performs isomerization for pour-point control and smoke-point control, both of which are important in high-quality jet fuel.

Hydrocracking employs high pressure, high temperature, and a catalyst. In the process, high molecular weight aromatic feedstocks are converted into lower boiling products under a wide range of very high pressures (1,000 to 2,000 psi) and fairly high temperatures (400 to 815°C; 750 to 1,500°F), in the presence of hydrogen and special catalysts. Another important role of hydrogen in the hydrocracking process is to reduce tar formation and prevent buildup of coke on the catalyst. Hydrogenation also serves to convert sulfur and nitrogen compounds present in the feedstock to hydrogen sulfide and ammonia.

In the first stage of the process (Figure 3.3), the preheated feedstock is mixed with recycled hydrogen and sent to the first-stage reactor, where catalysts convert sulfur and nitrogen compounds to hydrogen sulfide and ammonia. Limited hydrocracking also occurs.

3.3 HYDROGEN PRODUCTION

Hydrogen in molecular form can be produced from many different sources, and in many ways such as (1) from any hydrocarbon derivative by the steam reforming process, (2) from water by electrolysis, a well-known electrochemical process, or from (3) from various biological materials (Conte et al., 2001).

Briefly, most hydrogen produced in refineries involves the steam-methane reforming process, which is a mature production process in which high-temperature steam (700 to 1,000°C, 1,290 to 1,830°F) is used to produce hydrogen from a methane source, such as natural gas. In the electrochemical process, hydrogen is separated from a hydrogen-nitrogen-carbon dioxide mixture ($H_2/N_2/CO_2$) by applying a direct current to a proton-conducting membrane. The production of hydrogen from biomass typically involves gasification of the biomass by means of a controlled partial-oxidation process involving heat, steam, and oxygen to convert biomass to hydrogen and other products, without combustion (Chapter 5). The goal of each type of process is the production of a high-purity

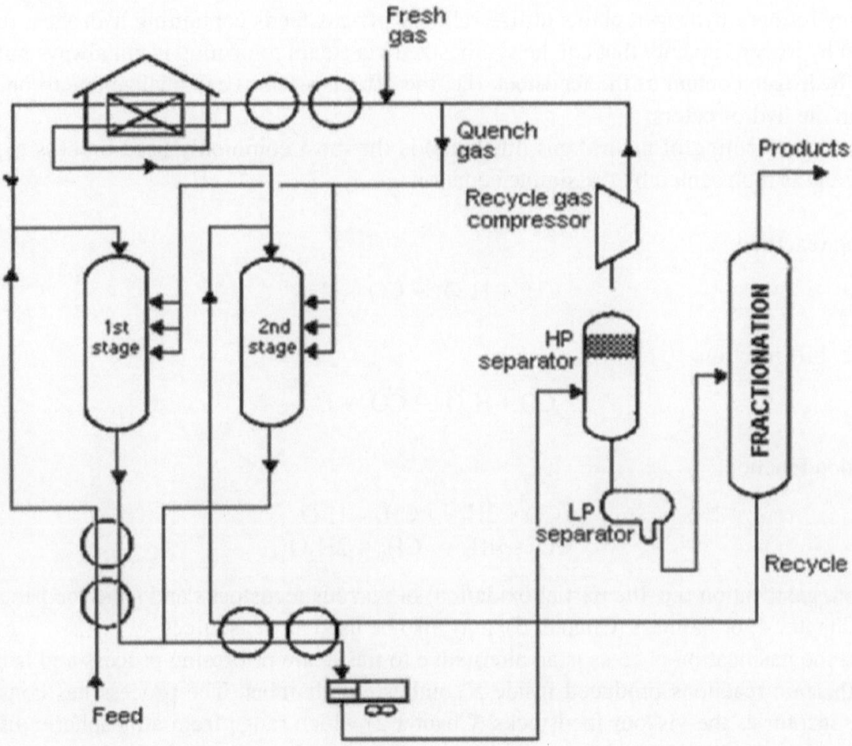

FIGURE 3.3 A Single-Stage or Two-Stage (Optional) Hydrocracking Unit.
Source:
OSHA Technical Manual, Section IV, Chapter 2: Petroleum Refining Processes. www.osha.gov/dts/osta/otm/otm_iv/otm_iv_2.html.

hydrogen stream (>95% v/v hydrogen) that can be achieved from a low purity (on the order of 30 to 50% v/v hydrogen) feed by means of a dedicated via a separation process (Chapter 6).

Typically, hydrogen is produced in a refinery feedstocks refinery by the steam-methane reforming process (commonly referred to by the acronym 'the SMR process') using natural gas for feedstock. In the early refineries, hydrogen for the hydroprocesses was provided as a result of catalytic reforming processes in which dehydrogenation is a major chemical reaction and, as a consequence, hydrogen gas is produced. The low-boiling products from the catalytic reformer contain a high ratio of hydrogen to methane so the stream is freed from ethane and/or propane to get a high concentration of hydrogen in the stream.

More specifically, within the modern refinery system, hydrogen is most typically produced through the steam-methane reforming process (Cruz and Oliveira Junior, 2008). In fact, many existing refinery hydrogen plants use a conventional process, which produces a medium-purity hydrogen product (94% to 97% v/v hydrogen) by removing the carbon dioxide in an absorption system and methanation of any remaining carbon oxides. Since the 1980s, most hydrogen plants are built with the pressure swing adsorption (PSA) technology to recover and purify the hydrogen to purities above 99.9%. Thus,

Natural gas → desulfurization
Desulfurized natural gas → steam reforming
Steam reformed desulfurized gas stream → shift conversion
Product from shift conversion → pressure swing adsorption → hydrogen.

Since many refinery hydrogen plants utilize refinery off-gas feeds containing hydrogen, the actual maximum hydrogen capacity that can be synthesized via steam reforming is not always guaranteed since the hydrogen content of the feedstock (i.e. the off-gas stream) can change due to operational changes in the hydrotreaters.

The steam reforming of natural gas (methane) is the most commonly used process to produce hydrogen and is represented by the simple equation:

Reforming reaction:

$$CH_4 + H_2O \rightarrow CO + 3H_2$$

Water gas shift reaction:

$$CO + H_2O \rightarrow CO_2 + H_2$$

Methanation reaction:

$$CO + 3H_2 \rightarrow CH_4 + H_2O$$
$$CO_2 + 4H_2 \rightarrow CH_2 + 2H_2O$$

In addition, gasification (i.e. the partial oxidation) of viscous feedstocks and solid feedstocks, such as coal (Chapter 4) or biomass (Chapter 5), may also be used as the source.

In fact, the gasification process is an alternative to the steam reforming process and is based on very exothermic reactions produced inside a combustion chamber. The process has considerable flexibility insofar as the viscous feedstocks (Chapter 2) which range from atmospheric distillation residue, vacuum distillation residue, residual 'tar' from a solvent de-asphalting process, residual 'tar' from a visbreaking process, and refinery coke can be used as feedstocks to a gasifier.

In fact, the gasification process is a way to convert viscous fossil fuel-related feedstocks into either a combustible gas or a synthesis gas for subsequent utilization. The gasification process can be operated not only with refinery generated viscous feedstocks (such as process residua) but fired not only with coal but also with biomass, various types of waste, and mixed feedstocks (Chapter 5; Speight, 2014b). Thus, in the process, an unconventional feedstock oil may also be used as the source of hydrogen. In the process, the feedstock is reacted at high pressure (1,300 to 1,800 psi) with pure oxygen (that is supplied in strictly controlled quantities) for partial oxidation to carbon monoxide and hydrogen. In the purification step, the carbon dioxide that is produced in the shift reaction is removed by absorption in a basic solvent such as potassium carbonate (K_2CO_3) solution.

Generally, as refineries have evolved, there have been two forms of hydrogen production in a refinery and these are (1) hydrogen production using steam methane reformers (SMR) or other processes, and (2) hydrogen production as a by-product of other chemical processes although other processes are recognized (Table 3.1). Natural gas is used almost exclusively as feedstock for on-site hydrogen production in steam-methane reforming units in many countries.

Historically, the hydrogen produced during catalytic reforming processes as a by-product of the production of the aromatic compounds used in naphtha and in solvents satisfied the hydrogen requirement of many refineries. However, as reforming processes changed from fixed-bed to cyclic to continuous regeneration, process pressures have dropped and hydrogen production per barrel of reformate has tended to increase. However, hydrogen production as a by-product is not always adequate to the needs of the refinery and other processes are necessary. Thus, hydrogen production by steam reforming or by partial oxidation of residua has also been used, particularly where heavy refinery feedstocks are available. Steam reforming is the dominant method for hydrogen production and is usually combined with pressure swing adsorption (PSA) to purify the hydrogen to greater than 99% by volume (Bandermann and Harder, 1982).

The gasification of distillation residuum (or a process residuum) and coke to produce hydrogen and/or power may become an attractive option for refiners (Fleshman, 1997; Gross and Wolff,

2000). The premise is that the gasification section of a refinery will be the disposal unit (often referred to as the 'garbage can') for residues from the deasphalting process, high-sulfur coke, as well as other refinery wastes is worthy of consideration.

Several other processes are available for the production of the additional hydrogen that is necessary for the various heavy feedstock hydroprocessing sequences, and it is the purpose of the present chapter to present a general description of these processes. In general, most of the external hydrogen is manufactured by steam-methane reforming or by oxidation processes. Other processes such as ammonia dissociation, steam-methanol interaction, or electrolysis are also available for hydrogen production, but economic factors and feedstock availability assist in the choice between processing alternatives. In general, although there may be other uses for hydrogen, a refinery uses hydrogen to lower the sulfur content of fuels.

However, as refineries and refinery feedstocks have evolved during the last four decades, the demand for hydrogen has increased and reforming processes are no longer capable of providing the quantities of hydrogen necessary for feedstock hydrogenation. Within the refinery, other processes are used as sources of hydrogen. Thus, the recovery of hydrogen from the by-products of the coking units, visbreaker units, and catalytic cracking units is also practiced in some refineries.

In coking units and visbreaker units, viscous feedstocks are converted to refinery feedstocks: coke, oil, low-boiling hydrocarbon derivatives (liquefied petroleum gas and naphtha). Depending on the process, hydrogen is present in a wide range of concentrations. Since coking processes need gas for heating purposes, adsorption processes are best suited to recover the hydrogen because they feature a very clean hydrogen product and an off-gas suitable as fuel.

Catalytic cracking is the most important process step for the production of low-boiling products from gas oil and increasingly from vacuum gas oil and heavy feedstocks. In catalytic cracking, the molecular mass of the main fraction of the feed is lowered, while another part is converted to coke that is deposited on the hot catalyst. The catalyst is regenerated in one or two stages by burning the coke off with air that also provides the energy for the endothermic cracking process. In the process, paraffins and naphthenes are cracked to olefins and alkanes with shorter chain length, mono-aromatic compounds are dealkylated without ring cleavage, di-aromatics and poly-aromatics are dealkylated and converted to coke. Hydrogen is formed in the last type of reaction, whereas the first two reactions produce low-boiling hydrocarbon derivatives and, therefore, require hydrogen. Thus, a catalytic cracker can be operated in such a manner that enough hydrogen for subsequent processes is formed.

In reforming processes, naphtha fractions are reformed as blend stocks to improve the quality of gasoline (Speight, 2000, 2014a, 2017). The important reactions occurring during this process are the dehydrogenation of naphthenes to aromatics, which is an endothermic reaction that is favored by low pressures, and the reaction temperature lies in the range of 300 to 450°C (570 to 840°F). The reaction is performed on platinum catalysts, with other metals, e.g., rhenium, as promoters.

The trend to increase the number of hydrogenation processes (hydrotreating processes and/or hydrocracking processes) in refineries (Dolbear, 1998) coupled with the need to process the heavier oils, which require substantial quantities of hydrogen for upgrading, has resulted in vastly increased demands for this gas.

In the modern refinery, hydrogen is required in large volumes driven by the need for ultra-pure (i.e., ultra-low sulfur) products and the incursion of high-sulfur feedstocks into refinery operations where hydrogen is necessary for a variety of applications in several different scales of supply.

The processes described in this chapter are those gasification processes that are often referred to as 'the garbage disposal units' of the refinery. Hydrogen is produced for use in other parts of the refinery as well as for energy, and it is often produced from process by-products that may not be of any use elsewhere. Such by-products might be the highly aromatic, heteroatom, and metal-containing reject from a deasphalting unit or from a mild hydrocracking process. However attractive this may seem, there will be a need to incorporate a gas cleaning operation to remove any environmentally objectionable components from the hydrogen gas.

After production, hydrogen needs to be purified before use, and there are several methods by which purification can be accomplished, including processes such as pressure swing adsorption

TABLE 3.4
Hydrogen Purification Processes

Physical processes
Adsorption
Pressure swing adsorption
Temperature swing adsorption
Vacuum adsorption
Low-temperature separation
Cryogenic distillation
Low-temperature adsorption
Membrane separation
Inorganic membrane
Metal membrane
Carbon molecular sieve
Organic membrane
Polymer membrane
Chemical processes
Metal hydride separation
Catalysis

(PSA), cryogenic distillation, or membrane separation. These methods are covered in more detail elsewhere in this text (Chapter 6). However, in the current context, a brief mention of the methods is necessary.

When the hydrogen content of the refinery gas is greater than 50% v/v, the gas should first be considered for hydrogen recovery, using a membrane (Grashoff et al., 1983; Brüschke, 1995, 2003) or a pressure swing adsorption unit (Mokhatab et al., 2006; Speight, 2014a, 2019). The tail gas or reject gas that will still contain a substantial amount of hydrogen can then be used as steam reformer feedstock. Generally, the feedstock purification process uses three different refinery gas streams to produce hydrogen. First, high-pressure hydrocracker purge gas is purified in a membrane unit that produces hydrogen at medium pressure and is combined with medium pressure off-gas that is first purified in a pressure swing adsorption unit. Finally, low-pressure off-gas is compressed, mixed with reject gases from the membrane and pressure swing adsorption units, and used as steam reformer feed.

Various processes are available to purify the hydrogen stream but since the product streams are available in a wide variety of composition, flows, and pressures, the best method of purification will vary (Table 3.4). These processes are presented below (alphabetically and not in any order of preference) to maintain the context of this chapter. In addition, there are several factors that must also be taken into consideration in the selection of a purification method. These are: (1) hydrogen recovery, (2) product purity, (3) pressure profile, (4) reliability, and last but certainly not least (5) the cost. These factors are not considered here but are presented in more detail elsewhere in this text (Chapter 5) since the emphasis at this point is on the technical aspects of the purification processes which are (1) cryogenic separation, (2) use of a desiccant separation system, (3) the use of a membrane system, including the use of metal membranes and polymer membranes, (4) pressure swing adsorption, and (5) wet scrubbing. Each of these technologies will be presented in detail in a later chapter (Chapter 5).

3.4 FEEDSTOCKS

There are four main sources for the commercial production of hydrogen, and they are (1) natural gas, (2) refinery feedstocks, (3) coal or coke, and (4) electrolysis, which account for substantial portions of global hydrogen production.

Using the coal-based technology as an example, supercritical water, i.e., water at a temperature and pressure that are at or above the critical values (374.3°C/705.7°F and 304.6 psi), is used and provides a homogeneous and rapid reaction so that the chemical energy of coal is directly and efficiently converted into hydrogen (Du et al., 2021; Sun et al., 2021).

For example, the splitting of water by electrolysis of water can be used to produce high-purity hydrogen.

$$2H_2O + \text{electricity} \rightarrow 2H_2 + O_2$$

Alkaline electrolyzers are suited for stationary applications and are available at operating pressures up to 375 psi.

Electrolyte:

$$4H_2O \rightarrow 4H+ + 4OH^-$$

Cathode:

$$4\,H+ + 4e^- \rightarrow 2H_2$$

Anode:

$$4OH^- \rightarrow O_2 + 2H_2O + 4e^-$$

Thus:

$$2H_2O \rightarrow O_2 + 2H_2$$

Typically, commercial electrolyzer units consist of a number of electrolytic cells arranged in a cell stack.

However, fossil fuel sources have remained the dominant source of industrial hydrogen.

Natural gas typically contains more than 90% methane and ethane with only minor amounts of propane and higher boiling hydrocarbon derivatives (Mokhatab et al., 2006; Speight, 2019). Natural gas may (or most likely will) contain traces of carbon dioxide with some nitrogen and other impurities. Purification of natural gas, before reforming, is usually relatively straightforward. Traces of sulfur must be removed to avoid poisoning the reformer catalyst; zinc oxide treatment in combination with hydrogenation is usually adequate.

Refinery gas, containing a substantial amount of hydrogen, can be an attractive steam reformer feedstock since it is produced as a by-product. Processing of refinery gas will depend on its composition, particularly the levels of olefins and of propane and heavier hydrocarbons. Olefins, which can cause problems by forming coke in the reformer, are converted to saturated compounds in the hydrogenation unit. Higher boiling hydrocarbons in refinery gas can also form coke, either on the primary reformer catalyst or in the preheater. If there are more than minimal amounts of C_3 and higher compounds, a promoted reformer catalyst should be considered in order to avoid carbon deposits.

Refinery gas from different sources vary in suitability as hydrogen plant feed. Catalytic reformer off-gas, for example, is saturated, very low in sulfur, and often has high hydrogen content. The process gases from a coking unit or from a fluid catalytic cracking unit are much less desirable because of the content of unsaturated constituents. In addition to olefins, these gases contain substantial amounts of sulfur that must be removed before the gas is used as feedstock. These gases are also generally unsuitable for direct hydrogen recovery since the hydrogen content is usually too low. Hydrotreater off-gas lies in the middle of the range. It is saturated, so it is readily used as hydrogen plant feed. Content of hydrogen and heavier hydrocarbons depends to a large extent on the upstream pressure. Sulfur removal will generally be required.

However, before the feedstock is introduced to a process, there is a need for application of a strict feedstock purification protocol. Prolonging catalyst life in hydrogen production processes is attributable to effective feedstock purification, particularly sulfur removal. A typical natural gas or other low-boiling hydrocarbon feedstock contains traces of hydrogen sulfide and organic sulfur.

In addition, the evolution of refineries to include biomass-type (renewable) feedstocks (including the various types of carbonaceous waste, sometimes referred to as bio-waste) is operational in some refineries that are driven by changes to legislation and hydroprocessing renewable feedstocks is becoming more commonplace. While refineries have been processing fossil fuel-derived feedstocks since the late 19th century, converting biomass-derived feedstocks into saleable hydrocarbon products is relatively recent and introduces a different set of challenges such as (1) a number of exothermic reactions along with a high consumption of hydrogen, (2) the risk of corrosion due to a high acid number and the presence of oxygenated compounds, (3) the eventual catalyst deactivation due to the presence of contaminants such as silicon and phosphorus, (4) the impact on the density and cold flow properties of the end product, which might then require additional processing, (5) the ability to secure a reliable supply of renewable feedstocks of sufficient quality.

All of these challenges require consideration of (1) the specific needs of each refinery, (2) the chemical and physical properties of the available feedstocks, (3) the quality of the end product, (4) the availability of hydrogen, (5) the ease of integration of the biomass-type feedstocks with the existing refinery units, and (6) the type of catalyst to be employed.

Thus, hydroprocessing biomass-type feedstocks is complex and choosing the optimal catalyst technology requires consideration of the factors outlined above. The choices are critical because they have a direct bearing on (1) the process design, (2) the choice of the reactor, (3) the process parameters, and, last but certainly not least, (4) the type of catalyst to be employed.

To be more specific, transforming biomass-type feedstocks into a particular type of product demands the application of specific catalysts. For example, specifically designed guard beds are required to handle the high levels of contaminants such as phosphorus, sodium, and silicon present in some renewable feedstocks. Selecting the right hydrotreating catalyst is critical for choosing the optimal reaction pathway (hydrodeoxygenation, HDO, or decarboxylation) because the catalyst has a major effect on the distribution and yields of the product.

3.5 PROCESS CHEMISTRY

Hydrotreating and hydrocracking are processes that are designed to remove specific impurities (which act as catalyst poisons) such as sulfur, nitrogen, oxygen, halogens, and metals to produce a variety of products. Hydrotreating also is used to upgrade the quality of the products by, for example, converting olefin derivatives to alkane (paraffin) derivatives. On the other hand, hydrocracking processes were developed for converting refractory feedstocks to naphtha and middle distillates (such as kerosene).

Briefly, and by way of introduction, the so-called 'mild hydrocracking process' uses operating conditions similar to those of a vacuum gas oil desulfurizer to convert vacuum gas oil to significant yields of lower boiling products. The reaction parameters may also be used as pretreatment parameters for viscous feedstocks.

The hydrotreating processes typically use a catalyst in the presence of large amounts of hydrogen under high pressure and temperature to react the feedstock constituents and impurities with hydrogen.

To remove sulfur compounds, it is necessary to hydrogenate the feedstock to convert the organic sulfur to hydrogen that is then reacted with zinc oxide (ZnO) at approximately 370°C (700°F) that results in the optimal use of the zinc oxide as well as ensuring complete hydrogenation. Thus, assuming efficient purification of the feedstock in which all objectionable contaminants are removed, the chemistry of hydrogen production can be defined (Figure 3.4).

In the steam reforming process, low-boiling hydrocarbon derivatives such as methane are reacted with steam to form hydrogen.

$$CH_4 + H_2O \rightarrow 3H_2 + CO$$

Steam reforming

$$CH_4 + H_2O = CO + 3H_2$$

Dry reforming*

$$CO_2 + CH_4 = 2CO + 2H_2$$

Partial oxidation**

$$2CH_4 + O_2 = 2CO + 4H_2$$

Methanol-steam reforming

$$CH_3OH + H_2O = CO_2 + 2H_2$$

Methanol-oxy reforming

$$2CH_3OH + H_2 = 2CO_2 + 4H_2$$

***Also referred to as 'carbon dioxide reforming'**
****Also referred to as 'oxy-reforming**

FIGURE 3.4 The Chemistry of Hydrogen Production.

H is the heat of reaction. A more general form of the equation that shows the chemical balance for higher-boiling hydrocarbons is:

$$C_nH_m + nH_2O \rightarrow (n + m/2)H_2 + nCO$$

The reaction is typically carried out at a temperature on the order of 815°C (1,500°F) using a nickel catalyst that is contained in the tubes of a reforming furnace.

$$CH_4 \rightarrow 2H_2 + C$$

$$C + H_2O \rightarrow CO + H_2$$

When higher boiling feedstocks are used, the carbon is not removed fast enough and builds up, thereby requiring catalyst regeneration or replacement. Carbon buildup on the catalyst (when high-boiling feedstocks are employed) can be avoided by the addition of alkali compounds, such as potash, to the catalyst, thereby encouraging or promoting the carbon-steam reaction.

However, even with a catalyst that is promoted by the presence of alkali, feedstock cracking limits the process to hydrocarbon products with a boiling point less than 180°C (350°F). Natural gas (CH_4), propane ($CH_3CH_2CH_3$), butane (C_4H_{10}, typically $CH_3CH_2CH_2CH_3$), and low-boiling naphtha are most suitable. Pre-reforming, a process that uses an adiabatic catalyst bed operating at a lower temperature, can be used as a pretreatment to allow heavier feedstocks to be used with a lower potential for carbon deposition (coke formation) on the catalyst.

After reformation, the carbon monoxide in the gas is reacted with steam to form additional hydrogen (the 'water-gas shift reaction'):

$$CO + H_2O \rightarrow CO_2 + H_2$$

This leaves a mixture consisting primarily of hydrogen and carbon monoxide that is removed by conversion to methane:

$$CO + 3H_2O \rightarrow CH_4 + H_2O$$

$$CO_2 + 4H_2 \rightarrow CH_4 + 2H_2O$$

The critical variables for steam reforming processes are (1) temperature, (2) pressure, and (3) the steam/hydrocarbon ratio. Steam reforming is an equilibrium reaction, and conversion of the hydrocarbon feedstock is favored by high temperature, which in turn requires higher fuel use. Because of the volume increase in the reaction, conversion is also favored by low pressure, which conflicts with the need to supply the hydrogen at high pressure. In practice, materials of construction limit temperature and pressure.

On the other hand, and in contrast to reforming, the shift conversion reaction is favored by low temperature. The gas from the reformer is reacted over an iron oxide catalyst at 315 to 370°C (600 to 700°F) with the lower limit being dictated by the activity of the catalyst at low temperature.

Hydrogen can also be produced by the partial oxidation (often referred to as POX or P_{OX}) of hydrocarbon feedstocks, in which the feedstock is oxidized in a limited or controlled supply of oxygen.

$$2CH_4 + O_2 \rightarrow CO + 4H_2$$

The water gas-shift reaction also occurs, and a mixture of carbon monoxide and carbon dioxide are produced in addition to hydrogen. Thus:

$$CO + H_2O \rightleftharpoons CO_2 + H_2$$

With the development of industrial processes (particularly the refinery processes) that required hydrogen, a more efficient method of hydrogen production was needed, and as a result, the water gas-shift was combined with the gasification of coal to produce hydrogen.

Gasification:

$$\text{Coal} + \text{steam} \rightarrow CO_2 + H_2$$

Water has shift:

$$CO + H_2O \rightarrow CO_2 + H_2$$

Pressure swing adsorption:

$$CO_2 + H_2 \rightarrow H_2$$

The catalyst tube materials do not limit the reaction temperatures in partial oxidation processes and higher temperatures may be used to enhance the conversion of methane to hydrogen. Indeed, much of the design and operation of hydrogen plants involves protecting the reforming catalyst and the catalyst tubes because of the extreme temperatures and the sensitivity of the catalyst. In fact, minor variations in feedstock composition or operating conditions can have significant effects on the life of the catalyst or the reformer itself. This is particularly true of changes in molecular weight of the feed gas, or poor distribution of heat to the catalyst tubes.

Since the high temperature takes the place of a catalyst, partial oxidation is not limited to the lower-boiling feedstocks that are required for steam reforming. Partial oxidation processes were first considered for hydrogen production because of expected shortages of lower boiling feedstocks and the need to have available a disposal method for higher-boiling, high-sulfur streams such as asphalt or refinery feedstock coke.

In the catalytic partial oxidation process (also known as the auto-thermal reforming process), a low-boiling hydrocarbon feedstock is reacted with oxygen and the resulting hot mixture is passed over a reforming catalyst. The use of a catalyst allows for the use of lower temperatures than in non-catalytic partial oxidation, which causes a reduction in oxygen demand.

The feedstock requirements for catalytic partial oxidation processes are similar to the feedstock requirements for steam reforming, and low-boiling hydrocarbon derivatives from refinery gas to naphtha are preferred. The oxygen substitutes for much of the steam in preventing coking, and a lower steam/carbon ratio is required. In addition, because a large excess of steam is not required, the catalytic partial oxidation reaction produces more carbon monoxide and less hydrogen than steam reforming. Thus, the process is more suited to situations where carbon monoxide is the more desirable product, such as, for example, as synthesis gas for chemical feedstocks.

3.6 COMMERCIAL PROCESSES

Commercially, there are variants which are categorized according to the flow geometry and typically are (1) the fluidized bed gasifier, in which the feedstock is suspended in the gas flow and mixed with that undergoing gasification and (2) the moving bed gasifier, also called a fixed-bed gasifier, in which fresh feedstock can be fed from the top of the reactor while a gas stream that is much lower in flowrate in comparison to a fluidized-bed operating under similar solids flowrate, is fed from the bottom and (3) the entrained bed gasifier which has high feedstock conversion on short residence (Table 3.5). These solid particles slowly move down the reactor and are withdrawn from the bottom of the bed. A demetallization reactor may be used in conjunction with any of the three reactors because of the need to prevent contamination of the bed by feedstocks containing metal-containing constituents.

In the case of a fluidized-bed gasifier, the feedstock is suspended in the gas flow and mixed with that undergoing gasification. In the moving bed gasifier, also called a fixed-bed gasifier, gases flow relatively slowly upward through the bed of feedstock. Both concurrent and counter-current technologies are available, but the former is more common.

Depending on the properties of the naphtha feedstock (as measured by the paraffin, olefin, naphthene, and aromatic content) and catalysts used, reformate can be produced with very high concentrations of toluene, benzene, xylene, and other aromatics that are useful components when gasoline is produced by blending different stocks as well as for the manufacture of petrochemicals. Hydrogen, a significant by-product, is separated from reformate for recycling and use in other processes (such as the platforming process, the powerforming process, the ultraforming process, and the thermofor catalytic reforming process (Parkash, 2003; Gary et al., 2007; Speight, 2014a; Hsu and Robinson, 2017; Speight, 2017).

TABLE 3.5
Characteristics of the Different Types of Gasifiers

Gasifier Type	Fuel Properties
Fixed/moving-bed	Particle size: 1 to 10 cm
	Mechanically stable fuel particles
	Unblocked passage of gas through the bed)
	Pellets or briquettes preferred
	Updraft configuration
	More tolerant of biomass moisture content (up to 40 to 50% w/w)
	Drying occurs as biomass moves down the gasifier
Fluidized bed	Ash melting temperature of fuel
	Higher limit for operating temperature.
	Fuel particle size is relatively small
	Ensures good contact with the bed material
	Good fuel flexibility due to the high thermal inertia of the bed.
Entrained bed	Fuel particle size: <50 micrometers.
	Pulverized for high fuel conversion in short residence times
	Low moisture content
	Ash melting behavior can influence reactor/process design.

Hydrogen chloride from the hydrogenation of chlorine compounds may form an acid or ammonium chloride salt.

In spite of the use of low-quality hydrogen (that contains hydrocarbon gases on the order of up to 40% v/v), a high-purity hydrogen stream (95%–99% by volume hydrogen) is required for hydrodesulfurization, hydrogenation, hydrocracking, and petrochemical processes (Parkash, 2003; Gary et al., 2007; Speight, 2014a; Hsu and Robinson, 2017; Speight, 2017).

There are many different commercial catalytic reforming processes, including Platforming, Powerforming, Ultraforming, and Thermofor catalytic reforming (Parkash, 2003; Gary et al., 2007; Speight, 2011, 2014a; Hsu and Robinson, 2017; Speight, 2017). Using the Platforming process as the example, in the first step of the preparation of the feedstock (naphtha), the removal of any impurities that would cause catalyst degradation is performed. The feedstock is then mixed with hydrogen, vaporized, and passed through a series of alternating furnace and fixed-bed reactors containing the platinum catalyst. After this, the reactor effluent is cooled and sent to a separator, where the hydrogen-rich gas stream is removed and exits the top of the separator for recycling. The liquid product from the bottom of the separator is sent to a fractionator called a stabilizer (butanizer), and the bottom product (reformate) is sent to storage, while butane derivatives and lower boiling gases pass overhead and are sent to the saturated gas plant.

Safe catalyst handling is very important and care must be taken not to break or crush the catalyst when loading the beds, as the small fines will plug up the reformer screens. Also, caution is advised because hydrogen chloride from the hydrogenation of chlorine compounds may form an acid or ammonium chloride salt.

3.6.1 Autothermal Reforming

The autothermal reforming process (often referred to as the ATR process) is a process for producing synthesis gas, composed of hydrogen, carbon monoxide, and carbon dioxide. In the process, the reaction takes place in a single chamber in which the methane is subjected to an exothermic partial oxidation process. The process is a combination of the steam reforming process and the partial oxidation process, and both reactions occur themselves in one compact reactor. The main difference between the autothermal reforming process and the steam-methane reforming process is that the latter process (i.e., the steam-methane reforming process) does not use or require oxygen.

The autothermal process is generally operated between temperatures on the order of 950 to 1,050°C (1,740 to 1,920°F), a pressure on the order of 450 to 750 psi, a steam-to-carbon molar ratio on the order of 0.5 to 1.5 and an oxygen-to-carbon molar ratio on the order of 0.6 to 1.0 (Lamb et al., 2020).

In the process, the organic feedstock and steam are mixed directly with oxygen and air in the reformer, which is typically a refractory lined vessel that contains the catalyst, together with an injector located at the top of the vessel (Aasberg-Petersen et al., 2002; Hagh, 2004). The partial oxidation reactions occur in the combustion zone of the reactor, and the gaseous mixture then flows through a catalyst bed where the actual reforming reactions occur. The heat that is generated in the combustion zone (from exothermic partial oxidation reactions) is utilized in the reforming zone, and therefore, there is the potential for the process to be in complete heat balance.

More specifically, the autothermal reforming reactor consists of three zones: (1) the burner, where the feed streams are mixed in a turbulent diffusion flame, (2) the combustion zone, where partial oxidation reactions produce a mixture of carbon monoxide and hydrogen, and (3) the catalytic zone, where the gases leaving the combustion zone attain thermodynamic equilibrium. Key elements in the reactor are the burner and the catalyst bed – the burner provides mixing of the feed streams and the natural gas is converted in a turbulent diffusion flame.

$$CH_4 + 3/2O_2 \rightarrow CO + 2H_2O$$

When carbon dioxide is present in the feed, the H_2/CO ratio produced is on the order of 1:1, but when the process employs steam, the H_2/CO ratio produced is 2.5:1.

$$2CH_4 + O_2 + CO_2 \rightarrow 3H_2 + 3CO + H_2O$$
$$4CH_4 + O_2 + 2H_2O \rightarrow 10H_2 + 4CO$$

The risk of soot formation in an ATR reactor depends on a number of parameters including feed gas composition, temperature, pressure, and especially burner design. Soot precursors may be formed in the combustion chamber during operation, and it is essential that the design of the burner, catalyst and reactor is such that the precursors are destroyed by the catalyst bed to avoid soot formation.

In addition, the process offers (1) flexible operation, including short startup periods and fast load changes, as well as (2) the potential for soot-free operation, which, however, is feedstock dependent.

The optimum conditions for hydrogen production require (i) high temperature at the exit of the reforming reactor on the order of 800 to 900°C (1,470 to 1,650°F), (ii) a high excess of steam with a molar steam-to-carbon ratio on the order of 2.5 to 3, and (iii) relatively low pressures, typically less than 450 psi.

3.6.2 Combined Reforming

Combined reforming incorporates the combination of both steam reforming and autothermal reforming (Wang et al., 2004; Rahbari et al., 2018). The off-gas from the steam reformer is then sent to an oxygen-fired secondary reactor, the autothermal reforming reactor, where any hydrocarbon derivatives in the gas stream are converted to synthesis gas by partial oxidation followed by steam reforming.

Partial oxidation:

$$CH_4 + 0.5\ O_2 \rightarrow 2H2 + CO$$

or (for completeness):

$$2CH_4 + O_2 \rightarrow 4H2 + 2CO$$

Steam reforming:

$$CH_4 + H_2O \rightarrow 3H_2 + CO$$

Another configuration requires the feedstock to be split into two streams, which are then fed in parallel to the steam reforming and autothermal reactors.

3.6.3 Dry Reforming

The dry reforming process consists of an endothermic reaction between methane and carbon dioxide for the production of a mixture of carbon monoxide and hydrogen. The process is not used to the same extent as the steam reforming process, as the predominant use of the main use of the dry reforming process is for processes that require a high proportion of carbon monoxide in the synthesis gas (Bao and Yu, 2018). The process can be represented by the simple equation:

$$CH_4 + CO_2 \leftrightarrows 2CO + 2H_2$$

High temperatures and low pressures favor the dry reforming reaction since higher reactant conversion and high product yield can be achieved at higher reaction temperatures, particularly in a temperature range of 650 to 1,050°C (1,200 to 1,920°F) and at a pressure close to atmospheric pressure.

Since coking quickly deactivates nickel (Ni) catalysts, rhodium (Rh) and ruthenium (Ru) catalysts are used in most dry reforming applications. However, while nickel-based catalysts are suitable for this reaction and commercially available, the major drawback is that nickel catalysts also catalyze carbon deposition by the decomposition of methane or by the Boudouard reaction, which leads to potential catalyst breakdown and catalyst deactivation as well as reactor clogging.

$$CH_4 \rightarrow C + 2H_2$$

$$2CO \rightarrow C + CO_2$$

A number of methods for reducing the amount of carbon formed on nickel catalysts during reforming reactions have been found, such as the addition of alkali and alkaline-earth metal dopants, such as potassium (K), sodium (Na), magnesium oxide (MgO), calcium oxide (CaO), or lanthanide/actinide dopants, which result in a more carbon resistant catalyst, in most cases due to the promotion of carbon dioxide dissociation and, hence, more surface oxygen for reaction with carbon.

3.6.4 Steam-Methane Reforming

Steam-methane reforming (sometimes referred to as steam reforming, the SMR process) (Chadeesingh, 2011; Muradov, 2015), while not truly the subject of this chapter, is worthy of note because of the production of synthesis gas. The process is carried out by passing a preheated mixture comprising essentially methane and steam through one or more catalyst-filled tubes (Figure 3.5.). Since the reaction is endothermic, heat must be provided in order to effect the conversion and is achieved by the use of burners located adjacent to the tubes. The products of the process are a mixture of hydrogen and carbon monoxide, as well as some carbon dioxide.

Thus, the process is the method by which natural gas or another methane stream, such as a biogas stream or a landfill gas stream (Chapter 5), is reacted with steam in the presence of a catalyst to produce hydrogen and carbon dioxide. When starting with natural gas (as the source of methane) that includes sulfur or other impurities, a pretreatment cleanup step is necessary to remove the impurities. Typically, the process produces a hydrogen-rich gas that is on the order of 70 to 75% v/v hydrogen on a dry mass basis, along with smaller amounts of methane (2 to 6% v/v), carbon monoxide (7 to 10% v/v), and carbon dioxide (6 to 14% v/v) (Hirschenhofer et al., 2000). Thus:

$$CH_4 + H_2O \rightarrow CO + 3H_2$$

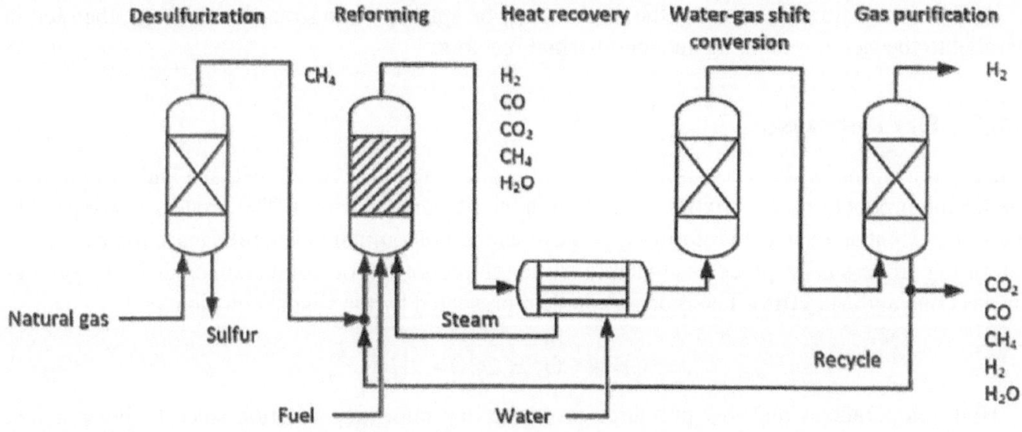

FIGURE 3.5 Hydrogen Production by the Steam Reforming Process.

An alternative to this process, utilizing carbon dioxide and thus potentially mitigating its environmentally harmful emissions, is dry methane reforming:

$$CH_4 + CO_2 \rightarrow 2CO + 2H_2$$

Typically, the reaction is conducted in a series of multi-tubular packed-bed reactors which consist of a series of long and narrow tubes that are situated within the combustion chamber of a furnace that provides the necessary energy to maintain the reactor at a constant temperature during the process. The design and configuration of the furnace may vary but are typically categorized into: (1) a top-fired furnace, (2) a bottom-fired furnace, and (3) a side-fired furnace.

Inside the tubes, a mixture of methane and steam comes into contact with the catalyst (typically a catalyst based on nickel) that has a high surface area-to-volume ratio that is necessary because of the potential diffusion limitations due to the high operating temperature of the furnace. The shape of the catalyst particles is also a deciding factor, and the typical shapes employed are (1) spoked wheels, (2) gear wheels, and (3) rings with holes, which have a low pressure drop that is advantageous for this particular use (Reimert et al., 2011).

Also, a primary reformer is used to effect conversion of methane on the order of 90 to 92% v/v. Here, the hydrocarbon feed is partially reacted with steam over a nickel-alumina catalyst to produce a synthesis gas with an H_2/CO ratio of approximately 3:1. This is done in a fired tube furnace at 900°C (1,650°F) at a pressure of 220 to 450 psi.

The deposition of carbon on the catalyst can be a major problem that occurs when a nickel-based catalyst is employed in the primary reformer (Rostrup-Nielsen, 1984; Alstrup, 1988, Rostrup-Nielsen, 1993). To bypass this potential issue, a successful technique is to use a steam/carbon ratio in the feed gas that does not allow the formation of carbon, which unfortunately can result in a lowering of the efficiency of the process. A second approach utilizes sulfur passivation (Rostrup-Nielsen, 1984; Udengaard et al., 1992). A third approach is to use Group VIII metals (Figure 3.6) that do not form carbides, such as, for example, platinum (Pt).

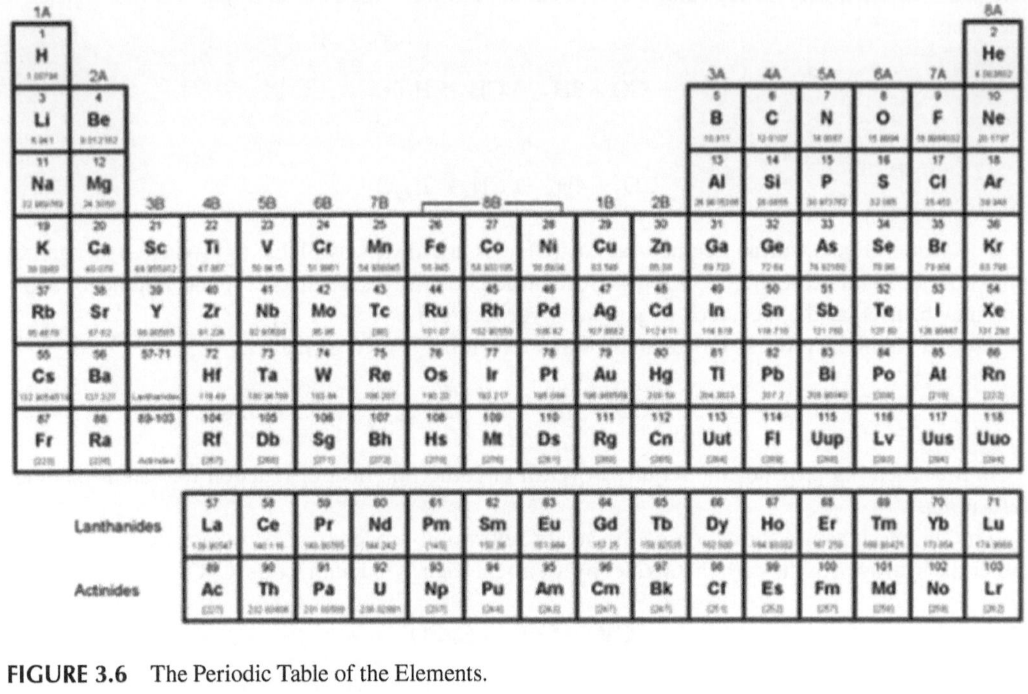

FIGURE 3.6 The Periodic Table of the Elements.

In addition, a major challenge in the development of the steam-methane reforming process is the energy-intensive nature of the process due to the high endothermic character of the reactions. The trend in development thus is one that seeks higher energy efficiency.

The steam-methane reforming process is the benchmark process that has been employed over a period of several decades for hydrogen production and involves reforming natural gas in a continuous catalytic process in which the major reaction is the formation of carbon monoxide and hydrogen from methane and steam.

$$CH_4 + H_2O \rightarrow CO + 3H_2$$

Higher molecular weight feedstocks can also be reformed to hydrogen, using propane as the example:

$$C_3H_8 + 3H_2O \rightarrow 3CO + 7H_2$$

In general form, the equation can be written as:

$$C_nH_m + nH_2O \rightarrow nCO + (0.5m + n)H_2$$

In the process, the feedstock is first desulfurized by passage through activated carbon, which may be preceded by caustic and water washes and the desulfurized product is mixed with steam and passed over a nickel-based catalyst (730 to 845°C, 1,350 to 1,550°F and 400 psi, 2,758 kPa). Effluent gases are cooled by the addition of steam or condensate to approximately 370°C (700°F), at which point carbon monoxide reacts with steam in the presence of iron oxide in a shift converter to produce carbon dioxide and hydrogen.

$$CO + H_2O = CO_2 + H_2$$

Since the presence of any carbon monoxide or carbon dioxide in the hydrogen stream can interfere with the chemistry of the catalytic application, a third stage is used to convert these gases to methane.

$$CO + 3H_2 \rightarrow CH_4 + H_2O$$

$$CO_2 + 4H_2 \rightarrow CH_4 + 2H_2O$$

For many refiners, sulfur-free natural gas (CH_4) is not always available to produce hydrogen by this process. In that case, higher-boiling hydrocarbon derivatives (such as propane, butane, or naphtha) may be used as the feedstock to generate hydrogen (*q.v.*).

The net chemical process for steam methane reforming is then given by:

$$CH_4 + 2H_2O \rightarrow CO_2 + 4H_2 \qquad \Delta H_{298K} = +165.2 \text{ kJ/mol}$$

Indirect heating provides the required overall endothermic heat of reaction for steam-methane reforming.

The thermodynamic limitation of the steam reforming process can be overcome by removing either the hydrogen or the carbon dioxide (by a sorbent such as calcium oxide, CaO) as it is produced:

$$CaO + CO_2 \rightarrow CaCO_3$$

The use of a sorbent enables lower reaction temperatures, and, in the case of calcium oxide, the sorbent can be regenerated to the oxide form by the calcination reaction:

$$CaCO_3 \rightarrow CaO + CO_2$$

The steam-methane reforming process would be an ideal hydrogen production process if it were not for the fact that large quantities of natural gas, a valuable resource, are required as both feed gas and combustion fuel. For each mole of methane reformed, more than one mole of carbon dioxide is co-produced and must be disposed of. This can be a major issue as it results in the same amount of greenhouse gas emission as would be expected from the direct combustion of natural gas or methane. In fact, the production of hydrogen as a clean burning fuel by way of steam reforming of methane and other fossil-based hydrocarbon fuels is not in environmental balance if, in the process, carbon dioxide and carbon monoxide are generated and released into the atmosphere, although alternate scenarios are available (Gaudernack, 1996). Moreover, as the reforming process is not totally efficient, some of the energy value of the hydrocarbon fuel is lost by conversion to hydrogen but with no tangible environmental benefit, such as a reduction in the emission of greenhouse gases. Despite these apparent shortcomings, the process has the following advantages: (1) it produces 4 moles of hydrogen for each mole of methane consumed, (2) feedstocks for the process (methane and water) are readily available, (3) the process is adaptable to a wide range of hydrocarbon feedstocks, (4) it operates at low pressures, less than 450 psi, (5) it requires a low steam/carbon ratio (2.5 to 3), (6) it has good utilization of input energy (reaching 93%), and (7) it uses catalysts that are stable and resist poisoning, and (8) it has good process kinetics.

The feedstock handling system needs to include a surge drum, feed pump, vaporizer (usually steam-heated), followed by further heating before desulfurization.

The reformer will generally use a potash-promoted catalyst to avoid coke buildup from cracking of the heavier feedstock. If liquefied petroleum gas is to be used only occasionally, it is often possible to use a methane-type catalyst at a higher steam/carbon ratio to avoid coking. Naphtha will require a promoted catalyst unless a pre-former is used.

3.6.5 Steam-Naphtha Reforming

The steam reforming reaction is the most widespread process for the generation of hydrogen-rich synthesis gas from low-boiling hydrocarbon derivatives, particularly methane. The feed materials such as natural gas, liquid gas or naphtha are endothermically converted with water steam into synthesis gas in catalytic tube reactors. An option to the use of methane as the feedstock is the use of other low-boiling hydrocarbon derivatives, such as naphtha.

The process is a continuous process that is used for the production of hydrogen from liquid hydrocarbon derivatives and is, in fact, similar to steam-methane reforming (Murata, 1997; Muradov, 1997, 1998, 2000; Brandmair et al., 2003; Find et al., 2003). Following pretreatment to remove sulfur compounds, the feedstock is mixed with steam and taken to the reforming furnace (675 to 815°C, 1,250 to 1,500°F, 300 psi), where hydrogen is produced.

In the process, the naphtha feedstock is passed over a catalyst bed in the presence of hydrogen to remove any sulfur impurities, after which the desulfurized feedstock is mixed with hydrogen (on the order of five molecules of hydrogen to one of hydrocarbon) and heated to a temperature of 500 to 540°C (930 to 1,000°F). The gaseous mixture passes downward through catalyst pellets in a series of three or more reactor vessels that operate at less than 100 psi. After leaving the final reactor, the product is condensed to a liquid, separated from the hydrogen stream, and passed to a fractionating column, where the low-boiling hydrocarbon products are removed by distillation. The hydrogen leaving the product separator is compressed and returned to the reactor system.

During the reforming process, some carbon may be deposited on the catalyst, causing a gradual deterioration of the product yield pattern. In many cases, the units are continuous regeneration

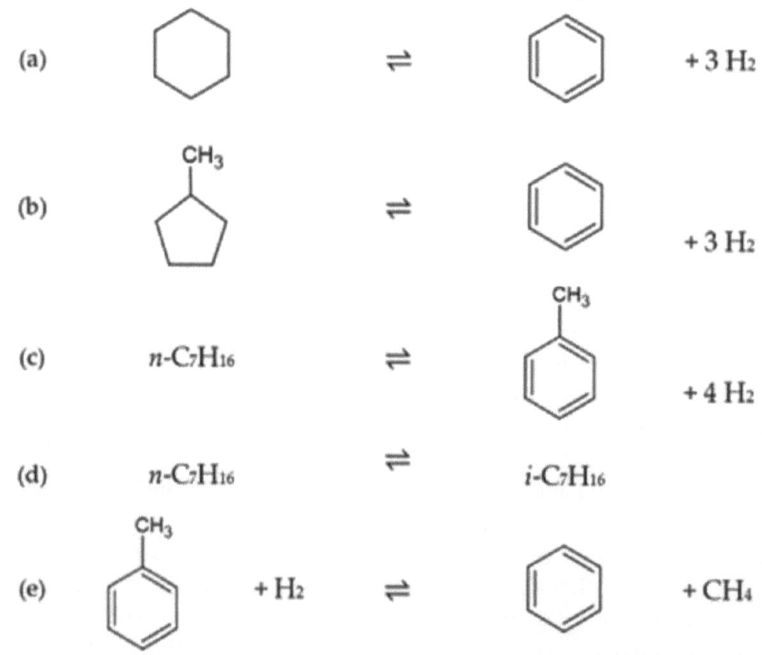

FIGURE 3.7 Examples of the Reactions that Occur in the Steam-Naphtha Reforming Process.

systems, which employ a moving bed of catalyst particles that is gradually withdrawn from the reactor system and passed through a regenerator vessel, where the carbon is removed and the catalyst rejuvenated for reintroduction to the reactor system.

However, the catalytic naphtha reforming reactions are sensitive to the reaction parameters (i.e. the temperature and the pressure) and the hydrogen to hydrocarbon ratio (H_2/feedstock and the effect on the various reforming reactions can differ (Melo and Morlanés, 2005).

A simplified classification of the main reactions that take place during naphtha reforming is: (1) naphthene derivatives are dehydrogenated to aromatic derivatives, (2) alkyl cyclopentane derivatives are dehydro-isomerized to aromatic derivatives, (3) alkane derivatives are dehydrocyclized to aromatic derivatives, (4) n-alkane derivatives are isomerized to branched alkane derivatives, and (5) cracking reactions leading to lower molecular weight products that are based on the constituents of the naphtha feedstock (Figure 3.7).

Each reaction may not be affected in the same way by the operating conditions, while reactions that form aromatic derivatives are endothermic and exothermic reactions such as paraffin isomerization to alkene derivatives (olefin derivatives) and hydrocracking reactions and hydrogenolysis reactions will be facilitated by lower temperatures, and in the latter case (i.e., the hydrogenolysis reactions), higher hydrogen partial pressure.

3.6.6 Hypro Process

The hypro process is a continuous catalytic method that is designed to convert natural gas to hydrogen:

$$CH4 \rightarrow C + 2H2$$

The process was developed by the UOP company (the Universal Oil Products company) with the aim of providing hydrogen for refinery processes.

In this process, light hydrocarbons are decomposed in a fluidized bed over a nickel catalyst at 800 to 1,100°C (1470 to 2,010°F). The catalyst is continuously regenerated in a second fluidized bed by burning the resulting carbon black.

As a follow-on to the hypro process, the process concepts for methane pyrolysis can be divided into three categories, which are (1) thermal decomposition, (2) plasma decomposition, and (3) catalytic decomposition.

3.6.7 Partial Oxidation

The partial oxidation process (also referred to as (the POX process or the P_{OX} process) is the manner in which the feedstock (such as methane or a suitable hydrocarbonaceous fuel) reacts exothermically in the presence of a small amount of air (Vernon et al., 1990; Rostrup-Nielsen, 2002; Zhu et al., 2004). Because incomplete combustion occurs, a gas containing hydrogen and carbon monoxide is produced.

The reactions occur when a less-than-stoichiometric hydrocarbon-air mixture is partially combusted in a reformer, as is often the case when the higher molecular weight hydrocarbon derivatives are employed as the feedstock.

$$C_nH_m + (2n + m)/2O_2 \rightarrow nCO + (m/2)\ H_2O$$

As a noteworthy mention here, hydrogen can be produced from coal by means of any one of several gasification processes (e.g., fixed-bed gasification, fluidized-bed gasification, or entrained flow gasification) (Chapter 4). In practice, high-temperature entrained flow processes are favored to maximize the conversion to gas, thus avoiding the formation of significant amounts of char, tar, and phenol derivatives.

A typical reaction for the process involves the carbon in the coal being converted to carbon monoxide and hydrogen:

$$C(coal) + H_2O + heat \rightarrow CO + H_2$$

Since this reaction is endothermic, additional heat is required, as with the steam-methane reforming process.

The carbon monoxide is converted to hydrogen and carbon dioxide by means of the water-gas shift reaction, described in the equation:

$$CO + H_2O \rightarrow CO_2 + H_2$$

The production of hydrogen from coal is a commercially mature process, but it is more complex than the production of hydrogen from natural gas. For coal or any carbonaceous feedstock (Chapter 4), the reaction can be represented simply as (understanding that the reaction, in reality, is extremely complex):

$$[CH]_{feedstock} + O_2 \rightarrow CO + H_2$$

In the process, the feedstock, which may include steam, is mixed directly with oxygen that is introduced by an injector that is located near the top of the reaction vessel. The combustion occurs in a pre-combustion chamber in the presence of a small amount of air (i.e. the reaction is a partial combustion reaction) and converted into carbon monoxide and hydrogen. Because partial oxidation is an exothermic reaction, some of the heat of combustion is released which can bring the temperature of the gas to a temperature on the order of 870°C (1,°F600°F). However, the temperature of the gas stream needs to be lowered before entering the combustion chamber otherwise the density of the gas is too low to have good volumetric efficiency. A thermal partial oxidation reactor is similar to an autothermal reactor except that a catalyst is not employed.

Both partial oxidation and reforming reactions occur in the combustion zone below the burner.

The principal advantage of the partial oxidation process is the ability of the system to accommodate almost any carbonaceous feedstock, which can comprise very high molecular-weight organic constituents such as petroleum residua and petroleum coke (Gunardson and Abrardo, 1999; Speight, 2014a). High temperatures on the order of 1300°C (2370°F), are required to achieve near complete reaction, which necessitates the consumption of some of the hydrogen and a greater than stoichiometric consumption of oxygen, i.e. oxygen-rich conditions. If the goal is to produce hydrogen, the partial oxidation process is often used in combination with steam reforming, using the heat of the partial oxidation for the endothermic steam reforming.

A possible means of improving the efficiency of synthesis gas production is by use of the catalytic partial oxidation technology (also referred to as the CPOX or CP_{OX} technology), which has the potential to offer several advantages over steam reforming and thermal particle oxidation, particularly higher energy efficiency (Enger et al., 2008). Furthermore, a hydrogen/carbon monoxide ratio on the order of 2.0, i.e. the ideal stoichiometric ratio for the Fischer-Tropsch synthesis of methanol (methyl alcohol, CH_3OH), is produced by this technology, which can occur by either of two routes: (1) the direct route and (2) the indirect route.

Compared with conventional synthesis gas production methods, the 'direct route' would reduce the amount of catalyst used and the conversion occurs by a mechanism involving a surface reaction on the catalyst.

$$2CH_4 + O_2 \rightarrow 2CO + 4H_2$$

On the other hand, the 'indirect route' involves the total combustion of methane (to carbon dioxide and water) followed by steam reforming and the water-gas shift reaction:

Total combustion:

$$CH_4 + 2O_2 \rightarrow CO_2 + 2H_2O$$

Water gas shift reaction:

$$CO + H_2O \rightarrow CO_2 + H_2$$

3.6.8 Pyrolysis Processes

Pyrolysis can be used to produce hydrogen and carbon from methane or from natural gas by direct contact pyrolysis. In the process, methane or natural gas is bubbled through a bed of either low-melting-point metals (e.g., lead or tin), granular or catalytic materials (e.g., silicon carbide, α-alumina, NiMo/γ-alumina), or a mechanical mixture of molten metal and solid media (Serban et al., 2003).

Based on the temperature range, pyrolysis processes are divided into three temperature ranges, which are (1) low-temperature processes, 500°C, 930°F, (2) medium temperature processes, 500 to 800°C, 930 to 1,470°F and high-temperature processes, >800°C, >1,470°F). In addition, fast pyrolysis is one of the latest processes for the transformation of organic material into products with high energy content (Uemura et al., 1999; Weimer at el., 2000; Dahl and Weimer, 2001):

$$C_nH_m \rightarrow nC + (m/2)H_2$$

If a hydrocarbon feedstock such as natural gas (methane) is to be used for hydrogen production by direct decomposition, then the process that is optimized to yield hydrogen production may not be suitable for production of high-quality carbon black by-product intended for the industrial rubber market. Moreover, it appears that the carbon produced from high-temperature (850 to 950°C; 1,560 to 1,740°F) direct thermal decomposition of methane is soot-like material with a high tendency for catalyst deactivation (Murata, 1997). Thus, if the object of methane decomposition is hydrogen production, the carbon by-product may not be marketable as high-quality carbon black for rubber and tire applications. The production of hydrogen by the decomposition of hydrogen sulfide is another option that might be employed (Clark and Wassink, 1990; Clark, 1995; Zaman and Chakma, 1995; Donini, 1996; Luinstra, 1996).

At temperatures less than 1,500°C (2730°F), the thermodynamic equilibrium is unfavorable toward hydrogen formation. However, in the presence of catalysts such as platinum-cobalt (at 1,000°C; 1,830°F), disulfides of molybdenum or tungsten Mo or W at 800°C (1,470°F) (Kotera et al., 1976), or other transition metal sulfides supported on alumina (at 500 to 800°C; 930 to 1470°F), decomposition of hydrogen sulfide proceeds rapidly (Kiuchi, 1982; Bishara et al., 1987; Al-Shamma and Naman, 1989; Clark and Wassink, 1990; Megalofonos and Papayannakos, 1997; Arild, 2000; Raissi, 2001). In the temperature range of about 800 to 1,500°C (1470 to 2730°F), thermolysis of hydrogen sulfide can be treated simply.

$$H_2S \rightarrow H_2 + 1/xS_x$$

In this equation, $x = 2$.

At temperatures in excess of 1,000°C (1,830°F), there is a limited advantage to using catalysts since the thermal reaction proceeds to equilibrium very rapidly (Clark and Wassink, 1990). The hydrogen yield can be doubled by preferential removal of either H2 or sulfur from the reaction environment, thereby shifting the equilibrium. The reaction products must be quenched quickly after leaving the reactor to prevent the occurrence of reversible reactions.

3.6.9 Recovery from Refinery Gas

Crude oil-related gases, particularly gas streams categorized as 'refinery gas' (which are a combination of gas streams from various processes, such as distillation, cracking, reforming, or other upgrading processes), are the product gases produced when crude oil is refined. The gas streams are a category of saturated and unsaturated gaseous hydrocarbons, predominantly in the C_1 to C_4 carbon number range (even as high as the C_6 hydrocarbon derivatives, including olefin derivatives), that may also (depending upon the process) contain hydrogen. More specifically, refinery gas is the non-condensable gas obtained during the production of products from various processes (such as cracking processes) in refineries and also includes gases that are returned from the petrochemical industry. The gas stream may also be designated as 'fuel gas' that is to be combusted separately or in any combination with any type of gas.

Some gas streams may also contain inorganic constituents, such as nitrogen, hydrogen sulfide, carbon monoxide, and carbon dioxide. As such, refinery gases (unless produced as a salable product that must meet specifications prior to sale) are often of unknown or variable composition and toxic. Some streams also contain varying amounts of other chemicals including ammonia, mercaptan derivatives, 1,3-butadiene, and/or benzene. The site-restricted refinery gas streams (i.e., those gases that are not produced as saleable products) often serve as fuels consumed on-site, as intermediates for purification and recovery of various gaseous products, or as feedstocks for isomerization and alkylation processes within a facility.

In the current context, these gas streams may also be a source of hydrogen. However, the composition of each refinery gas stream is not always consistent, which means that different processes may be needed to separate the components of the gas stream. Nevertheless, recovering hydrogen from a refinery gas stream can contribute to the hydrogen demand in a refinery.

However, hydrogen losses through purging off-gas to a gas stream to be used as a refinery fuel gas can often result in significant loss of low-boiling (C_3 hydrocarbon) valuable derivatives. The same as hydrogen, light ends are much more valuable than methane, and their use as fuel gas should be minimized. In terms of the refinery off-gases that contain hydrogen, the interaction between hydrogen recovery and light-ends recovery can be an advantage insofar as removing either hydrogen or the light ends will make the remaining fraction more concentrated and, therefore, easier to upgrade.

The basic configuration for hydrogen recovery from refinery gases involves a two-stage partial condensation process, with post purification via pressure swing adsorption (Dragomir et al., 2010). For example,

The typical gas processing (gas treating) configuration for the recovery of hydrogen from a refinery gas streams involves a two-stage partial condensation process followed by purification using pressure swing adsorption (Chapter 6) (Dragomir et al., 2010). The major steps in this process involve (1) compressing and pretreating the crude refinery gas stream, (2) chilling the gas stream to an intermediate temperature on the order of –51 to –85°C (-60 to 120°F), (3) separation of the partially condensed stream in a flash drum followed by (4) expansion of the liquid stream through a Joule-Thompson valve to generate refrigeration, (5) followed by a wash column which can (as an option) be substituted by another flash drum.

The crude liquefied petroleum gas stream is collected at the bottom of the column and a methane-rich vapor exits at the top of the column. The vapor from the flash drum is further cooled before being fed to another flash drum where a hydrogen-rich stream and a methane-rich liquid are produced. The hydrogen-rich gas is then sent to the pressure swing adsorption units for further purification (Kohl and Nielsen, 1997).

3.6.10 Electrolysis

Electrolysis is a promising option for the production of carbon-free hydrogen from renewable sources (such as biomass) and is the process by which water molecules can be converted into hydrogen and oxygen using electricity (Ivy, 2004; Swalla, 2008; Chisholm and Cronin, 2016). In the process, hydrogen is produced by an electrolysis process in which electricity is passed through two electrodes in water, causing the water molecule to decompose thereby producing hydrogen at the cathode and oxygen at the anode.

The two most common types of electrolyzer unit for the process are (1) an alkaline unit and (2) a solid polymer membrane electrolyte, which is often referred to as a PEM electrolyte. The electrolysis reaction produces pure hydrogen and, as a by-product, pure oxygen.

The alkaline unit involves the use of an aqueous solution of potassium hydroxide (KOH), which is used because of its high conductivity and is referred to as an alkaline electrolyzer, which can be either unipolar or bipolar. The unipolar electrolyzer has electrodes connected in parallel, and a membrane is placed between the cathode and anode, which separates the hydrogen and oxygen as the gases are produced but allows the transfer of ions.

The other type of electrolysis unit is the solid polymer electrolyte (SPE) electrolyzer, which is also referred to as a proton exchange membrane (PEM) electrolyzer, in which the electrolyte is a solid ion conducting membrane as opposed to the aqueous solution in the alkaline electrolyzers. The membrane allows the hydrogen ion (H+ ion) to transfer from the anode side of the membrane to the cathode side, where molecular hydrogen is formed. The solid polymer electrolyte membrane also serves to separate the hydrogen and oxygen gases, as oxygen is produced at the anode on one side of the membrane and hydrogen is produced on the opposite side of the membrane.

The chemistry, although seemingly complex, can be represented in the following equations:

At the anode:

$$H_2O \rightarrow \tfrac{1}{2}O_2 + 2H+ + 2e^- \text{ (acidic conditions)}$$

$$2OH^- \rightarrow \tfrac{1}{2}O_2 + 2H_2O + 2e^- \text{ (alkaline conditions)}$$

At the cathode:

$$2H+ + 2e- \rightarrow H_2$$

$$2H_2O + 2e^- \rightarrow H_2 + 2OH^-$$

Total reaction:

$$2H_2O \rightarrow O_2 + 2H_2$$

To facilitate the practical use of water splitting for the generation of hydrogen, the design of efficient catalysts plays a major role in improving the efficiency of the process. Despite the significant progress in understanding the electrocatalytic processes, several challenges remain for the ultimate commercial large-scale production of hydrogen by water splitting electrolysis. Also, more effective electrocatalyst screening strategies are needed to establish the standard evaluation protocol for effective comparisons of the performances of the different catalysts.

3.7 CATALYSTS

As a promising means of refinery processing, hydrogen has emerged as a major player in the refinery. However, a key challenge to the use of hydrogen is the efficient production of hydrogen to meet the commercial-scale demand for hydrogen in the modern refinery.

Catalytic operations include hydrogenation, steam reforming, shift conversion, and methanation. Catalysts will play a key role in the generation of hydrogen by various processes (Epron and Duprez, 2020).

3.7.1 Reforming Catalysts

The reforming catalyst is usually supplied as nickel oxide that, during start-up, is heated in a stream of inert gas, then steam. When the catalyst is near the normal operating temperature, hydrogen or a low-boiling hydrocarbon is added to reduce the nickel oxide to metallic nickel.

The high temperatures (up to a temperature on the order of 870°C, 1,600°F), and the nature of the reforming reaction require that the reforming catalyst be used inside the radiant tubes of a reforming furnace. The active agent in the reforming catalyst is nickel, and normally the reaction is controlled both by diffusion and by heat transfer. The catalyst life is limited as much by physical breakdown as by deactivation.

Sulfur is the main catalyst poison, and the catalyst poisoning is theoretically reversible with the catalyst being restored to near full activity by steaming. However, in practice, the deactivation may cause the catalyst to overheat and coke, to the point that it must be replaced. Reforming catalysts are also sensitive to poisoning by heavy metals, although these are rarely present in low-boiling hydrocarbon feedstocks and in naphtha feedstocks.

Coke deposition on the reforming catalyst and the ensuing gloss of catalyst activity is the most characteristic issue that must be assessed and mitigated.

While methane-rich streams such as natural gas or refinery gas are the most common feeds to hydrogen plants, there is often a requirement for a variety of reasons to process a variety of higher boiling feedstocks, such as liquefied petroleum gas and naphtha. Feedstock variations may also be inadvertent due, for example, to changes in refinery off-gas composition from another unit or because of variations in naphtha composition due to feedstock variance to the naphtha unit.

Thus, when using higher boiling feedstocks in a hydrogen plant, coke deposition on the reformer catalyst becomes a major issue. Coking is most likely in the reformer unit at the point where both temperature and hydrocarbon content are high enough. In this region, hydrocarbons decompose and form coke faster than the coke is removed by reaction with steam or hydrogen. When catalyst deactivation occurs, there is a simultaneous temperature increase with a concomitant increase in coke formation and deposition. In other zones where the hydrocarbon-to-hydrogen ratio is lower, there is less risk of coking.

Coking depends to a large extent on the balance between catalyst activity and heat input, with the more active catalysts producing higher yields of hydrogen at lower temperatures, thereby reducing the risk of coking. A uniform input of heat is important in this region of the reformer since any catalyst voids or variations in catalyst activity can produce localized hot spots leading to coke formation and/or reformer failure.

Coke formation result in hotspots in the reformer that increase pressure drop, reduce feedstock (methane) conversion, leading eventually to reformer failure. Coking may be partially mitigated by increasing the steam/feedstock ratio to change the reaction conditions, but the most effective solution may be to replace the reformer catalyst with one designed for higher boiling feedstocks.

A *standard* steam-methane reforming catalyst uses nickel on an alpha-alumina ceramic carrier that is acidic in nature. Promotion of hydrocarbon cracking with such a catalyst leads to coke formation from higher boiling feedstocks. Some catalyst formulations use a magnesia/alumina (MgO/Al_2O_3) support that is less acidic than alpha-alumina, which reduces cracking on the support and allows higher boiling feedstocks (such as liquefied petroleum gas) to be used.

Further mitigation of the tendency for coke formation can be achieved by adding an alkali promoter, typically some form of potash (KOH), to the catalyst. Besides reducing the acidity of the carrier, the promoter catalyzes the reaction of steam and carbon. While carbon continues to be formed, it is removed faster than it can build up. This approach can be used with naphtha feedstocks boiling point up to approximately 180°C (350°F). Under the conditions in a reformer, potash is volatile, and it is incorporated into the catalyst as a more complex compound that slowly hydrolyzes to release potassium hydroxide (KOH). Alkali-promoted catalyst allows for the use of a wide range of feedstocks but, in addition to possible potash migration, which can be minimized by proper design and operation, the catalyst is also somewhat less active than conventional catalyst.

Another option to reduce coking in steam reformers is to use a *pre-reformer* in which a fixed-bed of catalyst, operating at a lower temperature, upstream of the fired reformer is used. In a pre-reformer, adiabatic steam-hydrocarbon reforming is performed outside the fired reformer in a vessel containing high nickel catalyst. The heat required for the endothermic reaction is provided by hot flue gas from the reformer convection section. Since the feed to the fired reformer is now partially reformed, the steam methane reformer can operate at an increased feed rate and produce 8 to 10% additional hydrogen at the same reformer load. An additional advantage of the pre-reformer is that it facilitates higher mixed feed preheat temperatures and maintains relatively constant operating conditions within the fired reformer regardless of variable refinery off gas feed conditions. Inlet temperatures are selected so that there is minimal risk of coking and the gas leaving the pre-reformer contains only steam, hydrogen, carbon monoxide, carbon dioxide, and methane. This allows a standard methane catalyst to be used in the fired reformer and this approach has been used with feedstocks up to low-boiling kerosene (boiling range on the order of 150 to 275°C (300 and 525°F). Since the gas leaving the pre-reformer poses reduced risk of coking, it can compensate to some extent for variations in catalyst activity and heat flux in the primary reformer.

3.7.2 Shift Conversion Catalysts

An important reaction in a steam reforming plant is the reaction known as the shift conversion reaction:

$$CO + H_2O \rightarrow CO_2 + H_2$$

Two basic types of shift catalysts are used in steam reforming plants: iron/chrome high-emperature shift catalysts, and copper/zinc low temperature shift catalysts.

High-temperature shift catalysts operate in a temperature range on the order of 315 to 430°C (600 to 800°F) and consist primarily of magnetite (Fe_3O_4) with three-valent chromium oxide (Cr_2O_3) added as a stabilizer. Metallic iron is a common catalyst for Fischer-Tropsch reactions, and hydrocarbons will be produced.

Low-temperature shift catalysts operate at temperatures on the order of 205 to 230°C (400 to 450°F). The low-temperature shift catalyst is primarily used in wet scrubbing plants that use methanation for final purification.

3.7.3 Methanation Catalysts

In wet scrubbing plants, the final hydrogen purification procedure involves methanation in which the carbon monoxide and carbon dioxide are converted to methane:

$$CO + 3H_2O \rightarrow CH_4 + H_2O$$

$$CO_2 + 4H_2 \rightarrow CH_4 + 2H_2O$$

The active agent is nickel, on an alumina (Al_2O_3) carrier.

3.8 HYDROGEN MANAGEMENT AND DISTRIBUTION

Many existing refinery hydrogen plants use a conventional process, which produces a medium-purity (94% to 97% v/v) hydrogen product by removing the carbon dioxide in an absorption system and methanation of any remaining carbon oxides. Since the 1980s, most hydrogen plants are built with pressure swing adsorption (PSA) technology to recover and purify the hydrogen to purities above 99.9%. Since many refinery hydrogen plants utilize refinery off-gas feeds containing hydrogen, the actual maximum hydrogen capacity that can be synthesized via steam reforming is not certain since the hydrogen content of the off-gas can change due to operational changes in the hydrotreaters.

3.8.1 Hydrogen Management

Hydrogen management has become a priority in current refinery operations and when planning to produce lower sulfur gasoline and diesel fuels (Chapter 7) (Zagoria et al., 2003; Davis and Patel, 2004; Luckwal and Mandal, 2009). Along with increased hydrogen consumption for deeper hydrotreating, additional hydrogen is needed for processing heavier and higher sulfur crude slates. In many refineries, hydroprocessing capacity and the associated hydrogen network are limiting refinery throughput and operating margins.

Improved hydrogen utilization and expanded or new sources for refinery hydrogen, and hydrogen purity optimization are now required to meet the needs of the future transportation fuel market and the drive toward higher refinery profitability. Many refineries developing hydrogen management programs fit into the two general categories of either a catalytic reformer supplied network or an on-purpose hydrogen supply.

Some refineries depend solely on catalytic reformer(s) as their source of hydrogen for hydrotreating. Often, they are semi-regenerative reformers where off-gas hydrogen quantity, purity, and availability change with feed naphtha quality, as octane requirements change seasonally, and when the reformer catalyst progresses from start-of-run to end-of-run conditions and then goes offline for regeneration. Typically, during some portions of the year, refinery margins are reduced as a result of hydrogen shortages.

3.8.2 Hydrogen Distribution

Currently, hydrogen is distributed through three methods: (1) pipeline, (2) high-pressure tube trailers, and (3) liquefied hydrogen tankers.

Typically, the pipelines are located near large refinery feedstocks, refineries, and chemical plants in Illinois, California, and the Gulf Coast.

Multiple hydrotreating units compete for hydrogen – either by selectively reducing throughput, managing intermediate tankage logistics, or running the catalytic reformer sub-optimally just to satisfy downstream hydrogen requirements. Part of the operating year still runs in hydrogen surplus, and the network may be operated with relatively low hydrogen utilization (consumption/production) at 70 to 80%. Catalytic reformer off-gas hydrogen supply may swing from 75 to 85% v/v hydrogen purity. Hydrogen purity upgrade can be achieved through some hydrotreaters by absorbing high-boiling hydrocarbon derivatives. But without supplemental hydrogen purification, critical control of hydrogen partial pressure in hydroprocessing reactors is difficult, which can affect catalyst life, charge rates, and/or gasoline yields.

More complex refineries, especially those with hydrocracking units, also have on-purpose hydrogen production, typically with a steam methane reformer that utilizes refinery off gas and supplemental natural gas as feedstock. The steam methane reformer plant provides the swing hydrogen requirements at higher purities (92 to more than 99% hydrogen) and serves a hydrogen network configured with several purity and pressure levels. Multiple purities and existing purification units allow for more optimized hydroprocessing operation by controlling hydrogen partial pressure for maximum benefit. Typical hydrogen utilization is on the order of 85% to 95%.

REFERENCES

Aasadnia, M., Mehrpooya, M., and Ghorbani, B. 2021. A Novel Integrated Structure for Hydrogen Purification Using the Cryogenic Method. *Journal of Cleaner Production*, 278: 123872. https://www.sciencedirect.com/science/article/abs/pii/S0959652620339172.

Aasberg-Petersen, K., Christensen, T.S., Nielsen, C.S., and Dybkjær, I. 2002. Recent Developments in Autothermal Reforming and Pre-reforming for Synthesis Gas Production in GTL Applications. Preprints. *Division of Fuel Chemistry, Journal of the American Chemical Society*, 47(1): 96–97.

Al-Shamma, L.M., and Naman, S.A. 1989. Kinetic Study for Thermal Production of Hydrogen from Hydrogen Sulfide by Heterogeneous Catalysis of Vanadium Sulfide in a Flow System. *International Journal of Hydrogen Energy*, 14(3): 173–179.

Arild, V. 2000. Production of Hydrogen and Carbon with a Carbon Black Catalyst. PCT Int. Appl. No. 0021878.

Balasubramanian, B., Ortiz, A.L., Kaytakoglu, S., and Harrison, D.P. 1999. Hydrogen from Methane in a Single-Step Process. *Chemical Engineering Science*, 54(15–16): 3543–3552.

Bandermann, F., and Harder, K.B. 1982. Production of Hydrogen via Thermal Decomposition of Hydrogen Sulfide and Separation of Hydrogen and Hydrogen Sulfide by Pressure Swing Adsorption. *International Journal of Hydrogen Energy*, 7(6): 471–475.

Bao, Z., and Yu, F. 2018. Catalytic Conversion of Biogas to Syngas via Dry Reforming Process. *Advances in Bioenergy*, 3: 43–76. Chapter 2.

Bernardo, G., Araújo, T., Lopes, T.D.S., Sousa, J., and Mendes, A. 2020. Recent Advances in Membrane Technologies for Hydrogen Purification. *International Journal of Hydrogen Energy*, 45(12): 7313–7338.

Bezler, J. 2003. Optimized Hydrogen Production – A Key Process Becoming Increasingly Important in Refineries. Proceedings. DGMK Conference on Innovation in the Manufacture and Use of Hydrogen. Dresden. October 15–17. Page 65.

Bishara, A., Salman, O.S., Khraishi, N., and Marafi, A. 1987. Thermochemical Decomposition of Hydrogen Sulfide by Solar Energy. *International Journal of Hydrogen Energy*, 12(10): 679–685.

Brandmair, M., Find, J., and Lercher, J.A. 2003. Combined Autothermal Reforming and Hydrogenolysis of Alkanes. Proceedings. DGMK Conference on Innovation in the Manufacture and Use of Hydrogen. Dresden. October 15–17. Page 273–280.

Brüschke, H. 1995. Industrial Application of Membrane Separation Processes. *Pure and Applied Chemistry*, 345, 67(6): 993–1002.

Brüschke, H. 2003. Separation of Hydrogen from Dilute Streams (e.g. Using Membranes). Proceedings. DGMK Conference on Innovation in the Manufacture and Use of Hydrogen. Dresden. October 15–17. Page 47.

Chadeesingh, R. 2011. The Fischer-Tropsch Process. In *The Biofuels Handbook* J.G. Speight (Editor). Royal Society of Chemistry, London. Part 3, Chapter 5.

Chisholm, G., and Cronin, L. 2016. Hydrogen From Water Electrolysis. In *Storing Energy: With Special Reference to Renewable Energy Sources*. Elsevier, Amsterdam. Page 315–343.

Clark, P.D., and Wassink, B. 1990. A Review of Methods for the Conversion of Hydrogen Sulfide to Sulfur and Hydrogen. *Alberta Sulfur Research Quarterly Bulletin*, 26(2/3/4): 1.

Clark, P.D., Dowling, N.I., Hyne, J.B., and Moon, D.L. 1995. Production of Hydrogen and Sulfur from Hydrogen Sulfide in Refineries and Gas Processing Plants. *Quarterly Bulletin*, 32(1): 11–28.

Conte, M., Iacobazzi, A., Ronchetti, M., and Vellone, R. Different Hydrogen Production Methods. *Journal of Power Sources*, 100: 171–187.

Cruz, F.E., and de Oliveira Junior, S. 2008. Petroleum Refinery Hydrogen Production Unit: Exergy and Production Cost Evaluation. *International Journal of Thermodynamics*, 11(4): 187–193.

Dahl, J., and Weimer, A.W. 2001. Preprints. *Division of Fuel Chemistry, Journal of the American Chemical Society*, 221.

Dalrymple, D.A., Trofe, T.W., and Leppin, D. Gas Industry Assesses New Ways to Remove Small Amounts of Hydrogen Sulfide. *Oil and Gas Journal*, May, 23.

Davis, R.A., and Patel, N.M. 2004. *Petroleum Technology Quarterly*, 2004(Spring): 28–35.

Demirbaş, A. 2005. Recovery of Chemicals and Gasoline-Range Fuels from Plastic Wastes via Pyrolysis. *Energy Sources*, 27(14): 1313–1319.

Demirbaş, A. 2006. Hydrogen from Various Biomass Species via Pyrolysis and Steam Gasification Processes. *Energy Sources*, 28(3): 245–252.

Dolbear, G.E. 1998. Hydrocracking: Reactions, Catalysts, and Processes. In *Petroleum Chemistry and Refining*. J.G. Speight (Editor). Taylor & Francis, Washington, DC. Chapter 7.

Donini, J.C. 1996. Separation and Processing of Hydrogen Sulfide in the Fossil Fuel Industry. Minimum Effluent Mills Symposium. Page 357–363.

Dragomir, R., Drnevich, R.F., Morrow, J., Papavassiliou, V., Panuccio, G., and Watwe, R. 2010. Technologies for Enhancing Refinery Gas Value. Proceedings. AIChE 2010 Spring Meeting. San Antonio, TX. November 7–12.

Du, Z., Liu, C., Zhai, J., Guo, X., Xiong, Y., Su, W., and He, G. 2021. A Review of Hydrogen Purification Technologies for Fuel Cell Vehicles. *Catalysts*, 11(3): 393. https://www.mdpi.com/2073-4344/11/3/393.

Ehwald, H., Kürschner, U., Smejkal, Q., and Lieske, H. 2003. Investigation of Different Catalysts for Autothermal Reforming of i-Octane. Proceedings. DGMK Conference on Innovation in the Manufacture and Use of Hydrogen. Dresden. October 15–17. Page 345.

Epron, F., and Duprez, D. 2020. Hydrogen Production by Catalytic Processes. In *Current Trends and Future Developments on* (Bio-) Membranes. Membrane Systems for Hydrogen Production. Chapter 3, Page 57–89. https://www.sciencedirect.com/science/article/abs/pii/B9780128171103000035.

Find, J., Nagaoka, K., and Lercher, J.A. 2003. Steam Reforming of Light Alkanes in Micro-structured Reactors. Proceedings. DGMK Conference on Innovation in the Manufacture and Use of Hydrogen. Dresden. October 15–17. Page 257.

Fleshman, J.D. 1997. In *Handbook of Petroleum Refining Processes*, 2nd Edition. R.A. Meyers (Editor). McGraw-Hill, New York. Chapter 6.2.

Gary, J.H., Handwerk, G.E., and Kaiser, M.J. 2007. *Petroleum Refining: Technology and Economics*, 5th Edition. CRC Press, Taylor & Francis Group, Boca Raton, FL.

Gaudernack, B. 1996. Hydrogen from Natural Gas without Release of Carbon Dioxide into the Atmosphere. Hydrogen Energy Prog. Proceedings. 11th World Hydrogen Energy Conference. 1. Page 511–523.

Grashoff, G.J., Pilkington, C.E., and Corti, C.W. 1983. A Review of the Technology Emphasizing the Current Status of Palladium Membrane Diffusion. *Platinum Metals Review*, 27(4): 157–169.

Gross, M., and Wolff, J. 2000. Gasification of Residue as a Source of Hydrogen for the Refining Industry in India. Proceedings. Gasification.

Gunardson, H.H., and Abrardo, J.M. 1999. Produce CO-Rich Synthesis Gas. *Hydrocarbon Processing*, April: 87–93.

Hagh, B.F. 2004. Comparison of Autothermal Reforming for Hydrocarbon Fuels. Preprints. *Division of Fuel Chemistry, Journal of the American Chemical Society*, 49(1): 144–147.

Hamm, J.B., Ambrosi, A., Griebeler, J.G., Marcilio, N.R., Tessaro, I.C., and Pollo, L.D. 2017. Recent Advances in the Development of Supported Carbon Membranes for Gas Separation. *International Journal of Hydrogen Energy*, 42(39): 24830–24845.

Hirschenhofer, J.H., Stauffer, B.D., Engleman, R.R., and Klett, M.G. 2000. *Fuel Cell Handbook*, 4th Edition. Federal Energy Technology Center, United States Department of Energy. B/T Books, Orinda, CA.

Hsu, C.S., and Robinson, P.R. (Editors) 2017. *Handbook of Petroleum Technology*. Springer AG, Cham.

Hufton, J.R., Mayorga, S., and Sircar, S. 1999. Sorption-Enhanced Reaction Process for Hydrogen Production. *AIChE Journal*, 45(2): 248–256.

Ivy, J. 2004. Summary of Electrolytic Hydrogen Production Milestone Completion Report. Report NREL/MP-560-36734. National Renewable Energy Laboratory, Golden, Colorado. https://www.osti.gov/biblio/15009552.

Kiuchi, H. 1982. Recovery of Hydrogen from Hydrogen Sulfide with Metals and Metal Sulfides. *International Journal of Hydrogen Energy*, 7(6).

Kohl, A.L., and Nielsen, R.B. 1997. *Gas Purification*, 5th Edition. Gulf Professional Publishers, Elsevier, Amsterdam.

Kotera, Y., Todo, N., and Fukuda, K. 1976. Process for Production of Hydrogen and Sulfur from Hydrogen Sulfide as Raw Material. U.S. Patent No. 3,962,409. June 8.

Lacroix, M., Marrakchi, H., Calais, C., Breysse, M., and Forquy, C. 1991. Catalytic Properties of Transition Metal Sulfides for the Dehydrogenation of Sulfur Containing Molecules. *Studies in Surface Science and Catalysis*, 59: 277–285.

Kee, H.K., Choi, H.Y., Choi, K.H., Park, J.H., and Lee, T.H. 2004. Hydrogen Separation Using the Electrochemical Method. *Journal of Power Sources*, 132(1–2): 92–98.

Lamb, J.J., Hillestad, M., Rytter, E., Bock, R., Nordgård, A.S.R., Lien, K.M., Burheim, O.S., and Pollet, B.G. 2020. Traditional Routes for Hydrogen Production and Carbon Conversion. In *Hydrogen, Biomass and Bioenergy: Integration Pathways for Renewable Energy Applications*. Academic Press Inc., New York. Chapter 3, Page 21–53.

Luckwal, K., and Mandal, K.K. 2009. Improve Hydrogen Management of Your Refinery. *Hydrocarbon Processing*, 88(2): 55–61.

Luinstra, E. 1996. Hydrogen from Hydrogen Sulfide – A Review of the Leading Processes. Proceedings. 7th Sulfur Recovery Conference. Gas Research Institute, Chicago, IL. Page 149–165.

Megalofonos, S.K., and Papayannakos, N.G. 1997. Kinetics of Catalytic Reaction of Methane and Hydrogen Sulfide over MoS2. *Journal of Applied Catalysis A: General*, 65(1–2): 249–258.

Melo, F., and Morlanés, N. 2005. Naphtha Steam Reforming for Hydrogen Production. *Catalysis Today*, 107–108: 458–466.

Miller, G.Q., and Penner, D.W. 2003. Meeting Future Needs for Hydrogen – Possibilities and Challenges. Proceedings. DGMK Conference on Innovation in the Manufacture and Use of Hydrogen. Dresden. October 15–17. Page 7.

Mokhatab, S., Poe, W.A., and Speight, J.G. 2006. *Handbook of Natural Gas Transmission and Processing*. Elsevier, Amsterdam.

Muradov, N.Z. 1998. CO2-Free Production of Hydrogen by Catalytic Pyrolysis of Hydrocarbon Fuel. *Energy and Fuels*, 12(1): 41–48.

Muradov, N.Z. 2000. Thermocatalytic Carbon Dioxide-Free Production of Hydrogen from Hydrocarbon Fuels. Proceedings. Hydrogen Program Review. NREL/CP-570-28890.

Muradov, N.Z. 2003. Emission-Free Fuel Reformers for Mobile and Portable Fuel Cell Applications. *Journal of Power Sources*, 118(1–2): 320–324.

Muradov, N.Z. 2015. Low-Carbon Production of Hydrogen from Fossil Fuels. In *Compendium of Hydrogen Energy - Hydrogen Production and Purification. Woodhead Publishing Series in Energy*. Elsevier Group, Sawston. Page 489–522.

Murata, K., Ushijima, H., and Fujita, K. 1997. Process for Producing Hydrogen from Hydrocarbon. United States Patent 5,650,132.

Nagaoka, K., Jentys, A., and Lecher, J.A. 2003. Autothermal Reforming of Methane over Mono- and Bi-metal Catalysts Prepared from Hydrotalcite-Like Precursors. Proceedings. DGMK Conference on Innovation in the Manufacture and Use of Hydrogen. Dresden. October 15–17. Page 171.

Parkash, S. 2003. *Refining Processes Handbook*. Gulf Professional Publishing, Elsevier, Amsterdam.

Rahbari, A., Ramdin, M., van den Broeke, L.J.P., and Vlugt, T.J.H. 2018. Combined Steam Reforming of Methane and Formic Acid to Produce Syngas with an Adjustable H2:CO Ratio. *Industrial and Engineering Chemistry Research*, 57(31): 10663–10674.

Raissi, A.T. 2001. Technoeconomic Analysis of Area II Hydrogen Production. Part 1. Proceedings. US DOE Hydrogen Program Review Meeting, Baltimore, MD.

Ranke, H., and Schödel, N. 2003. Hydrogen Production Technology – Status and New Developments. Proceedings. DGMK Conference on Innovation in the Manufacture and Use of Hydrogen. Dresden. October 15–17. Page 19.

Reimert, R., Friedemann, M., Renner, H.-J., Boll, W., Supp, E., Brejc, M., Liebner, W., and Schaub, G. 2011. *Gas Production 2: Processes. Ullmann's Encyclopedia of Industrial Chemistry*. John Wiley & Sons Inc., Hoboken, NJ.

Rostrup-Nielsen, J.R. 1984. Sulfur-Passivated Nickel Catalysts for Carbon-Free Steam Reforming of Methane. *Journal of Catalysis*, 85(1): 31–43.

Rostrup-Nielsen, J.R. 1993. Production of Synthesis Gas. *Catalysis Today*, 19(4): 305–324.
Rostrup-Nielsen, J.R. 2002. Syngas in Perspective. *Catalysis Today*, 71(3–4): 243–247.
Sasikumar, B., Arthanareeswaran, G., and Ismail, A. 2018. Recent Progress in Ionic Liquid Membranes for Gas Separation. *Journal of Molecular Liquids*, 266: 330–341.
Schorer, L., Schmitz, S., and Weber, A. 2019. Membrane Based Purification of Hydrogen System (MEMPHYS). *International Journal of Hydrogen Energy*, 44(25): 12708–12714.
Scherzer, J., and Gruia, A.J. *Hydrocracking Science and Technology.* Marcel Dekker Inc., New York.
Serban, M., Lewis, M.A., Marshall, C.L., and Doctor, R.D. 2003. Hydrogen Production by Direct Contact Pyrolysis of Natural Gas. *Energy and Fuels*, 17(3): 705–713.
Speight, J.G. 2000. *The Desulfurization of Heavy Oils and Residua*, 2nd Edition. Marcel Dekker Inc., New York.
Speight, J.G. (Editor) 2011. *The Biofuels Handbook.* Royal Society of Chemistry, London.
Speight, J.G. 2013. *The Chemistry and Technology of Coal*, 3rd Edition. CRC Press, Taylor & Francis Group, Boca Raton, FL.
Speight, J.G. 2014a. *The Chemistry and Technology of Petroleum*, 5th Edition. CRC Press, Taylor & Francis Group, Boca Raton, FL.
Speight, J.G. 2014b. *Gasification of Unconventional Feedstocks.* Gulf Professional Publishing Company, Elsevier, Oxford.
Speight, J.G. 2016. Hydrogen in Refineries. In *Hydrogen Science and Engineering: Materials, Processes, Systems, and Technology.* D. Stolten and B. Emonts (Editors). Wiley-VCH Verlag GmbH & Co., Weinheim. Chapter 1, Page 3–18.
Speight, J.G. 2017. *Handbook of Petroleum Refining.* CRC Press, Taylor & Francis Group, Boca Raton, FL.
Speight, J.G. 2019. *Natural Gas: A Basic Handbook*, 2nd Edition. Gulf Publishing Company, Elsevier, Cambridge, MA.
Speight, J.G. 2024. *Petroleum Refining Technology Series Volume 4: Hydrotreating and Hydrocracking in Petroleum Refining.* CRC Press, Taylor & Francis Group, Boca Raton, FL.
Speight, J.G. 2024. *Petroleum Refining Technology Series Volume 6: Product Improvement Processes.* CRC Press, Taylor & Francis Group, Boca Raton, FL.
Sun, J., Feng, H., Xu, J., Jin, H., and Guo, L. 2021. Investigation of the Conversion Mechanism for Hydrogen Production by Coal Gasification in Supercritical Water. *International Journal of Hydrogen Energy*, 46(17): 10205–10215.
Swalla, D.R. 2008. Feasibility Study of Hydrogen Production from Existing Nuclear Power Plants Using Alkaline Electrolysis. Final Technical Report October 1, 2006–October 30, 2008. General Electric Global Research Center, Niskayuna Which Is a Town within Schenectady County, New York. https://www.osti.gov/servlets/purl/945378.
Uemura, Y., Ohe, H., Ohzuno, Y., and Hatate, Y. 1999. Carbon and Hydrogen from Hydrocarbon Pyrolysis. Proceedings. International Conference Solid Waste Technology Management. Page 15: 5E/25-5E/30.
Vauk, D., Di Zanno, P., Neri, B., Allevi, C., Visconti, A., and Rosanio, L. 2008. What Are Possible Hydrogen Sources for Refinery Expansion? *Hydrocarbon Processing*, 87(2): 69–76.
Vernon, P.D.F., Green, M.L.H., Cheetham, A.K., and Ashcroft, A.T. 1990. Partial Oxidation of Methane to Synthesis Gas. *Catalysis Letters*, 6(2): 181–186.
Wang, W., Stagg-Williams, S.M., Noronha, F.B., Mattos, L.V., and Passos, F.B. 2004. Preprints. *Division of Fuel Chemistry, Journal of the American Chemical Society*, 49(1): 133.
Wang, L., Jin, Z., Chen, X., Su, Y., and Huang, X. 2023. The Origin and Occurrent of Natural Hydrogen. *Energies*, 16(5): 2400. https://www.h2knowledgecentre.com/content/journal4495#:~:text=The%20primary%20sources%20of%20the,adsorbed%20and%20trapped%20in%20inclusions.
Weimer, A.W., Dahl, J., Tamburini, J., Lewandowski, A., Pitts, R., Bingham, C., and Glatzmaier, G.C. 2000. Thermal Dissociation of Methane Using a Solar Coupled Aerosol Flow Reactor. Proceedings. Hydrogen Program Review. NREL/CP-570-28890.
Zagoria, A., Huychke, R., and Boulter, P.H. 2003. Refinery Hydrogen Management – The Big Picture. Proceedings. DGMK Conference on Innovation in the Manufacture and Use of Hydrogen. Dresden. October 15–17. Page 95.
Zaman, J., and Chakma 1995. Production of Hydrogen and Sulfur from Hydrogen Sulfide. *Fuel Processing Technology*, 41(2): 159–198.
Zhu, Q., Zhao, X., and Deng, Y. 2004. Advances in the Partial Oxidation of Methane to Synthesis Gas. *Journal of Natural Gas Chemistry*, 13: 191–203.NCES

4 Hydrogen Production from Non-Typical Sources

4.1 INTRODUCTION

In addition to the steam-methane reforming reaction and other reforming reactions as presented earlier in this text that are used for hydrogen production (Chapter 3) there are other methods, usually categorized as non-typical sources of hydrogen production.

In the context of this book, the production of hydrogen from non-typical sources includes the gasification of a variety of carbonaceous feedstocks (1) coal, (2) refinery coke, (3) crude oil residua, particularly distillation residua as well as the residua also includes residua from other non-distillation processes, (4) tar sand bitumen, and (5) the refinery by-products that are loosely termed 'asphalt', 'tar', and 'pitch' (Conte et al., 2001).

Gasification is a process that converts organic carbonaceous feedstocks (i.e., feedstocks containing organic carbon) or a hydrocarbonaceous feedstock (i.e., feedstocks containing organic carbon and hydrogen) into a variety of gaseous products (which are feedstock dependent) such as carbon monoxide, carbon dioxide, and hydrogen by reacting the feedstock at high temperatures (>700°C, 1,290°F) by a partial oxidation reaction (without combustion) using a controlled amount of oxygen and/or steam. The resulting gas mixture (often loosely referred to as 'synthesis gas' or 'syngas' or even 'producer gas') is itself a fuel.

Thus, gasification is a process that converts organic carbonaceous feedstocks into carbon monoxide, carbon dioxide, and hydrogen by reacting the feedstock at high temperatures (>700°C, 1,290°F), without combustion, with a controlled amount of oxygen and/or steam. The product gas mixture (often referred to as 'synthesis gas' and sometime 'producer gas') is a fuel (Lee et al., 2007; Speight, 2013a, 2014b, 2020).

For many decades coal has been the primary feedstock for gasification units but recent concerns about the use of fossil fuels and the resulting environmental pollutants, irrespective of the various gas cleaning processes and environmental cleanup efforts at gasification plants, there is a move to feedstocks other than coal for gasification processes (Speight, 2013a, 2013b, 2014b). In addition, there are other methods by which hydrogen can be produced including thermochemical processes that are used to convert a variety of feedstocks into gas or liquids from which hydrogen can be separated (Chapter 5).

More pertinent in the present context, the gasification process can also use carbonaceous feedstocks other than those listed above (i.e., coal, refinery coke, crude oil residua, refinery by-products that are loosely termed 'asphalt', ;'tar', and 'pitch', and tar sand bitumen) such as biomass, carbonaceous waste, and other similar bio-wastes (Chapter 5) thereby allowing the various feedstocks such as crude oil processing residues, coal, biomass and other carbonaceous wastes can be used to their fullest potential. In fact, the refining industry has seen fit to use the gasification of distillation residua as a source of hydrogen for the past several decades (Speight, 2014a, 2017).

The advantage of the gasification process when a carbonaceous feedstock (a feedstock containing carbon) or hydrocarbonaceous feedstock (a feedstock containing carbon and hydrogen) is employed is that the product of focus – synthesis gas – is potentially more useful as an energy source and results in an overall cleaner process. The production of synthesis gas is a more efficient production of an energy source than, say, the direct combustion of the original feedstock because synthesis gas can be (1) combusted at higher temperatures, (2) used in fuel cells, (3) used to produce methanol, (4)

DOI: 10.1201/9781003186694-4

used as a source of hydrogen, and (5) particularly because the synthesis gas can be converted via the Fischer-Tropsch process into a range of synthesis liquid fuels suitable for use in gasoline engines or in diesel engines (Chapter 4, Chapter 5).

This chapter presents a description of the various materials that can be used as feedstocks for a gasification processes to produce a variety of gases, including hydrocarbon gases and particularly synthesis gas (a mixture of carbon monoxide and hydrogen) and hydrogen.

4.2 NON-TYPICAL SOURCES OF HYDROGEN

Gasification processes can accept a variety of feedstocks, but the reactor must be selected based on the feedstock properties and the anticipate behavior of the feedstock in the gasification process, especially when coal is considered as the feedstock for a gasifier. Furthermore, because of the historical use of coal for gasification purposes (Speight, 2013a), the inclusion of coal among the gasification feedstocks in this section is warranted.

Overall, use of gasification technology (Speight, 2013a, 2014b) with the necessary gas cleanup options can have a smaller environmental footprint and lesser effect on the environment than landfill operations or combustion of the biomass or organic waste. In fact, there are strong indications that gasification is a technically viable option for waste conversion, including residual waste from separate collection of municipal solid waste. The process can meet existing emission limits and can have a significant effect on the reduction of landfill disposal using known gasification technologies to produce a synthesis gas product than can be employed as a fuel, a petrochemical feedstock as well as for use as a feedstock to the Fischer-Tropsch process (Arena, 2012; Fabry et al., 2013; Speight, 2014b; Luque and Speight, 2015).

However, it is compared to a typical fossil fuel (if there is such a fuel), the complex ligno-cellulosic structure of biomass is more difficult to gasify or combust. The nature of the mineral impurities in conjunction with the presence of various inorganic species, as well as sulfur and nitrogen containing compounds, adversely impacts the benign thermal processing of the oxygenated hydrocarbon structure of the biomass. While combustion of biomass feedstocks results in fuel-bound nitrogen and sulfur (being converted to NOx and SOx, respectively) steam gasification involves thermal treatment under a reducing atmosphere resulting in fuel-bound nitrogen release as molecular nitrogen (N_2) and fuel-bound sulfur conversion to hydrogen sulfide (H_2S) that is more easily removed by means of adsorption beds. Unlike combustion, the gasification process is more energy intensive. Careful engineering of the process is necessary if the result is to produce rather than consume a significant amount of energy or power as a result of the thermal treatment (Butterman and Castaldi, 2008).

4.2.1 COAL

The original method of gas production from coal was produced by carbonizing the coal in horizontal retorts which was an intermittent process that required cyclic operation for periodic charging and cleaning. Later advancements were vertical retorts that were continuously operated, and intermittent vertical retorts that employed 'steaming', or introduction of steam toward the end of the carbonizing process. The steam served to increase the recovery of the remaining coal gas (mostly carbon monoxide and hydrogen), and also increased the quality of the coke produced. The gas produced in this manner was referred to as 'water gas'. However, for many decades since the 19th Century, coal has been the primary feedstock for gasification units Century (Speight, 2013a). The plentiful supply and the properties of coal made it an ideal feedstock for gasification units.

Coal occurs in different forms or types and the harder coal, such as anthracite coal, can be regarded as organic metamorphic rocks because of a higher degree of maturation (Speight, 2013a). Variations in the nature of the source material and local or regional variations in the coalification processes cause the vegetal matter to evolve differently. Thus, various classification systems

TABLE 4.1
The Various Types of Coal

Coal Type	Description
Lignite	Also referred to as brown coal and is the lowest rank* of coal that is used almost exclusively as fuel for steam-electric power generation.
	Jet coal is a compact form of lignite that is sometimes polished and has been used as an ornamental stone since the Iron Age.
Sub-bituminous coal	Exhibits properties ranging from those of lignite to those of bituminous coal.
	Used primarily as fuel for steam-electric power generation.
Bituminous coal	A dense coal, usually black, sometimes dark brown, often with well-defined bands of bright and dull material.
	Used primarily as fuel in steam-electric power generation, with substantial quantities.
	Also used for heat and power applications in manufacturing and to produce coke.
Anthracite	The highest rank coal and is a hard, glossy, black coal.
	Used primarily for residential and commercial space heating.

*Rank refers to the relative amount of carbon in the coal (Speight, 2013a).

exist to define the different types of coal. Thus, as geological processes increase their effect over time, the coal precursors are transformed over geological time to produce the various types of coal (Table 4.1).

Chemically, coal is a hydrogen-deficient hydrocarbon with an atomic hydrogen-to-carbon ratio near 0.8, as compared to crude oil hydrocarbon derivatives, which have an atomic hydrogen-to-carbon ratio approximately equal to 2, and methane (CH_4) that has an atomic carbon-to-hydrogen ratio equal to 4. For this reason, any process used to convert coal to alternative fuels must add hydrogen or redistribute the hydrogen in the original coal to produce hydrogen-rich products and coke (Speight, 2013a).

In the gasification process, which involved heating coal at a high temperature in the absence or relative absence (i.e., partial oxidation), and the products of the gasification process, often referred to simply as 'coal gas' or 'town gas' which is a flammable gas that was used as a fuel for domestic and industrial heating and lighting systems and supplied to the user via a piped distribution system.

The term 'town gas' is a more general term referring to manufactured gaseous fuels produced for sale to consumers and municipalities. However, hydrogen is produced by the reaction of coal with oxygen in a less than stoichiometric ratio (at 1200 to 1350°C/2190 to 2460°F) to yield a mixture of carbon monoxide and hydrogen in approximately equal volumes.

However, in addition to synthesis gas (the mixtures of carbon monoxide and hydrogen), methane, and other hydrocarbon derivatives are also produced depending on the reaction parameters.

4.2.2 Refinery Coke

Refinery coke is (often referred to as 'coke', 'pet coke', or petcoke) the insoluble carbon-rich, often honeycomblike, carbonaceous material that is produced from crude oil and crude oil fractions during thermal processing and also during catalytic processing. More particularly, refinery coke is the residue left by the destructive distillation (i.e., thermal cracking such as the delayed coking process) of crude oil residua. More simply, refinery coke (in this context) is the carbonization product of high-boiling crude oil fractions obtained in a refinery (Speight, 2014, 2017).

The coke formed in catalytic cracking operations is usually non-recoverable because of the materials deposited on the catalyst during the process and such coke is often employed as fuel for the process (Speight, 2014a, 2017). It is often characterized as a solid material with a honeycomb-type

TABLE 4.2
Examples of the Approximate Composition of Refinery Coke

Element	Amount, % w/w
Carbon	98–99.5
Hydrogen	0.1
Nitrogen	<0.2
Sulfur	<0.5
Mineral matter	0.01–0.7

of appearance having high carbon content (95%+ w/w) with some hydrogen and, depending on the process, as well as sulfur and nitrogen (Table 4.2). The color of the coke varies from gray to black, and the material is insoluble in organic solvents.

As an additional note, calcined coke is the product from the calcining process in which the green coke (i.e., raw coke directly out of the coking unit) is raised to high temperature without melting under restricted supply of ambient oxygen, generally for the purpose of removing impurities or volatile substances and/or to incur thermal decomposition.

Typically, the composition of refinery coke varies with the source of the crude oil, but in general, large amounts of high-molecular-weight condensed aromatic ring derivatives (rich in carbon but correspondingly poor in hydrogen) make up a high proportion of the coke. The solubility of refinery coke in carbon disulfide has been reported to be as high as 50 to 80%, but this is in fact often a misinterpretation and is due to a soluble product adsorbed on the coke.

Indeed, the flexibility of the gasification technology permits the refinery to handle any kind of refinery residue, including crude oil coke, tank bottoms and refinery sludge and makes available a range of value-added products, electricity, steam, hydrogen and various feedstocks for petrochemical processes (Speight, 2013a, 2014, 2017)

With respect to gasification, no other technology processing low value refinery residues can come close to the emission levels achievable with gasification (Speight, 2014a, 2017) and is projected to be a major part of the refinery of the future (Speight, 2011b).

Gasification is also a method for converting crude oil coke and other refinery non-volatile waste streams (often referred to as refinery residuals and include but not limited to atmospheric residuum, vacuum residuum, visbreaker tar, and deasphalter pitch) into power, steam and hydrogen for use in the production of cleaner transportation fuels. And for the gasification of coal and other carbonaceous feedstocks, the main requirement for a feedstock to a gasification unit is that the feedstock contains both hydrogen and carbon (Speight, 2013a, 2014b).

Gasification of coke can provide high-purity hydrogen for a variety of uses within the refinery such as (1) sulfur removal, (2) nitrogen removal, as well as removal of other impurities from intermediate to finished product streams and in hydrocracking operations for the conversion of heavy distillates and oils into light products, naphtha, kerosene, and diesel fuel (Speight, 2014a, 2017). Hydrocracking and severe hydrotreating require hydrogen which is at least 99% v/v pure, while less severe hydrotreating can gas stream containing hydrogen on the order of 90% v/v purity.

During gasification some soot (typically 99%+ carbon) is produced, which ends up in the quench water. The soot is transferred to the feedstock by contacting, in sequence, the quench water blowdown with naphtha, and then the naphtha-soot slurry with a fraction of the feed. The soot mixed with the feed is recycled to the gasifier, thus achieving 100% conversion of carbon to gas.

4.2.3 Crude Oil Residua

A crude oil resid (residuum, pl. residua, resids) is the non-volatile residue obtained from crude oil after the nondestructive atmospheric distillation or the nondestructive vacuum distillation of the crude oil has removed all of the volatile constituents. The temperature of the distillation is usually maintained below 350°C (660°F) since the rate of thermal decomposition of crude oil constituents is minimal below this temperature but the rate of thermal decomposition of crude oil constituents is substantial above 350°C (660°F) (Speight, 2014a, 2017).

Typically, resids are non-volatile near-black or black, viscous products that are obtained by distillation of crude oil under atmospheric pressure (atmospheric residuum) or under reduced pressure (vacuum residuum). A resid may be liquid at room temperature (typically, an atmospheric resid) or almost solid (typically, a generally vacuum resid) depending upon the nature of the crude oil feedstock (Speight, 2014a, 2017). When a residuum is obtained from a crude oil and thermal decomposition has commenced, it is more usual to refer to this product as 'pitch', although this term is usually applied to the non-volatile product from coal tar. The differences between the parent crude oil and the (atmospheric and vacuum) residua are due to the relative amounts of various constituents present and which are removed from or remain in the non-volatile residuum by virtue of their relative volatility.

The chemical composition of a residuum from an asphaltic crude oil is complex. Physical methods of fractionation usually indicate high proportions of asphaltene constituents (heptane-insoluble constituents) and resin constituents (simply: heptane-soluble materials but propane-insoluble materials), even in amounts up to 50% (or higher) of the residuum. In addition, the presence of ash-forming metallic constituents, including such organometallic compounds as those of vanadium and nickel, is also a distinguishing feature of residua and the heavier oils. Furthermore, the less volatile the residuum, the greater the concentration of sulfur and metals in the residuum and the greater the deterioration in physical properties (Speight, 2014a, 2017).

4.2.4 Tar Sand Bitumen

The 'tar sand bitumen' is used interchangeably with the term 'oil sand bitumen' (in Canada) and the word 'bitumen' (also, on occasion, incorrectly referred to as 'native asphalt' and 'extra heavy crude oil') includes a wide variety of reddish brown to black materials of semisolid, viscous to brittle character that can exist in nature with no mineral impurity or with a mineral matter content that exceeds 50% w/w (Speight, 2009, 2013c, 2014a, 2017). Naturally occurring bitumen is frequently found filling pores and crevices of sandstone, limestone, or argillaceous sediments such as the deposits referred to as tar sand (which are sediments containing, made of, or resembling clay or clay-containing sediments), in which case the organic and associated mineral matrix is known as 'rock asphalt' (Speight, 2014a, 2017).

Thus, bitumen is a naturally occurring material that is found in deposits where the permeability of the deposit is low and, therefore, the passage of fluids through the deposit is restricted and can only be achieved by (for example) the prior application of fracturing techniques. In terms of the composition, tar sand bitumen is a high-boiling material with little, if any, material boiling below 350°C (660°F) and the boiling range may approximate the boiling range of an atmospheric residuum (or in some cases, the boiling range of a vacuum residuum).

For clarification and legal purposes, *tar sands* have been defined in the United States through FE 76-4 (United States congress, 1976) as:

> the several rock types that contain an extremely viscous hydrocarbon which is not recoverable in its natural state by conventional oil well production methods including currently used enhanced recovery techniques. The hydrocarbon-bearing rocks are variously known as bitumen-rocks oil, impregnated rocks, oil sands, and rock asphalt.

In the crude oil industry, the expression 'tar sand' is typically used to describe sandstone reservoirs that are impregnated with a heavy, viscous black crude oil that cannot be retrieved through a well by conventional production techniques (FE 76-4, United States congress, 1976).

4.2.5 Asphalt, Tar, and Pitch

Asphalt does not occur naturally but is manufactured from crude oil and is a black or brown material that has a consistency varying from a viscous liquid to a glassy solid (Speight, 2014a, 2017). To a point, asphalt can resemble bitumen (isolated form tar sand formation), hence the tendency to refer to bitumen (incorrectly) as *native asphalt.* It is recommended that there be differentiation between asphalt (manufactured) and bitumen (naturally occurring) other than by use of the qualifying terms *crude oil* and *native* since the origins of the materials may be reflected in the resulting physicochemical properties of the two types of materials. It is also necessary to distinguish between the asphalt which originates from crude oil by refining and the product in which the source of the asphalt is a material other than crude oil, e.g., *Wurtzilite asphalt* (Speight, 2014a). In the absence of a qualifying word, it should be assumed that the word *asphalt* (with or without qualifiers such as *cutback, solvent,* and *blown,* which indicate the process used to produce the asphalt) refers to the product manufactured from crude oil.

When the asphalt is produced simply by distillation of an asphaltic crude oil, the product can be referred to as 'residual asphalt' or 'straight-run asphalt'. For example, if the asphalt is prepared by solvent extraction of residua or by a low-boiling light hydrocarbon (propane is the typical example) or if blown (oxidized) or otherwise treated, the term should be modified accordingly to qualify the product (e.g., solvent asphalt, propane asphalt, blown asphalt). The resulting product (now referred to as 'asphalt') is also used for roofs, coatings, floor tiles, soundproofing, waterproofing, and other building-construction elements and in a number of industrial products, such as batteries.

Tar is a product of the destructive distillation of many bituminous or other organic materials and is a brown to black, oily, viscous liquid to semi-solid material. However, *tar* is most commonly produced from *bituminous coal* and is generally understood to refer to the product from coal, although it is advisable to specify *coal tar* if there is the possibility of ambiguity. The most important factor in determining the yield and character of the coal tar is the carbonizing temperature. Three general temperature ranges are recognized, and the products have acquired the designations: *low-temperature tar* (approximately 450 to 700°C; 540 to 1,290°F); *mid-temperature tar* (approximately 700 to 900°C; 1290 to 1,650°F); and *high-temperature tar* (approximately 900 to 1,200°C; 1,650 to 2,190°F). Tar released during the early stages of the decomposition of the organic material is called *primary tar* since it represents a product that has been recovered without the secondary alteration that results from prolonged residence of the vapor in the heated zone. In the chemical-process industries, pitch is the black or dark brown residue obtained by distilling coal tar, wood tar, fats, fatty acids, or fatty oils.

Any of the above derivatives can be used as a gasification feedstock. The properties of asphalt change markedly during the aging process (oxidation in service) to the point where the asphalt fails to perform the task for which it was designed. In some cases, the asphalt is recovered and reprocessed for additional use or it may be sent to a gasifier.

4.3 THE GASIFICATION PROCESS

The term 'gasification process' refers to a thermal process (also referred to as a 'partial oxidation' process or a 'Pox process', POX or P_{OX}) that is employed for the conversion of fossil-based organic carbonaceous materials at high temperatures (>700°C, 1,290°F), without combustion, with a controlled amount of oxygen (hence, the term 'partial oxidation process') and/or steam into carbon monoxide, hydrogen, and carbon dioxide.

In the process, a carbonaceous (organic) feedstock is converted into gaseous products at a high temperature (typically in excess of 700°C, 1,290°F) with a controlled amount of oxygen (i.e., without combustion,) and/or steam into carbon monoxide, hydrogen, and carbon dioxide.

The process can accept a variety of feedstocks but the process parameters and the type of reactor are selected based on the feedstock properties and the anticipated behavior of the feedstock in the process. Furthermore, because of the historical use of the gasification of coal to produce gas for lighting purposes and for heating purposes (Speight, 2013a, 2021), it is the feedstock against which the suitability of all other feedstocks is measured.

The typical gasification system consists of several process plants including (1) a feedstock preparation area, (2) the type of gasifier, (3) a gas cleaning section, and (4) a sulfur recovery unit, as well as (v) downstream process options that are dependent on the nature of the products. Also, the gas generated from biomass by the gasification process can be differentiated in direct processes (autothermal processes) and indirect processes (allothermal processes).

4.3.1 Gasifiers

A gasification reactor differs from a combustor in that the amount of air or oxygen available inside the gasification reactor is carefully controlled so that only a relatively small portion of the feedstock burns completely. In addition, the gasifier contains separate sections for gasification and combustion. The gasification section consists of three parts: (1) the riser, (2) the settling chamber, and (3) the downcomer whereas the combustion section contains only one part, the combustor.

Four types of gasifier are currently available for commercial use: (1) the counter-current fixed-bed gasifier, (2) co-current fixed-bed gasifier, (3) the fluidized-bed gasifier, and (4) the entrained flow gasifier (Speight, 2013a; Luque and Speight, 2015; Speight, 2020, 2021). The typical gasification system incorporated into the refinery consists of several process plants including (1) feed preparation, (2) the gasifier, (3) an air separation unit (ASU), (4) synthesis gas cleanup, (5) sulfur recovery unit (SRU), and (6) downstream process options such as Fischer-Tropsch synthesis (FTS) and methanol synthesis (MTS), depending on the desired product slate.

In a fixed-bed process, the feedstock is supported by a grate and combustion gases (steam, air, oxygen, etc.) pass through the supported feedstock whereupon the hot gases that are produced gases exit from the top of the reactor. The reactor heat is supplied internally or from an outside source, but some feedstocks cannot be used in an unmodified fixed-bed reactor. Due to the liquid-like behavior, the fluidized-beds are very well mixed, which effectively eliminates the concentration and temperature gradients inside the reactor. The process is also relatively simple and reliable to operate as the bed acts as a large thermal reservoir that resists rapid changes in temperature and operation conditions. The disadvantages of the process include the need for recirculation of the entrained solids carried out from the reactor with the fluid, and the nonuniform residence time of the solids that can cause poor conversion levels. The abrasion of the particles can also contribute to serious erosion of pipes and vessels inside the reactor (Kunii and Levenspiel, 1991) – another reason why feedstock pretreatment (Chapter 2) is essential.

The fixed-bed gasifier is further classified as (1) an updraft gasifier and (2) a downdraft gasifiers. In the former type of gasifier (i.e., the updraft gasifier), the feedstock is supplied from the top, while the gasifying agent (air or oxygen) is supplied from the bottom (counter-current). In the latter type of gasifier (i.e., the downdraft gasifier), the feedstock and the gasifying agent (air or oxygen) are introduced from the top (co-current). For the updraft gasifiers, the sequence of the feedstock passage is drying, pyrolysis, and reduction, finally arriving at the combustion zone, with synthesis gas drawn out from the top. In the case of the downdraft gasifier, both the feedstock and oxygen or air are supplied in the drying zone, going the through pyrolysis, combustion, and reduction, with synthesis gas drawn out from the bottom (Dogru et al., 2002). On the other hand, the counter-current fixed-bed (updraft) gasifier consists of a fixed-bed of the carbonaceous fuel through which the gasification agent (steam, oxygen and/or air) flows in counter-current configuration. In addition, an updraft gasifier has four well-defined zones which (1) the partial combustion zone, (2) the reduction zone, (3) the pyrolysis zone, and (4) the drying zone.

The product gas exits the gasification unit together with the products of pyrolysis from the pyrolysis zone and steam from the drying zone. The resulting combustible gas is rich in higher molecular weight hydrocarbon derivatives (tars) and, therefore, has a higher calorific value, which makes updraft gasifiers more suitable where the gas is to be used for heat production. The mineral ash is either removed dry or as a slag. In fixed-bed gasifiers the feedstock is fed to the reactor from its top, through an opening on the reactor's head and moves downwards by gravity. Depending on the way that the gasification medium (air, oxygen or steam) is introduced in the reactor, fixed-bed systems are divided into updraft gasifiers and downdraft gasifiers.

The co-current fixed-bed (downdraft) gasifier is similar to the counter-current type of gasifier except that the gasification agent gas flows in co-current configuration with the fuel (downwards, hence the name down draft gasifier) which give rise to the term 'co-current gasifier' because the air moves in the same direction as that of fuel (i.e., in a downward direction). A downdraft gasifier is so designed that tar product, which are produced in the pyrolysis zone, travel through the combustion zone, where the tar is decomposed into lower molecular weight products or burned. As a result, the mixture of gases in the exit stream is relatively clean. The position of the combustion zone is thus a critical element in the downdraft gasifier – the main advantage being that it produces gas with low tar content. The operating temperature varies from a minimum of 900°C (1,650°F) to a maximum of 1,000 to 1,050°C (1830 to 1,920°F).

On the other hand, in a cross-draft gasifier, the air enters from one side of the gasifier reactor and leaves from the other. Also, a cross-draft gasifier does not need a grate insofar as the product ash falls to the bottom and does not come in the way of normal operation.

In the fluidized-bed gasifier, the feedstock is fluidized in oxygen (or air) and steam and the carbonaceous feedstock is brought into an inert bed of fluidized material (such as sand or char) – the air is distributed through nozzles located at the bottom of the bed. The fluidized-bed reactors are diverse but characterized by the fluid behavior of the catalyst (Dry (2001). Typically, the feedstock is introduced into the fluidized system either above-bed or directly into the bed, depending upon the size and density of the feedstock and how it is affected by the bed velocities. During normal operation, the bed media is maintained at a temperature between 550 and 1,000°C (1,020 and 1,830°F). When the feedstock is introduced under such temperature conditions, the drying and pyrolysis steps proceed rapidly, driving off all gaseous portions of the fuel at relatively low temperature. Any char that remains is oxidized within the bed to provide the heat source for the drying and devolatilizing reactions to continue.

Fluidized-bed gasifiers produce a uniformly high (800 to 1,000°C, 1,470 to 183°F) bed temperature. A fluidized-bed gasifier works as a hot bed of sand particles agitated constantly by air. Fluidized-bed gasifiers are most useful for fuels that form highly corrosive ash that would damage the walls of slagging gasifiers. The ash is removed dry or as high molecular weight agglomerated materials – a disadvantage of some carbonaceous feedstocks is that they generally contain high levels of mineral constituents that can produce corrosive ash.

Compared with the fixed-bed gasifiers, in the fluidized-bed gasifier, the sequence of processes (drying, pyrolysis, oxidation and reduction) is not obvious at a certain point of the gasifier since they take place in the entire reactor thus resulting to a more homogeneous type of reaction. This means the existence of more constant and lower temperatures inside the reactor, where no 'hot spots' are observed. Due to the lower operating temperatures, ash does not melt, so it is easier to remove from the reactor. At the same time, sulfur and chloride contaminants can be absorbed in the inert bed material thus eliminating the fouling hazard and reducing the maintenance costs. Also, fluidized-bed gasifiers are much less subject to feedstock quality than fixed-bed systems, and can even operate with mixed (blended) feedstocks.

In the entrained flow gasifier, the dry pulverized solid feedstock, an atomized liquid, or a liquid slurry is gasified with oxygen – air is also a less frequent oxidizer – in co-current flow. The high temperature and high pressure in the gasifier also lead to a higher throughput of the feedstock but the thermal efficiency is somewhat lower as the gas must be cooled before it can be sent to a gas processing facility. All entrained flow gasifiers remove a major part of the ash as a slag as the operating temperature is well above the ash fusion temperature. The operating temperatures are on the order of 1,200 to 1,600°C (2,190 to 2,910°F) and the pressure is 300 to 1,200 psi. Entrained-flow gasifiers are suitable for use with carbonaceous feedstocks as long as the feedstock (low moisture) and has a low content of mineral matter (manifested as a low ash in the product). Due to the short residence time (0.5 to 4.0 seconds), high temperatures are required for such gasifiers and – as a result – entrained-flow gasifiers are that the gas contains only small quantities of tar.

The main feature of entrained-flow gasification is the requirement for a very fine feedstock particle size, even smaller than 0.1 mm which allows the feedstock easier to be carried out by the gasification medium. The feedstock can be introduced in the reactor either at a dry form or even as slurry in which the pulverized feedstock is mixed with water (or other suitable liquid carrier. The retention time of the feedstock inside the reactor is only a few seconds and thus in order to achieve high conversion rates, higher temperatures are applied (between 1,200 and 1,500°C, 2,190 and 2,730°F). At these temperature levels, the resulting ash melts, cools down and eventually accumulates as slag. In addition, the entrained-flow gasification process typically takes place at elevated pressure – for example, reaching an operating pressure on the order of 600 and 750 psi.

Entrained flow gasifiers are classified into two types: (1) the top-fed gasifiers and (2) the side-fed gasifiers. A top-fed gasifier is a vertical cylinder reactor where fine particles (pulverized) and the gasification agent are co-currently fed from the top in the form of a jet and the feedstock conversion is achieved by use of an inverted burner. The product gas is taken from the side of the lower section while slag is extracted from the bottom of the reactor. On the other hand, in a side-fed gasifier, the pulverized fed and the gasification agent are co-currently fed by nozzles installed in the lower reactor, resulting in an appropriate mixing of the feedstock and the gasification agent (air or oxygen). The gas is extracted from the top of the gasifier and the slag from the bottom. Both configurations are highly efficient, with a standard operating temperature (1,300 to 1,500°C, 2,370 to 2,730°F)]and pressure 300 to 1,000 psi.

Also, with regard to the entrained-flow reactor, it may be necessary to pulverize the feedstock (such as the case when the feedstock is a char). On the other hand, when the feedstock is sent into an entrained-flow gasifier, the feedstock can be in either the form of dry feedstock or slurry feedstock if, for example, bio-oil and a solid feedstock (including feedstock and feedstock char) are co-gasified. In general, dry-feed gasifiers have the advantage over slurry-feed gasifiers in that the former can be operated with lower oxygen consumption. Moreover, dry-feed gasifiers have an additional degree of freedom that makes it possible to optimize bio-synthesis gas production (Shen et al., 2012).

While there are many alternate uses for the synthesis gas produced by gasification, and a combination of products/utilities can be produced in addition to power. A major benefit of the integrated gasification combined cycle concept is that power can be produced with the lowest sulfur oxide (Sox) and nitrogen oxide (NOx) emissions of any liquid/solid feed power generation technology.

Finally, and by way of explanation, an indirect gasifier requires an external source of heat to be transferred to the feedstock to pyrolyze the feedstock under conditions of high severity, i.e., long residence time at high temperature. In the case of certain feedstocks (such as biomass), the process will provide a char stream using either slow or fast pyrolysis to temperatures of 600 to 750°C (1,110 to 1,380°F), in amounts representing between 12% and 25% w/w of the feedstock.

While upstream processing is typically influential in terms of the material handling, the choice of the operation parameters for the gasifier (i.e., temperature, gasifying agent, and catalysts) dictates the product gas composition and quality. As an example, the decomposition of some feedstocks may occur at a lower temperature than the decomposition of coal and therefore reactors that are compatible to the feedstock mixture are required (Speight, 2011; Brar et al., 2012; Speight, 2013a, 2013b). Furthermore, the characteristics of the feedstock and the type of gasifier type (as well as the operating parameters used in the gasifier) not only decide product gas composition but also dictate the amount of impurities to be handled downstream. Moreover, downstream processes need to be modified if coal is co-gasified with biomass.

Heavy metals (i.e. metals with a relatively high density, relatively high atomic weight, or relatively high atomic number) and impurities such as sulfur and mercury present in coal can make synthesis gas difficult to use and unhealthy for the environment. Alkali constituents that are typically present in some feedstocks can also cause corrosion problems high temperatures in downstream pipes.

However, first and foremost, the feedstock requires drying and size reduction before they can be fed into a gasifier. In addition, size reduction of the feedstock particles is needed to obtain a particle size that is appropriate to the reactor. However, drying is required to achieve moisture content suitable for gasification operations and feedstock densification may be necessary to prepare the feedstock pellets and to improve density and material flow in the feeder areas.

4.3.2 Fischer-Tropsch Synthesis

Bio-synthesis gas is used as a raw material in different thermochemical processes for the production of second-generation biofuels, both liquid, (such as methanol, ethanol, dimethyl ether (DME), and Fischer-Tropsch diesel) and gaseous (such as hydrogen and synthetic natural gas (SNG). In particular, the type and composition of the feedstock and the production process strongly influences the composition and heating value of the gaseous products (Sikarwar et al., 2016).

The synthesis reaction is dependent on a catalyst, mostly an iron or cobalt catalyst where the reaction takes place. There is either a low-temperature Fischer-Tropsch process (often represented as the LTFT process) or high temperature Fischer-Tropsch process (often represented as the HTFT process), with temperatures ranging between 200 to 240°C (390 to 465°F) for the low-temperature Fischer-Tropsch process and 300 to 350°C (570 to 660°F) for the high-temperature Fischer-Tropsch process. The high temperature Fischer-Tropsch process uses an iron-based catalyst, and the low-temperature Fischer-Tropsch process uses either an iron-based catalyst or a cobalt-based catalyst. The different catalysts include also nickel-based catalyst and ruthenium-based catalysts.

Finally, it is worthy of note that clean synthesis gas can also be used (1) as chemical building blocks to produce a broad range of chemicals using processes well established in the chemical and petrochemical industry), (2) as a fuel producer for highly efficient fuel cells (which run off the hydrogen made in a gasifier) or perhaps in the future, hydrogen turbines and fuel cell-turbine hybrid systems, and (3) as a source of hydrogen that can be separated from the gas stream and used as a fuel or as a feedstock for refineries (which use the hydrogen to upgrade crude oil products) (Speight, 2016, 2017, 2019a).

4.4 PROCESS CHEMISTRY

The gasification process consists of the conversion of the organic carbon in the carbonaceous feedstock into various components that make up the synthesis gas by reacting the feedstock at high temperatures (>700°C, 1,290°F), without combustion using a controlled amount of oxygen and/or steam (Marano, 2003; Lee, 2007; Lee et al., 2007; Higman and Van der Burgt, 2008; Sutikno and Turini, 2012; Speight, 2013a, 2014b; Luque and Speight, 2015; Speight, 2020). The products include gaseous hydrocarbon derivatives (methane, ethane, and propane), carbon monoxide, carbon dioxide, and hydrogen. Other products may also be produced depending upon the composition of the feedstock.

As might be anticipated because of the complexity of the composition of the various types of the feedstock, the gasification the chemistry of the feedstock gasification may consist of many overlapping processes such as (1) drying, (2) pyrolysis, and (3) partial oxidation.

The pyrolysis process can be categorized as (1) mild pyrolysis, (2) slow pyrolysis, or (3) fast pyrolysis. A simple way of representing the reaction is:

$$\text{Feedstock} \rightarrow H_2 + CO + CO_{2(gas)} + \text{hydrocarbons}_{(gas)} + \text{tar}_{(liquid)} + \text{char}_{(solid)}$$

Elevated temperatures, high heating rates and long volatile stage residence times are vital for synthesis gas production or hydrogen production. Since the liquid yield through fast pyrolysis is high, the gaseous yield is reduced. This approach may not be conducive to the production directly of usable products but can produce pyrolytic oils that require further upgrading owing to their high acidity.

It is important to distinguish the gasification process from a pyrolysis process. The main difference between pyrolysis and gasification is the absence of a gasifying agent in a pyrolysis process (Chapter 2, Chapter 4). Pyrolysis is a thermal degradation of organic compounds, at a range of temperatures on the order of 300 to 900°C (570 to 1650°F), under oxygen-deficient circumstances to produce various types of products such as gases (often referred to as synthesis gas), a liquid product (referred to as bio-oil), and a solid product (referred to as biochar).

On the other hand, the gasification process involves thermal cracking of solid or high boiling liquid carbonaceous material (such as a liquid feedstock or s semi-liquid feedstock) into a combustible gas mixture, mainly composed of hydrogen, carbon monoxide (CO), carbon dioxide (CO_2), and methane (CH_4) and other gases with some by-products (solid char or slag, oils, and water). The synthesis gas product has a chemical composition and properties that are largely affected by (1) the type and composition of the feedstock, (2) the operational conditions throughout pyrolysis and gasification such as reactor temperature, residence time, pressure, and (3) the reactor geometry.

In the current context, and in addition to the potential reactions that can occur in the gasifier, the gasification process is often considered to involve two distinct physico-chemical stages: (1) devolatilization of the feedstock to produce volatile matter and char, (2) followed by char gasification, which is complex and specific to the conditions of the reaction – both processes contribute to the complex kinetics of the gasification process (Sundaresan and Amundson, 1978; Luque and Speight, 2015).

4.4.1 General Aspects

At a temperature in excess of 500°C (930°F) the conversion of the feedstock to char and ash and char is completed. Furthermore, with an increase in heating rate, feedstock particles are heated more rapidly and are burned in a higher temperature region, but the increase in heating rate may have very little effect on the mechanism (Demirbaş, 2004; Tinaut et al., 2008; Irfan, 2009).

Most notable effects in the physical chemistry of the gasification process are those effects due to the chemical character of the feedstock as well as the physical composition of the feedstock (Speight, 2011a, 2013a, 2014a, 2014b, 2017). In more general terms of the character of the feedstock, gasification technologies generally require some initial processing of the feedstock with the type and degree of pretreatment a function of the type of feedstock and the operational parameters of the gasifier (Table 4.3, Table 4.4). This is especially true for a feedstock that can have several constituents which strongly influence the course and outcome of the conversion process (such as biomass, Chapter 5).

Another factor, often presented as a very general *rule of thumb*, is that optimum gas yields and gas quality are obtained at operating temperatures of approximately 595 to 650°C (1,100 to 1,200°F).

With some feedstocks, the higher the amounts of volatile material produced in the early stages of the process the higher the heat content of the product gas. However, in the case of some carbonaceous feedstocks, there is the potential for carbon dioxide to be an initial product of the process which will serve to lower the heat content of the product gas. Also, when the process temperature is not sufficiently high, the char oxidation reaction is suppressed and the overall heat content of the product gas is diminished. All such events serve to complicate the reaction rate and make the attempted derivation of a kinetic relationship that can be applied to all types of feedstock (and gasifier configurations) subject to serious question and doubt.

Pressure also plays a role in the type of product produced and is, as might be anticipated, dependent on the type of feedstock being processed and the gas product that is desired. In fact, some (or all) of the following processing steps will be required (1) pretreatment of the feedstock, (2) primary gasification of the feedstock, (3) secondary gasification of the carbonaceous residue from

TABLE 4.3
Examples of Properties that Influence the Gasification Process

Chemical Properties	**Elemental composition**	**C,H,N,O,S (% w/w)**
	Proximate analysis	Moisture
		Fixed carbon
		Volatile matter
		Mineral matter
Physical Properties	Density	
	Porosity	
	Particle size	
	Particle shape	
Thermal Properties	Heating value	
	Elemental composition	
	Types of minerals	Mineral ash
		Ash fusion temperature

TABLE 4.4
Characteristics of the Different Types of Gasifiers

Gasifier Type	**Feedstock Properties**
Fixed/moving bed	Particle size: 1 to 10 cm
	Mechanically stable fuel particles
	Unblocked passage of gas through the bed
	Pellets or briquettes preferred
Fluidized bed	Updraft configuration more tolerant to biomass moisture content
	Ash melting temperature of fuel: higher limit for operating temperature
	Fuel particle size relatively small to ensure good contact with bed material; typically <40 mm
	Good fuel flexibility due to high thermal inertia of the bed
Entrained bed	Fuel particle size: <50 micrometers
	Pulverized for high fuel conversion in short residence times
	Ash melting behavior can influence for reactor/process design
	Ash melting behavior can influence for reactor/process design

the primary gasifier; (4) removal of carbon dioxide, hydrogen sulfide, and other acid gases, (5) shift conversion for adjustment of the carbon monoxide/hydrogen mole ratio to the desired ratio, and (6) catalytic methanation of the carbon monoxide/hydrogen mixture to form methane.

While pretreatment of the feedstock for introduction into the gasifier is often considered to be a physical process in which the feedstock is prepared for the gasification process – typically as pellets or finely ground feedstock – there are chemical aspects that must also be considered (Chapter 2, Chapter 4). Co-gasification of different feedstocks, such as coal and with the residuum or coke from crude oil refining (often referred to as petcoke) offers a bridge between the depletion of crude oil stocks when coal, residua, and/or petcoke are used (Speight, 2014b).

The high reactivity of some feedstocks and the accompanying high production of volatile products suggest that some synergetic effects might occur in simultaneous thermochemical treatment of coal, residua, or petcoke, depending on the gasification conditions such as: (1) the feedstock type and origin, (2) reactor type, and (iii) process parameters (Penrose et al., 1999; Gray and Tomlinson, 2000; McLendon et al., 2004; Lapuerta et al., 2008; Fermoso et al., 2009; Shen et al., 2012; Khosravi and Khadse, 2013; Speight, 2013a, 2014a, 2014b; Luque and Speight, 2015; Speight, 2017).

For example, carbonaceous fuels are gasified in reactors, a variety of gasifiers such as the fixed-bed gasifier, the moving-bed gasifier, the fluidized-bed gasifier, the entrained-flow gasifier, and the molten bath gasifier have been developed that have differing feedstock requirements and are selected on the basis of the properties of the feedstock (Table 4.4) (Beenackers, 1999; Shen et al., 2012; Speight, 2014b; Luque and Speight, 2015).

4.4.2 Reactions

The gasification process is not a single step process, but involves a variety of chemical reactions (Table 4.5), and the is a complex thermal process that depends on the pyrolysis mechanism to generate gaseous precursors, which in the presence of reactive gases such as oxygen and steam convert the majority of the feedstock into a synthesis gas, which is mainly composed of carbon monoxide and hydrogen, and used to produce chemicals and liquid fuels over catalysts (Speight, 2016, 2017; 2019a). The distribution of weight and chemical composition of the products are also influenced by the prevailing conditions (i.e. temperature, heating rate, pressure, and residence time) and, last but by no means least, the nature of the feedstock (Speight, 2014a, 2014b, 2017).

If air is used for combustion, the product gas will have a heat content of approximately 150-300 Btu/ft^3 (depending on process design characteristics) and will contain undesirable constituents such as carbon dioxide, hydrogen sulfide, and nitrogen. The use of pure oxygen, although expensive, results in a product gas having a heat content on the order of 300 to 400 Btu/ft^3 with carbon dioxide and hydrogen sulfide as by-products (both of which can be removed from low or medium heat-content, low-Btu or medium-Btu) gas by any of several available processes (Speight, 2013a, 2014a, 2017).

If a high heat-content (high-Btu) gas (900 to 1,000 Btu/ft^3) is required, efforts must be made to increase the methane content of the gas. The reactions which generate methane are all exothermic and have negative values, but the reaction rates are relatively slow and catalysts may, therefore, be necessary to accelerate the reactions to acceptable commercial rates. Indeed, the overall reactivity of the feedstock and char may be subject to catalytic effects insofar as the mineral constituents of the feedstock (such as the mineral matter in the feedstock) may modify the reactivity by a direct

TABLE 4.5
Examples of Reactions that Occur During Gasification of a Carbonaceous Feedstock

$2C + O_2 \rightarrow 2CO$

$C + O2 \rightarrow CO_2$

$C + CO_2 \rightarrow 2CO$

$CO + H_2O \rightarrow CO_2 + H_2$ (shift reaction)

$C + H_2O \rightarrow CO + H_2$ (water gas reaction)

$C + 2H_2 \rightarrow CH_4$

$2H_2 + O_2 \rightarrow 2H_2O$

$CO + 2H2 \rightarrow CH_3OH$

$CO + 3H2 \rightarrow CH_4 + H_2O$ (methanation reaction)

$CO_2 + 4H_2 \rightarrow CH_4 + 2H_2O$

$C + 2H_2O \rightarrow 2H_2 + CO_2$

$2C + H_2 \rightarrow C_2H_2$

$CH_4 + 2H_2O \rightarrow CO_2 + 4H_2$

catalytic effect (Davidson, 1983; Baker and Rodriguez, 1990; Mims, 1991; Martinez-Alonso and Tascon, 1991).

If the feedstock begins to rapidly decompose with heat once its temperature rises above 240°C (465°F) and the breaks down into a combination of gases, liquids, and solid (char). In the process, the feedstock undergoes three steps in the conversation to gas – the first two steps, pyrolysis and combustion, occur very rapidly – all of which are highly dependent upon the properties of the feedstock (Table 4.3). However, there is also a preliminary step in which any moisture in the feedstock before it enters the pyrolysis step. All of the moisture needs to be (or will be) removed from the feedstock at some point (typically between 100 and 150°C, 212 and 300°F) in the higher temp processes. Moisture removal (often referred to as drying) is one of the major issues that has to be solved for successful gasification of the feedstock.

In the pyrolysis step, char is produced as the feedstock heats up and volatiles are released. In the combustion step, the volatile products and the char can react with oxygen to produce predominantly primarily carbon dioxide and carbon monoxide as well as heat that is required for subsequent gasification reactions. Finally, in the gasification step, the feedstock char reacts with steam to produce hydrogen (H_2) and carbon monoxide (CO).

Drying:

$$\text{Feedstock (wet)} \rightarrow \text{feedstock (dry)} + H_2O$$

Pyrolysis:

$$\text{Feedstock} \rightarrow \text{volatile products} + \text{char (coke)}$$

Combustion:

$$2C_{\text{volatiles + char}} + O_2 \rightarrow 2CO + H_2O$$

Gasification:

$$C_{\text{char}} + H_2O \rightarrow H_2 + CO$$

$$CO + H_2O \rightarrow H_2 + CO_2$$

Thus, in the initial stages of gasification, the rising temperature of the feedstock initiates devolatilization and the breaking of weaker chemical bonds to yield volatile tar, volatile oil, phenol derivatives, and hydrocarbon gases. These products generally react further in the gaseous phase to form hydrogen, carbon monoxide, and carbon dioxide. The char (fixed carbon) that remains after devolatilization reacts with oxygen, steam, carbon dioxide, and hydrogen.

In essence, the direction of the gasification process is subject to the constraints of thermodynamic equilibrium and variable reaction kinetics. The combustion reactions (reaction of the feedstock or char with oxygen) essentially go to completion. The thermodynamic equilibrium of the rest of the gasification reactions are relatively well defined and collectively have a major influence on thermal efficiency of the process as well as on the gas composition.

Typically, there may not be reliable kinetic information (for general application) on gasification reactions exists, partly because it is highly dependent on (1) the chemical nature of the feed, which varies significantly with respect to composition, mineral impurities, (2) feedstock reactivity, and (3) process parameters, such as temperature, pressure, and residence time. In addition, physical characteristics of the feedstock (or char) also play a role in phenomena such boundary layer diffusion, pore diffusion and ash layer diffusion which also influence the kinetic outcome. Also, some of the contaminants in the feedstock may have catalytic activity on some of the gasification reactions which can have further influence on the kinetic imprint of the gasification reactions.

For example, alkali metal salts are known to catalyze the steam gasification reaction of carbonaceous materials, hence the need for pretreatment of the feedstock to remove such metals (Chapter 2). The process is based on the concept that alkali metal salts (such as potassium carbonate (K_2CO_3), sodium carbonate (Na_2CO_3), potassium sulfide (K_2S), and sodium sulfide, Na_2S, as well as other metals-containing constituents) will catalyze the steam gasification of feedstocks. The order of catalytic activity of the metals on the gasification reaction is:

$$\text{Cesium (Cs)} > \text{rubidium (Rb)} > \text{potassium (K)} > \text{sodium (Na)} > \text{lithium (Li)}$$

Ruthenium-containing catalysts are used primarily in the production of ammonia. It has been shown that ruthenium catalysts provide five to 10 times higher reactivity rates than other catalysts. However, during the process, carbon is consumed, thereby reducing the effect of the ruthenium catalyst.

4.4.2.1 Water Gas Shift Reaction

The water gas shift reaction is used to increase the hydrogen yield and to lower the amount of carbon monoxide in a product gas, a WGS unit is employed. The water gas shift process is a well-established process, especially in large-scale industrial plants that produce hydrogen as well as for adjusting the carbon monoxide-hydrogen ratio in synthesis gas (Chadeesingh, 2011; Speight, 2013a). In addition, if a high-Btu gas is the desired product, the option is application of the water gas shift reaction, in which carbon monoxide is converted to hydrogen and carbon dioxide in the presence of a catalyst that can be represented by the simple equation. Thus:

$$CO + H_2O \rightarrow CO_2 + H_2$$

This reaction maximizes the amount of hydrogen in the product gas stream and the product gas is then scrubbed of particulate matter and sulfur is removed via physical absorption (Speight, 2013a, 2014a, 2017). The carbon dioxide is captured by physical absorption or a membrane and either vented or sequestered.

On the industrial scale, a water gas shift unit typically involves the use of one or more fixed-bed reactors. On a relatively small scale, a by-pass high-temperature water gas shift stage may be sufficient but for producing hydrogen, treatment of the feedstock gas stream may involve two or three stages at gradually lower inlet and outlet temperature to achieve a high conversion of the carbon monoxide.

Catalysts for the process (depending upon the temperature) are Co-Mo or Cu-Zn-based catalysts. The former catalyst (i.e. the Co-Mo catalyst) is resistant to the presence of sulfur components but Cu-Zn catalysts are very sensitive to sulfur poisoning and, therefore, sulfur removal is necessary (Liu et al., 2010).

4.4.2.2 Carbon Dioxide Gasification

The reaction of carbonaceous feedstocks with carbon dioxide produces carbon monoxide (*Boudouard reaction*) and (like the steam gasification reaction) is also an endothermic reaction:

$$C(s) + CO_2(g) \rightarrow 2CO(g)$$

This reaction may be valuable when biomass is the feedstock because of the potential to produce carbon dioxide directly from the biomass by the decompositon of carboxylic acid functions (Cao et al., 2019):

$$[\text{-C-}CO_2H] \rightarrow CO_2 + [\text{-C-CH}]$$

This gasification reaction is thermodynamically favored at high temperatures (>680°C, >1,255°F), which is also quite similar to the steam gasification. Also, the rate of the carbon dioxide gasification of a feedstock is different to the rate of the carbon dioxide gasification of carbon.

4.4.2.3 Hydrogasification

Hydrogasification is the gasification of feedstock, such as a char-type feedstock, in the presence of an atmosphere of hydrogen under pressure.

$$C_{char} + 2H_2 \rightarrow CH_4$$

The hydrogen-rich gas for hydrogasification can be manufactured from steam by using the char that leaves the hydrogasifier and the hydrogasification reaction is exothermic and is thermodynamically favored at relatively low temperature (<670°C, <1,240°F), unlike the endothermic both steam gasification and carbon dioxide gasification reactions. However, at low temperatures, the reaction rate is inevitably too slow. Therefore, a high temperature is always required for kinetic reasons, which in turn requires high pressure of hydrogen, which is also preferred from equilibrium considerations. This reaction can be catalyzed by salts such as potassium carbonate (K_2CO_3), nickel chloride ($NiCl_2$), iron chloride ($FeCl_2$), and iron sulfate ($FeSO_4$). However, use of a catalyst in feedstock gasification suffers from difficulty in recovering and reusing the catalyst and the potential for the spent catalyst becoming an environmental issue.

4.4.2.4 Methanation

A variety of metals have been used as catalysts for the methanation reaction. The most effective methanation catalysts, appear to be nickel and ruthenium, with nickel being the most widely (Cusumano et al., 1978):

Ruthenium (Ru) > nickel (Ni) > cobalt (Co) > iron (Fe) > molybdenum (Mo).

However, many catalysts are susceptible to sulfur poisoning and efforts must be taken to remove all hydrogen sulfide (H_2S) before the catalytic reaction starts. It is necessary to reduce the sulfur concentration in the feed gas to less than 0.5 ppm v/v in order to maintain adequate catalyst activity for a long period of time.

The methanation reaction is used to increase the methane content of the product gas, as needed for the production of high-Btu gas:

$$4H_2 + CO_2 \rightarrow CH_4 + 2H_2O$$

$$2CO \rightarrow C + CO_2$$

$$CO + H_2O \rightarrow CO_2 + H_2$$

Among these, the most dominant chemical reaction leading to methane is the first one. Therefore, if methanation is carried out over a catalyst with a bio-synthesis gas mixture of hydrogen and carbon monoxide, the desired hydrogen-carbon monoxide ratio of the feedstock gas is on the order of 3:1. The high proportion of water (vapor) produced is removed by condensation and recirculated as process water or steam.

4.5 GAS PRODUCTION AND PRODUCTS

The gasification of a carbonaceous feedstock (i.e. char produced from the feedstock such as char from the pyrolysis of a carbonaceous feedstock) is the conversion of the feedstock (by any one of a variety of processes) to produce gaseous products that are combustible as well as a wide range of chemical products from synthesis gas. The products from the gasification of the process may be of low, medium, or high heat-content (high-Btu) content as dictated by the process as well as by the ultimate use for the gas (Baker and Rodriguez, 1990; Probstein and Hicks, 1990; Lahaye and Ehrburger, 1991; Matsukata et al., 1992; Speight, 2013a).

There are many useful products from the gasification process, which include: (1) synthesis gas, an important aspect in the present context, (2) heat, (3) power, (4) biofuels, and (5) char, more often referred to as 'biochar'. The synthesis gas can be further processed by means of the Fischer-Tropsch process into methanol, dimethyl ether and other chemical feedstocks (Speight, 2014, 2017, 2019a).

Generally, biomass feedstocks are classified into four main groups: (1) woody biomass, (2) herbaceous biomass, (3) marine biomass, and (4) manure (Chapter 5) (Sikarwar et al., 2016). The gasifier is usually designed to generate a specific product (Chapter 3) but the feedstock material is an important parameter to specify and optimize where possible (Luque and Speight, 2015; Sikarwar et al., 2016, 2017).

4.5.1 Gas Production

Large scale gasifiers employ one of two types of fluidized bed configurations which are (1) bubbling fluidized bed and (2) circulating fluidized bed or (3) a combination of both indirectly fired to maintain the bed temperature below the fusion temperature of the ash.

A bubbling fluidized-bed gasifier consists of fine, inert particles of sand or alumina, which are selected based on their suitability of physical properties such as size, density and thermal characteristics. The dimension of the bed at some height above the distributor plate is increased to reduce the superficial gas velocity below the fluidization velocity to maintain inventory of solids and to act as a disengaging zone. A cyclone is used to trap the smaller size particle that exit the fluidized-bed to either return fines to the bed or to remove ash-rich fines from the system.

The feedstock is introduced either through a feed chute to the top of the bed or deep inside the bed. The deeper introduction of the feedstock into the bed of inert solids provides sufficient residence time for fines that would otherwise be entrained in the fluidizing gas.

The organic constituents of the feedstock may pyrolytically vaporize and are partially combusted in the bed. Exothermic combustion provides heat to maintain the bed at temperature and to volatilize additional feedstock. The bed needs to preheated to the start-up temperature using hydrocarbon resources such as natural gas, fuel oil, either by direct firing or by indirect heating. After the bed reaches the feedstock ignition temperature, the feedstock is slowly introduced into the bed to raise the bed temperature to the desired operating temperature which are normally in the range of 700 to 900°C (1290 to 1650°F). The bed temperature is governed by the desire to obtain complete devolatilization versus the need to maintain the bed temperature below the ash fusion temperature. The advantages of the fluidized-bed gasifiers are (1) yields a uniform product gas, (2) the ability to accept a wide range of fuel particles sizes, including fines, (3) a nearly uniform temperature distribution throughout the reactor which provides a high rate of heat transfer between inert material and the feedstock, which is beneficial for high conversion, but low tar production. There is the potential for a disadvantage that results in the formation of large bubbles at higher gas velocities and which bypass the bed reducing the high rate of heat /mass transfer significantly.

Autothermal gasifiers provide the necessary heat of reaction by means of partial oxidation within the gasification reactor. Thus, for synthesis gas production either pure oxygen (in entrained flow reactors) or mixtures of oxygen and steam (in fluidized-bed reactors) are used as gasification agent. The advantage of autothermal gasification is the direct heating of the reactants and therefore more efficient energy utilization.

Allothermal (or indirect) gasification is characterized by the separation of the processes of heat production and heat consumption (Pfeifer etal., 2009). The allothermal gasification process typically consists of two reactors, connected by an energy flow.

The transport of the heat can be achieved either by circulating a bed material or by heat exchangers. Allothermal gasifiers generally produce two gas streams: a medium calorific product gas (gasification reactor) with little or no nitrogen and a flue gas (combustion reactor). The production of nitrogen-free gas without the need of pure oxygen is one of the advantages over autothermal gasification processes. Another advantage is the complete carbon conversion and that there is no problematic waste produced. All carbon-containing streams from the product gas cleaning (such as dust and tar removal) can be recycled to the combustion zone and converted to heat, which is used for the gasification reactions (Tijmensen et al., 2002).

In a fixed-bed–fluidized, bed-entrained flow gasifier (Chapter 3), the differences in the design of the cause differences in composition of the gas which is also influenced other factors. Including feedstock composition, reactor type, and operating parameters (temperature, pressure, oxygen fuel ratio). During oxygen gasification, the combustion products (carbon dioxide and water) are in the product gas and take part in the chemical reactions, mainly in the water-gas shift reaction. On the other hand, a higher amount of hydrogen in the product gas can be found during steam gasification. The hydrogen that occurs in the product gas does not originate only from the fuel as in the case of oxygen gasification but also from the steam.

In entrained flow gasification, where the temperature is above 1,000°C (1,830°F), there is no tar produced, whereas in fluidized-bed gasification, where the temperature is below 1,000°C (1,830°F), tar is produced and has to be removed from the product gas.

4.5.2 Products

In all cases, the gas exiting the gasifier contains impurities that need to be removed for the advanced use of feedstock gasification gas. For example, if the gas is to be used a Fischer-Tropsch process, the gas must be ultra-clean. Any particulate matter in the gas exiting the gasifier can be removed by a hot gas filter, whereas the tar constituents and low-boiling hydrocarbon derivatives can be converted by reforming to synthesis gas. For tar removal, another possibility is scrubbing with an organic solvent.

However, the gas stream will also, more than likely, contain other impurities, such as hydrogen sulfide, carbonyl sulfide, ammonia, and alkali metals. The sulfur constitunets will further add complications to a gas reforming process by poisoning the catalyst. In addition to hydrocarbon derivatives, ammonia is partly converted to nitrogen in the reformer. The alkali metals, in turn, will condense in hot gas filtration taking place at around 500°C (930°F). The final conditioning of gas for synthesis purposes can be realized by conventional commercial technologies for acid gas removal, such as the Rectisol process, and the H2/CO ratio can be adjusted by a water gas shift reactor.

4.5.2.1 Synthesis Gas

Both nonrenewable and renewable energy sources are important for production of synthesis gas and hydrogen.

In most cases, synthesis gas has been (and in many cases, still is) produced from coal (gasification, carbonization), natural gas, and low-boiling hydrocarbon derivatives such as propane gas (steam reforming, partial oxidation, autothermal reforming, plasma reforming), crude oil fractions (dehydrocyclization and aromatization, oxidative steam reforming, pyrolytic decomposition), feedstock (gasification, steam reforming, biologic conversion), and water (electrolysis, photo-catalytic conversion, chemical and catalytic conversion) (Liu et al., 2010; Speight, 2011a, 2013a, 2013b, 2014, 2017, 2020). The relative competitiveness of different options depends on the economics of the given processes, which in turn depend on many factors such as the (1) suitability and availability of the feedstock, (2) efficiency of the catalysis, (3) scale of production, (4) required hydrogen purity, (5) economics of feedstock production and the processing steps.

For example, using the 'electrolysis process', by which water molecules are split directly into hydrogen and oxygen molecules using electricity and an electrolyzer device, the overall electrolysis reaction is:

$$2H_2O \rightarrow O_2 + 2H_2$$

The two most common types of electrolyzers are alkaline (such as a potassium hydroxide electrolyte) or a solid polymer membrane electrolyte (commonly referred to a PEM electrolyte). The electrolysis reaction produces pure oxygen as a by-product along with pure hydrogen. The oxygen can then be used for productive purposes such as enriching the oxygen content of greenhouses for food production.

Current commercial processes for synthesis gas and hydrogen production are largely dependent on fossil fuels both as the source of hydrogen and will continue to play an important role in the midterm (up to 50 years from now) (Speight, 2011b). Efficient separation of the hydrogen from the gaseous products is also a major issue that must be addressed. In this respect, pressure swing adsorption (PSA) is used in current industrial practice. Furthermore, several types of membranes are being developed that would, when incorporated into the separation process will enable more efficient gas separation (Ho and Sirkar, 1992).

The process for producing synthesis gas involves three individual components: (1) synthesis gas generation, (2) waste heat recovery, and (3) gas processing (Speight, 2013a, 2013b, 2014). Within each of the above three listed systems are several options – synthesis gas with a range of compositions ranging from high-purity hydrogen to high-purity carbon monoxide can be generated. Two major routes can be utilized for high purity gas production: (1) pressure swing adsorption and (2) utilization of a cold box, where separation is achieved by distillation at low temperatures. In fact, both processes can also be used in combination as well.

Synthesis gas is one of the key products of the gasification of a carbonaceous feedstock and is a vital source of environmentally fuels and chemicals (Chapter 5). Employing the Fischer Tropsch process other chemicals can be manufactured (Chadeesingh, 2011; Speight, 2019a). Hydrogen is produced from synthesis gas by means of the water gas shift reaction.

Contaminants such as particulates, tars, nitrogenous compounds such as ammonia (NH_3) and hydrogen cyanide (HCN), sulfur-containing inorganic compounds such as hydrogen sulfide (H_2S), carbonyl sulfide (COS), and carbon disulfide (CS_2). Halogens such as hydrogen chloride and traces of metals such as sodium (Na) and potassium (K) are present in varying quantities in synthesis gas produced from gasification.

The type of feedstock, operational conditions, and the gasifier type are the variables which determine the amount of tar produced. These contaminants in synthesis gas pose numerous technical and working problems and must be removed (Chapter 9). For example, hydrogen sulfide is responsible for equipment corrosion, tar causes fouling, and catalyst deactivation occurs due to the presence of tar, hydrogen sulfide, ammonia, hydrogen chloride, and trace metals.

Synthesis gas produced from various types of feedstocks and is often comparable in combustion efficiency to natural gas (Chadeesingh, 2011; Speight, 2020) which reduces the emissions of sulfur, nitrogen oxides, and mercury, resulting in a much cleaner fuel (Nordstrand et al., 2008; Lee et al., 2006; Sondreal et al., 2004, 2006; Yang et al., 2007; Wang et al., 2008). The resulting hydrogen gas can be used for electricity generation or as a transport fuel or in any aspects of refining that requires hydrogen (Speight, 2016). The gasification process also facilitates capture of carbon dioxide emissions from the combustion effluent.

The use of bio-synthesis gas offers the opportunity to furnish a broad range of environmentally clean fuels and chemicals and there has been steady growth in the traditional uses of synthesis gas. In fact, the major use of synthesis gas is in the manufacture of hydrogen for a growing number of purposes, especially in crude oil refineries (Speight, 2014a, 2016, 2017).

The hydroformylation of olefin derivatives (the Oxo reaction), a completely chemical use of synthesis gas, is the fourth largest use of carbon monoxide and hydrogen mixtures. A direct application of synthesis gas as fuel (and eventually also for chemicals) that promises to increase is its use for *integrated gasification combined cycle* (IGCC) units for the generation of electricity (and also chemicals), crude oil coke or viscous feedstocks (Holt, 2001). Finally, synthesis gas is the principal source of carbon monoxide, which is used in an expanding list of carbonylation reactions, which are of major industrial interest.

4.5.2.2 Low-Btu Gas

During the production of gas by oxidation with air, the oxygen is not separated from the air and, as a result, the gas product invariably has a low Btu content (low heat-content, 150 to 300 Btu/ft^3). Low heat-content gas contains several components, four of which are always major components present at levels of at least several percent; a fifth component, methane, is marginally a major component.

The nitrogen content of low heat-content gas ranges from somewhat less than 33% v/v to slightly more than 50% v/v and cannot be removed by any reasonable means; the presence of nitrogen at these levels makes the product gas *low heat-content* by definition. Nitrogen also strongly limits the applicability of the gas to chemical synthesis.

The two major combustible components are hydrogen and carbon monoxide; the H_2/CO ratio varies from approximately 2:3 to approximately 3:2. Any hydrogen sulfide present must be removed by one, or more, of several procedures (Mokhatab et al., 2006; Speight, 2014, 2019b).

4.5.2.3 Medium-Btu Gas

Medium Btu gas (medium heat content gas) has a heating value in the range 300 to 550 Btu/ft^3) and the composition is much like that of low heat-content gas, except that there is virtually no nitrogen. The primary combustible gases in medium heat-content gas are hydrogen and carbon monoxide. Medium heat-content gas is considerably more versatile than low heat-content gas; like low heat-content gas, medium heat-content gas may be used directly as a fuel to raise steam or used through a combined power cycle to drive a gas turbine, with the hot exhaust gases employed to raise steam, but medium heat-content gas is especially amenable to synthesize methane (by methanation), higher hydrocarbon derivatives (by Fischer-Tropsch synthesis), methanol, and a variety of synthetic chemicals.

In medium heat-content gas, the H_2/CO ratio varies from 2:3 C to 3:1 and the increased heating value correlates with higher methane and hydrogen contents as well as with lower carbon dioxide contents. Furthermore, the very nature of the gasification process used to produce the medium heat-content gas has a marked effect upon the ease of subsequent processing. For example, the CO_2-acceptor product is quite amenable to use for methane production because it has (1) the desired H_2/CO ratio just exceeding 3:1, (2) an initially high methane content, and (3) relatively low water and carbon dioxide contents. Other gases may require appreciable shift reaction and removal of large quantities of water and carbon dioxide prior to methanation.

4.5.2.4 High-Btu Gas

In the context of this book, substitute natural gas, also called synthetic natural gas) that is produced from feedstock gasification. Synthetic natural gas contains mainly methane and can be produced from a product gas stream that contains significant amounts of methane. This generally means that gasification should take place non-catalytically at approximately 900°C (1,650°F). At these temperatures, methane destruction generally is limited and the fluidized-bed gasification process is one of the obvious options.

However, to qualify as substitute natural gas, a product must contain at least 95% methane, giving an energy content (heat content) of synthetic natural gas on the order of 980 to 1080 Btu/ft^3). The commonly accepted approach to the synthesis of high heat-content gas is the catalytic reaction of hydrogen and carbon monoxide:

$$3H_2 + CO \rightarrow CH_4 + H_2O$$

To avoid catalyst poisoning, the feed gases for this reaction must be quite pure and, therefore, impurities in the product are rare. The large quantities of water produced are removed by condensation and recirculated as very pure water through the gasification system. The hydrogen is usually present in slight excess to ensure that the toxic carbon monoxide is reacted; this small quantity of hydrogen will lower the heat content to a small degree.

In addition, the methanation catalyst is troublesome and prone to poisoning by sulfur compounds and the decomposition of metals can destroy the catalyst. Hydrogasification may be thus employed to minimize the need for methanation:

$$[C]_{feedstock} + 2H2 \rightarrow CH_4$$

The product of hydrogasification is far from pure methane and additional methanation is required after hydrogen sulfide and other impurities are removed.

Besides the main components of carbon monoxide, hydrogen, carbon dioxide, methane and nitrogen, several trace substances are also present in the product gas. These include particulate matter) such as entrained ash, and bed material), sulfur compounds (hydrogen sulfide, carbonyl sulfide), alkali compounds (mainly chlorine compounds) and tars or higher molecular weight hydrocarbon derivatives that are prone to condensation at temperatures on the order of 300 to 350°C (570 to 660°F) and can, therefore, foul pipes and equipment. Removal of these substances is of particular importance in the production of synthetic natural gas since the methanation reaction uses catalysts that are highly sensitive to impurities in the gas. A number of different gas cleaning technologies exist within a refinery system or a gas processing plant that can be satisfactorily employed to remove the contaminants from various gas streams (Chapter 9) (Mokhatab et al., 2006; Speight, 2014a, 2017, 2019b)

4.5.3 Gas Upgrading

Raw biogas produced from digestion is approximately 60% v/v methane and 29% v/v carbon dioxide with trace elements of hydrogen sulfide and is inadequate for immediate use because of the corrosive nature of hydrogen sulfide. Organic substances that are typically found in waste include protein derivatives, carbohydrate derivatives, and fat derivatives, both in the forms of solid and solution.

4.6 OTHER PRODUCTS

The focus in the above sections has been on the gaseous products that are produced in the gasification process. However there are other products that require attention, particularly (1) liquid products and (2) solid products that may also find further use. Also worthy of mention is the ash that ism produced from the mineral matter that is present in the feedstock.

4.6.1 Liquid Products

The production of liquid fuels from a carbonaceous feedstock via gasification is often referred to as the *indirect liquefaction* of the feedstock (Speight, 2013a, 2014a, 2017).

The synthesis of hydrocarbon derivatives from carbon monoxide and hydrogen (synthesis gas) (the Fischer-Tropsch synthesis) is a procedure for the indirect liquefaction of various carbonaceous feedstocks (Speight, 2011a, 2011b). This process is the only liquefaction scheme currently in use on a relatively large commercial scale (for the production of liquid fuels from coal using the Fischer-Tropsch process (Singh, 1981).

In this process, temperatures in excess of 800°C (1470°F) and moderate pressures are employed to convert feedstock to produce synthesis gas:

$$[C]_{feedstock} + H_2O \rightarrow CO + H_2$$

The Fischer-Tropsch reaction is carried out at temperatures on the order of 200 to 350°C (390 to 60°F) and at pressures on the order of 75 to 4,000 psi. The hydrogen/carbon monoxide ratio is typically on the order of 2/2:1 or 2/5:1, since up to three volumes of hydrogen may be required to achieve the next stage of the liquids production, the synthesis gas must then be converted by means of the water-gas shift reaction) to the desired level of hydrogen:

$$CO + H_2O \rightarrow CO_2 + H_2$$

After this, the gaseous mix is purified and converted to a wide variety of hydrocarbon derivatives:

$$nCO + (2n + 1)H_2 \rightarrow C_nH_{2n+2} + nH_2O$$

4.6.2 Solid Products

The solid product (solid waste) of a gasification process is typically mineral ash which is the oxides of metal-containing constituents of the feedstock. The amount and type of solid waste produced is very much feedstock dependent.

Briefly, Coal gasification processes can produce three types of ash which are (1) fly ash, which includes char or unreacted feedstock, (2) bottom ash, and (3) slag, with most of the solid by-product ending in form of slag for high-temperature gasifiers. Non-slagging gasification produces coarse bottom ash and fine fly ash.

At the high temperature of the gasifier, most of the mineral matter of the feedstock is transformed and melted into slag, an inert glass-like material and, under such conditions, non-volatile metals and mineral compounds are bound together in molten form until the slag is cooled in a water bath at the bottom of the gasifier, or by natural heat loss at the bottom of an entrained-bed gasifier. Slag production is a function of mineral matter content of the feedstock – coal produces much more slag per unit weight than crude oil coke. Furthermore, as long as the operating temperature is above the fusion temperature of the ash, slag will be produced. The physical structure of the slag is sensitive to changes in operating temperature and pressure of the gasifier and a quick physical examination of the appearance of the slag can often be an indication of the efficiency of the conversion of feedstock carbon to gaseous product in the process.

Slag is comprised of black, glassy, silica-based materials and is also known as *frit*, which is a high density, vitreous, and abrasive material low in carbon and formed in various shapes from jagged and irregular pieces to rod and needle-like forms. Depending upon the gasifier process parameters and the feedstock properties, there may also be residual carbon char. Vitreous slag is much preferable to ash, because of its habit of encapsulating toxic constituents (such as heavy metals) into a stable, non-leachable material. Leachability data obtained from different gasifiers unequivocally shows that gasifier slag is highly non-leachable and can be classified as non-hazardous. Because of its particular properties and non-hazardous, non-toxic nature, slag is relatively easily marketed as a by-product for multiple advantageous uses, which may negate the need for its long-term disposal.

The physical and chemical properties of gasification slag are related to (1) the composition of the feedstock, (2) the method of recovering the molten ash from the gasifier, and (3) the proportion of devolatilized carbon particles (i.e. char) that are discharged with the slag. The rapid water-quench method of cooling the molten slag inhibits recrystallization, and results in the formation of a granular, amorphous material. Some of the differences in the properties of the slag may be attributed to the specific design and operating conditions prevailing in the gasifiers.

Char can be recycled back into the gasifier to increase carbon usage and has been used as a supplemental fuel source for use in a combustor. In terms of recycling char to the gasifier, a property that is important to fluidization is the effective particle density. If the char has a large internal void space, the density will be much less than that of the feedstock (especially coal) or char from slow carbonization of a carbonaceous feedstock.

4.7 COMMERCIAL PROCESSES

Hydrogen is an important commodity in the refining industry because of its use in hydrotreating processes, such as desulfurization, and in hydroconversion processes, such as hydrocracking (Chapter 1, Chapter 2) (Parkash, 2003; Gary et al., 2007; Speight, 2014; Hsu and Robinson, 2017; Speight, 2017, 2024). Part of the hydrogen required for the hydroprocesses is produced during the various reforming processes but that source is now insufficient for the hydrogen requirements of a modern refinery (Speight, 2000; Ancheyta and Speight, 2007; Speight, 2014, 2017). In addition, optimum hydrogen purity at the reactor inlet extends catalyst life by maintaining desulfurization kinetics at lower operating temperatures and reducing carbon laydown. Typical purity increases resulting from hydrogen purification equipment and/or increased hydrogen sulfide removal as well as tuning hydrogen circulation and purge rates, may extend catalyst life on the order of 25% higher than normal (Raissi, 2001; Vauk et al., 2008).

While the gasification of residua and coke to produce hydrogen and/or power may increase in use in refineries over the next two decades (Speight, 2011b), several other processes are available for the production of the additional hydrogen that is necessary for the various heavy feedstock hydroprocessing sequences (Speight, 2014) and it is the purpose of this section to present a general description of these processes.

4.7.1 Heavy Residue Gasification

Heavy residues are gasified and the produced gas is purified to fuel gas that is free of contaminants (Gross and Wolff, 2000). As an example, solvent deasphalter residuum (*deasphalter bottoms*) is gasified by partial oxidation method under pressure of about 570 psi and at a temperature between 1,300 and 1,500°C (2,370°F and 2,730°F). The gas from the waste heat boiler is then heat exchanged with the fuel gas and flows to the carbon scrubber, where unreacted carbon particles are removed from the generated gas by water scrubbing.

The gas from the carbon scrubber is further cooled by the fuel gas and boiler feed water and led into the sulfur compound removal section, where hydrogen sulfide (H2S) and carbonyl sulfide (COS) are removed from the gas to obtain clean fuel gas. This clean fuel gas is heated with the hot gas generated in the gasifier and finally supplied to the gas turbine at a temperature of 250 to 300°C (480 to 570°F).

In order to decrease the nitrogen oxide (NO_x) content in the flue gas, two methods can be applied. The first method is the injection of water into the gas turbine combustor. The second method is to selectively reduce the nitrogen oxide content by injecting ammonia gas in the presence of de-NO_x catalyst that is packed in a proper position of the heat recovery steam generator. The latter is more effective that the former to lower the nitrogen oxide emissions to the air.

4.7.2 Hybrid Gasification Process

In the Hybrid gasification process, the feedstock (such as a coal/residual oil slurry) is injected into the gasifier where pyrolysis occurs in the upper part of the reactor to produce gas and char. The char produced in the process is then partially oxidized to ash which is removed continuously from the bottom of the reactor.

As an example, coal and a vacuum residue were mixed to form a slurry which was then fed into the pressurized gasifier is thermally cracked at a temperature of 850 to 950°C (1,560 to 1,740°F) and is converted into gas, tar and char. The mixture oxygen and steam in the lower zone of the gasifier convert the char to gaseous products. The gas leaving the gasifier is quenched to a temperature of 450°C (840°F)in the fluidized-bed heat exchanger, and is then scrubbed to remove tar, dust and steam at around 200°C (390°F).

The ash that is discharged from the gasifier is cooled with steam and then discharged into the ash hopper after which it is burned with an incinerator to produce process steam. Coke deposited on the silica sand is removed in the incinerator.

4.7.3 Hydrocarbon Gasification

The gasification of hydrocarbon derivatives to produce hydrogen is a continuous, non-catalytic process that involves partial oxidation of the hydrocarbon feedstock (Breault, 2010). However, the term 'hydrocarbon gasification' can be misleading because the feedstocks may also be 'carbonaceous feedstocks' that contain elements other than carbon and hydrogen.

The process is adaptable to a variety of feedstocks and, in fact, hydrogen can be produced from coal or coke by a variety of gasification processes (such as the fixed-bed process, the fluidized-bed process, or the entrained flow process). In practice, the high-temperature entrained flow processes are favored to maximize conversion of the feedstock carbon to gas, thus avoiding the formation of significant amounts of char, residual tar, and derivatives of phenol.

In the process, air or oxygen (with steam or carbon dioxide) is used as the oxidant at 1,095 to 1,480°C (2,000 to 2,700°F). Any carbon produced (2 to 3% w/w of the feedstock) during the process is removed as a slurry in a carbon separator and pelletized for use either as a fuel or as raw material for carbon-based products.

A typical reaction for the process in which carbon is converted to carbon monoxide and hydrogen is:

$$C_{feedstock} + H_2O + heat \rightarrow CO + H_2$$

In the process, air or oxygen (with steam or carbon dioxide) is used as the oxidant at 1,095 to 1,480°C (2,000 to 2,700°F). Any carbon produced (2 to 3% w/w the feedstock) during the process is removed as a slurry in a carbon separator and pelletized for use either as a fuel or as raw material for carbon-based products.

Since this process is endothermic, additional heat is required, as with methane reforming. The carbon monoxide is converted to carbon dioxide and hydrogen through the water-gas shift reaction. Hydrogen production from coal is a commercially mature process (Chadeesingh, 2011; Speight, 2013), but it is more complex than the production of hydrogen from natural gas.

4.7.4 Hypro Process

The Hypro process is a continuous catalytic method that is employed for the manufacture of hydrogen from natural gas or the manufacture of hydrogen from refinery effluent gases, especially the decomposition of methane to hydrogen and carbon (Choudhary et al., 2003; Choudhary and Goodman, 2006):

$$CH4 \rightarrow C + 2H2$$

Steam reforming of methane represents the current method for hydrogen production in the refinery other popular methods of hydrogen production include autothermal reforming and partial oxidation are also employed. However, all these processes involve the formation of large amounts of unwanted (if hydrogen is the desired product) carbon oxides (CO_x, i.e. carbon monoxide where x = 1 and carbon dioxide where x = 2) as by-products.

Furthermore, hydrogen production routes, which do not require complex removal procedures for any carbon monoxide or carbon dioxide are desired for production of high-purity hydrogen. Thus, there is much interest in the catalytic decomposition of natural gas (whose major constituent is methane) for production of hydrogen. Since only hydrogen and carbon are formed in the decomposition process, separation of products is not an issue. For example, the high- and low-temperature water-gas shift reactions and carbon dioxide removal step (involved in the conventional methods) are completely eliminated. In addition, catalyst regeneration is extremely important for the practical application of the clean hydrogen production process.

4.7.5 Pyrolysis Processes

There are also pyrolysis processes that are available for the production of hydrogen and, specifically the interest has focused on the pyrolysis of methane (natural gas) and on the pyrolysis of hydrogen sulfide.

Natural gas is readily available and offers a relatively rich stream of methane with lower amounts of ethane, propane and butane also present. The thermocatalytic decompositon of natural gas hydrocarbon derivatives (c.f., Hypro process) offers an alternate method for the production of hydrogen (Uemura et al., 1999; Weimer at al., 2000; Dahl and Weimer, 2001):

$$C_nH_m \rightarrow nC + (m/2)H_2$$

The production of hydrogen by the decomposition of hydrogen sulfide has also been proposed (Clark and Wassink, 1990; Zaman and Chakma, 1995; Donini, 1996; Luinstra, 1996).

The decomposition of hydrogen sulfide is an endothermic process and equilibrium yields are poor (Clark et al., 1995). At temperatures less than 1,500°C (2,730°F), the thermodynamic equilibrium is unfavorable toward hydrogen formation. However, in the presence of catalysts such as platinum-cobalt (at 1,000°C; 1,830°F), disulfides of molybdenum (Mo) or tungsten (W) at 800°C (1,470°F) (Kotera et al., 1976), or other transition metal sulfides supported on alumina (at 500 to 800°C; 930 to 147°F), decomposition of hydrogen sulfide proceeds. In the temperature range of about 800 to 1,500°C (1,470 to 2,730°F), thermolysis of hydrogen sulfide can be treated simply:

$$H_2S \rightarrow H_2 + 1/xS_x \ \Delta H_{298K} = +34{,}300 \text{ Btu/lb}$$

In this equation, x = 2. Outside this temperature range, multiple equilibria may be present depending on temperature, pressure, and relative abundance of hydrogen and sulfur (Clark, 1990; Clark and Wassink, 1990).

In addition, the steam-iron process has been used for the production of hydrogen from coke. However, the process could not compete with the later on developed steam reforming of methane and the process fell into disuse. However, there is renewed interest in the development of the steam-iron process mainly focused on the use of renewable energy sources, such as biomass. The benefit of the steam-iron process compared to other thermochemical routes for biomass conversion is that hydrogen can be produced in a two-step redox cycle, without the need of any purification steps (such as pressure swing adsorption).

4.7.6 Shell Gasification Process

The Shell Gasification process (*partial oxidation process*) is a flexible process for generating synthesis gas, principally hydrogen and carbon monoxide, for the ultimate production of high-purity, high-pressure hydrogen, ammonia, methanol, fuel gas, town gas or reducing gas by reaction of gaseous or liquid hydrocarbon derivatives with oxygen, air or oxygen-enriched air. Traditionally, crude oil residua have been sold as marine bunker fuel or used on-site as furnace fuel. However, with changing legislation, refineries are under pressure to reduce both the emissions and the sulfur content of their products leaving some concern about the continued use of distillation residua as a fuel oil blend stocks (Breault, 2007).

The most important step in converting heavy residue to industrial gas is the partial oxidation of the oil using oxygen with the addition of steam. The gasification process takes place in an empty, refractory-lined reactor at temperatures of about 1,400°C (2,550°F) and pressures between 29 to 1140 psi. The chemical reactions in the gasification reactor proceed without catalyst to produce gas containing carbon amounting to some 0.5 to 2% by weight, based on the feedstock. The carbon is removed from the gas with water, extracted in most cases with feed oil from the water and returned to the feed oil. Steam is generated at 850 to 1565 psi and any surplus steam is used for energy production and heating purposes.

4.7.7 Steam-Naphtha Reforming

Liquid feedstocks, either liquefied petroleum gas or naphtha can also provide feedstocks for the steam-methane reforming process (Breault, 2010; Rostrup-Nielsen and Christiansen, 2011). The feed handling system needs to include a surge drum, feed pump, vaporizer (usually steam-heated) followed by further heating before desulfurization. A variety of naphtha-types in the gasoline boiling range may be employed, including feeds containing up to 35% aromatic constituents.

Many of these constituents are stable and will not be removed by zinc oxide therefore a hydrogenation unit will be required. In addition, as with refinery gas, olefin derivatives must also be hydrogenated if they are present. After the removal of the sulfur constituents, the feedstock is mixed with steam and taken to the reforming furnace (675 to 815°C, 1250 to 1500°F, 300 psi, where hydrogen is produced.

4.7.8 Texaco Gasification Process

The Texaco gasification process is a gasification process (partial oxidation process) for generating synthesis gas (Breault, 2010). In the process, the feedstock is injected into the reactor together with carbon dioxide, steam, or water into the gasifier. Thus, feedstock, together with the feedstock carbon slurry recovered in the carbon recovery section, is pressurized to a given pressure, mixed with high-pressure steam, and then blown into the gas generator through the burner together with oxygen. Therefore, solvent deasphalted residua or crude oil coke rejected from any of the types of coking units can be used as feedstock for this gasification process.

The produced gas from this gasification process can be used for the production of high-purity high-pressurized hydrogen, ammonia, and methanol. The heat that is recovered from the high-temperature gas is used for the generation of steam in the waste heat boiler. Alternatively the less expensive quench type configuration is preferred when high-pressure steam is not needed or when a high degree of shift is needed in the downstream carbon monoxide converter.

This process involves the partial oxidation of hydrocarbon derivatives to a gaseous mixture consisting of carbon monoxide and hydrogen:

$$C_xH_{2y} + x/2O_2 \rightarrow xCO + yH_2$$

$$C_xH_{2y} + xH_2O \rightarrow xCO + (x+y)H_2$$

The high-temperature gas leaving the reaction chamber of the gas generator enters the quenching chamber linked to the bottom of the gas generator and is quenched to 200 to 260°C (390 to 500°F) with water.

4.7.9 Recovery from Fuel Gas

The basic configuration for hydrogen recovery from refinery gases involves a two-stage partial condensation process, with post purification via pressure swing adsorption (Dragomir et al., 2010). The major steps in this process involve (1) compressing the crude refinery gas stream, (2) pretreating the crude gas stream after which (3) the gas stream is chilled to an intermediate temperature (on the order of –50 to –85°C, –60 to –120°F).

A methane rich vapor is obtained at the top of the column and is sent to a compressor unit and then to a pressure swing adsorption unit for further purification – the tail gas from this unit is compressed and returned to fuel together with the methane rich gas.

4.8 THE FUTURE

The future of synthesis gas production by gasification depends very much on the effect of gasification processes on the surrounding environment. It is these environmental effects and issues that will direct the success of gasification. In fact, there is the distinct possibility that within the foreseeable future the gasification process will increase in popularity in crude oil refineries – some refineries may even be known as gasification refineries (Speight, 2011b). A gasification refinery would have, as the center-piece of the refinery, gasification technology as is the case of the Sasol refinery in South Africa (Couvaras, 1997). The refinery would produce synthesis gas (from the carbonaceous feedstock) from which liquid fuels would be manufactured using the Fischer-Tropsch synthesis technology.

The manufacture of gas mixtures of carbon monoxide and hydrogen has been an important part of chemical technology for more than 100 years. Originally, such mixtures were obtained by the reaction of steam with incandescent coke and were known as *water gas*.

Partial oxidation processes using oxygen instead of steam also found wide application for synthesis gas manufacture, with the special feature that they could utilize low-value feedstocks such as viscous crude oil residues. In recent years, catalytic partial oxidation employing very short reaction times (milliseconds) at high temperatures (850 to 1,000°C, 1,560 to 1,830°F) is providing still another approach to synthesis gas manufacture (Hickman and Schmidt, 1993).

The conversion of the gaseous products of gasification processes to synthesis gas, a mixture of hydrogen (H_2) and carbon monoxide (CO), in a ratio appropriate to the application, needs additional steps, after purification (Chapter 5).

Gasification by means other than the conventional methods has also received some attention and has provided rationale for future processes (Rabovitser et al., 2010). In the process, a carbonaceous material and at least one oxygen carrier are introduced into a non-thermal plasma reactor at a temperature in the range of approximately 300°C to approximately 700°C (570 to approximately 1,290°F) and a pressure in a range from atmospheric pressure to approximate 1,030 psi and a non-thermal plasma discharge is generated within the non-thermal plasma reactor.

Almost all of the carbon is converted to fuel synthesis gas. The high operating temperatures (above 1,800°C, 3,270°F) allow for the breaking down of all tars, char and dioxins. For example, for waste material with high moisture content, a drier will be required.

In the plasma reactor two torches are used together with a gas (such as oxygen, helium or air) to generate plasma. The torches which extend into the plasma furnace are fitted with graphite electrodes. An electric current is passed through the electrodes and an electric arc is generated between the tip of the electrodes and the conducting receiver which is the slag at the bottom of the furnace. Because of the electrical resistivity across the system significant heat is required to strip away electrons from the molecules of the gas introduced resulting in an ionized superheated gas stream or plasma. This gas exits the torch at temperatures up to 10,000°C (18,000°F). Such high temperatures are used to break down the waste primarily into elemental gas and solid waste (slag) as well as reduce dioxins, sulfur oxides and carbon dioxide emissions. Unlike waste incineration, plasma gasification waste to energy technology has proven to have a benign impact on the environment.

The plasma furnace is the central component of the system where gasification and vitrification take place. Plasma torches are mounted at the bottom of the reactor, they provide high-temperature air (almost three times higher than traditional combustion temperatures) which allow for the gasification of the waste materials.

In the process, the carbonaceous feedstock and the oxygen carrier are exposed to the non-thermal plasma discharge, resulting in the formation of a product gas which comprises substantial amounts of hydrocarbon derivatives, such as methane, hydrogen and/or carbon monoxide. The products of the process are a fuel or gas known as synthesis gas and an inert vitreous material known as slag.

Furthermore, gasification and conversion of carbonaceous solid fuels to synthesis gas for application of power, liquid fuels and chemicals is practiced worldwide. Crude oil coke, coal, biomass, and refinery waste are major feedstocks for an on-site refinery gasification unit. The concept of blending of a variety of carbonaceous feedstocks (such as coal, biomass, or refinery waste) with a viscous feedstock of the coke from the thermal processing of the viscous feedstock is advantageous in order to obtain the highest value of products as compared to gasification of crude oil coke alone. Furthermore, based on gasifier type, co-gasification of carbonaceous feedstocks can be an advantageous and efficient process. In addition, the variety of upgrading and delivery options that are available for application to synthesis gas enable the establishment of an integrated energy supply system whereby synthesis gases can be upgraded, integrated, and delivered to a distributed network of energy conversion facilities, including power, combined heat and power, and combined cooling, heating and power (sometime referred to as *tri-generation*) as well as used as fuels for transportation applications.

As a final note, the production of chemicals from any feedstock is based on thermochemical conversion routes which are, in turn, based on feedstock gasification (Roddy and Manson-Whitton, 2012). The products are (1) a gas, which is the desired product and (2) a solid ash residue whose composition depends on the type of feedstock. Continuous gasification processes for various feedstocks have been under development since the early 1930s. Ideally, the gas produced would be a mixture of hydrogen and carbon monoxide, but, in practice, it also contains methane, carbon dioxide, and a range of contaminants.

A variety of gasification technologies is available across a range of sizes, from small updraft and downdraft gasifiers through a range of fluidized-bed gasifiers at an intermediate scale and on to larger entrained flow and plasma gasifiers (Brigwater, 2003a, 2003b; Roddy and Manson-Whitton, 2012). In an updraft gasifier, the oxidant is blown up through the fixed gasifier bed with the synthesis gas exiting at the top whereas in a downdraft gasifier, the oxidant is blown through the reactor in a downward direction with the synthesis gas exiting at the bottom.

Gasification processes tend to operate either above the ash melting temperature (typically 1,200°C, 2,190°F) or below the ash melting temperature (typically >1,000°C, >1,830°F). In the higher temperature processes, there is little methane or tar formation. The question of which gasification technology is the most appropriate depends on whether the priority is to (1) produce a very pure bio-synthesis gas, (2) accommodate a wide range of feedstock types, (3) avoid pre-processing of the feedstock, or (4) operate at a large scale, and produce chemical products from a variety of feedstocks.

The introduction of steam as a reactant influent during gasification has been shown to enhance the production of hydrogen during the gasification of a variety of fuels that include coal, biomass, and municipal solid waste (Minkova et al., 2000; Senneca, 2007. Introduction of carbon dioxide will enhance the production of carbon monoxide during high-temperature steam gasification while depressing the production of hydrogen and methane. Both steam and carbon dioxide have been shown to increase the char reactivity (Ye et al., 1998; Ochoa et al., 2001; Demirbaş, 2006) through a modification of the char pore structure and surface activity. Gasification using steam can result in an increased concentration of hydrogen in the synthesis gas with higher concentrations corresponding to higher gasification temperatures in the 700 to 1,100°C (1,290 to 2,110°F) range. If a methane fuel stream is the preferred product, careful monitoring of the process to operate below 600°C (1,110°F) would permit optimizing the production of methane while minimizing the production of hydrogen. A general characteristic of the fuel stream produced through gasification is a clean stream with a minimum of tar, soot and particulates. For example, the gasification of biomass results in a producer gas whose main components are carbon monoxide, hydrogen, and methane (Chapter 5). Other volatile components include carbon dioxide, acetaldehyde (CH_3CHO), acetic acid (CH_3CO_2H), phenol (C_6H_5OH), formaldehyde (HCHO), formic acid (HCO_2H), and acetone

CH_3COCH_3). If the gasification process is performed under high levels of nitrogen dilution, many of these species may not be detectable since they are present in the parts per million range (Butterman and Castaldi, 2008).

REFERENCES

Arena, U. 2012. Process and Technological Aspects of Municipal Solid Waste Gasification. A Review. *Waste Management*, 32(4): 625–639.

Baker, R.T.K., and Rodriguez, N.M. 1990. Coal Gasification. In *Fuel Science and Technology Handbook*. Marcel Dekker Inc., New York. Chapter 22.

Beenackers, A.A.C.M. 1999. Biomass Gasification in Moving Beds. A Review of European Technologies. *Renewable Energy*, 16(1–4): 1180–1186.

Butterman, H.C., and Castaldi, M.J. 2008. CO2 Enhanced Steam Gasification of Biomass Fuels. Proceedings. NAWTEC16. 16th Annual North American Waste-to-Energy Conference. Philadelphia, PA. May 19–21. https://asmedigitalcollection.asme.org/NAWTEC/proceedings-abstract/NAWTEC16/42932/157/328514.

Brage, C., Yu, Q., Chen, G., and Sjöström, K. 2000. Tar Evolution Profiles Obtained from Gasification of Biomass and Coal. *Biomass and Bioenergy*, 18(1): 87–91.

Brar, J.S., Singh, K., Wang, J., and Kumar, S. 2012. Cogasification of Coal and Biomass: A Review. *International Journal of Forestry Research*: 1–10.

Brigwater, A.V. (Editor) 2003a. *Pyrolysis and Gasification of Biomass and Waste*. CPL Press, Newbury, Berkshire.

Brigwater, A.V. 2003b. Renewable Fuels and Chemicals by Thermal Processing of Biomass. *Chemical Engineering Journal*, 91(2–3): 87–102.

Cao, J., Ding, S., Wang, H., Han, J., Ge, Q., and Zhu, X. 2019. Conversion of C2–4 Carboxylic Acids to Hydrocarbons on HZSM-5: Effect of Carbon Chain Length. *Industrial and Engineering Chemistry Research*, 58(24): 10307–10316. https://pubs.acs.org/doi/10.1021/acs.iecr.9b01755

Chadeesingh, R. 2011. The Fischer-Tropsch Process. In *The Biofuels Handbook*. J.G. Speight (Editor). The Royal Society of Chemistry, London. Part 3, Chapter 5, Page 476–517.

Chen, G., Sjöström, K., and Bjornbom, E. 1992. Pyrolysis/Gasification of Wood in a Pressurized Fluidized Bed Reactor. *Industrial and Engineering Chemistry Research*, 31(12): 2764–2768.

Conte, M., Iacobazzi, A., Ronchetti, M., and Vellone, R. Different Hydrogen Production Methods. *Journal of Power Sources*, 100: 171–187.

Couvaras, G. 1997. Sasol's Slurry Phase Distillate Process and Future Applications. Proceedings. Monetizing Stranded Gas Reserves Conference. Houston. December 1997.

Cusumano, J.A., Dalla Betta, R.A., and Levy, R.B. 1978. *Catalysis in Coal Conversion*. Academic Press Inc., New York.

Davidson, R.M. 1983. Mineral Effects in Coal Conversion. Report No. ICTIS/TR22. International Energy Agency, London.

Demirbaş, A. 2004. Effects of Temperature and Particle Size on Bio-char Yield from Pyrolysis of Agricultural Residues. *Journal of Analytical and Applied Pyrolysis*, 72(2): 243–248.

Demirbaş, A. 2006. Production and Characterization of Bio-chars from Biomass via Pyrolysis. *Energy Sources, Part A: Recovery, Utilization and Environmental Effects*, 28(5): 413–422.

Dogru, M., Howarth, C.R., Akay, G., Keskinler, B., and Malik, A.A. 2002. Gasification of Hazelnut Shells in a Downdraft Gasifier. *Energy*, 27(5): 415–427.

Dry, M.E. 1976. Advances in Fischer-Tropsch Chemistry. *Industrial and Engineering Chemistry Product Research and Development*, 15(4): 282–286.

Fabry, F., Rehmet, C., Rohani, V.-J., and Fulcheri, L. 2013. Waste Gasification by Thermal Plasma: A Review. *Waste and Biomass Valorization*, 4(3): 421–439.

Fermoso, J., Plaza, M.G., Arias, B., Pevida, C., Rubiera, F., and Pis, J.J. 2009. Co-gasification of Coal with Biomass and Petcoke in a High-Pressure Gasifier for Syngas Production. Proceedings. 1st Spanish National Conference on Advances in Materials Recycling and Eco-Energy. Madrid, Spain. November 12–13.

Gary, J.H., Handwerk, G.E., and Kaiser, M.J. 2007. *Petroleum Refining: Technology and Economics*, 5th Edition. CRC Press, Taylor & Francis Group, Boca Raton, FL.

Gray, D., and Tomlinson, G. 2000. Opportunities for Petroleum Coke Gasification Under Tighter Sulfur Limits for Transportation Fuels. Proceedings. Gasification Technologies Conference. San Francisco, CA. October 8–11.

Hanaoka, T., Inoue, S., Uno, S., Ogi, T., and Minowa, T. 2005. Effect of Woody Biomass Components on Air-Steam Gasification. *Biomass and Bioenergy*, 28(1): 69–76.

Hickman, D.A., and Schmidt, L.D. 1993. Syngas Formation by Direct Catalytic Oxidation of Methane. *Science*, 259(5093): 343–346.

Higman, C., and Van der Burgt, M. 2008. *Gasification*, 2nd Edition. Gulf Professional Publishing, Elsevier, Amsterdam.

Holt, N.A.H. 2001. *Integrated Gasification Combined Cycle Power Plants. Encyclopedia of Physical Science and Technology*, 3rd Edition. Academic Press Inc., New York.

Hsu, C.S., and Robinson, P.R. (Editors) 2017. *Handbook of Petroleum Technology*. Springer AG, Cham.

Irfan, M.F. 2009. Research Report: Pulverized Coal Pyrolysis & Gasification in N2/O2/CO2 Mixtures by Thermo-gravimetric Analysis. Novel Carbon Resource Sciences Newsletter, Kyushu University, Fukuoka. Volume 2. 27–33.

Khosravi, M., and Khadse, A. 2013. Gasification of Petcoke and Coal/Biomass Blend: A Review. *International Journal of Emerging Technology and Advanced Engineering*, 3(12): 167–173.

Kunii, D., and Levenspiel, O. 2013. *Fluidization Engineering*, 2nd Edition. Butterworth-Heinemann, Elsevier, Amsterdam.

Lahaye, J., and Ehrburger, P. (Editors) 1991. *Fundamental Issues in Control of Carbon Gasification Reactivity*. Kluwer Academic Publishers, Dordrecht.

Lapuerta, M., Hernández, J.J., Pazo, A., and López, J. 2008. Gasification and Co-gasification of Biomass Wastes: Effect of the Biomass Origin and the Gasifier Operating Conditions. *Fuel Processing Technology*, 89(9): 828–837.

Lee, S. 2007. Gasification of Coal. In *Handbook of Alternative Fuel Technologies*. S. Lee, J.G. Speight, and S. Loyalka (Editors). CRC Press, Taylor & Francis Group, Boca Raton, FL.

Lee, S., Speight, J.G., and Loyalka, S. 2007. *Handbook of Alternative Fuel Technologies*. CRC-Taylor & Francis Group, Boca Raton, FL.

Lee, S., and Shah, Y.T. 2013. *Biofuels and Bioenergy*. CRC Press, Taylor & Francis Group, Boca Raton, FL.

Liu, K., Song, C., and Subramani, V. (Editors) 2010. Hydrogen and Syngas Production and Purification Technologies. In *American Institute of Chemical Engineers*. John Wiley & Sons Inc., Hoboken, NJ.

Liu, G., Larson, E.D., Williams, R.H., Kreutz, T.G., and Guo, X. 2011. Making Fischer-Tropsch Fuels and Electricity from Coal and Biomass: Performance and Cost Analysis. *Energy and Fuels*, 25(1): 415–437.

Luque, R., and Speight, J.G. (Editors) 2015. *Gasification for Synthetic Fuel Production: Fundamentals, Processes, and Applications*. Woodhead Publishing, Elsevier, Cambridge.

Marano, J.J. 2003. Refinery Technology Profiles: Gasification and Supporting Technologies. Report prepared for the United States Department of Energy. National Energy Technology Laboratory. United States Energy Information Administration, Washington, DC. June.

Martinez-Alonso, A., and Tascon, J.M.D. 1991. *Fundamental Issues in Control of Carbon Gasification Reactivity*. J. Lahaye and P. Ehrburger (Editors). Kluwer Academic Publishers, Dordrecht.

Matsukata, M., Kikuchi, E., and Morita, Y. 1992. A New Classification of Alkali and Alkaline Earth Catalysts for Gasification of Carbon. *Fuel*, 71(7): 819–823.

McLendon, T.R., Lui, A.P., Pineault, R.L., Beer, S.K., and Richardson, S.W. 2004. High-Pressure Co-gasification of Coal and Biomass in a Fluidized Bed. *Biomass and Bioenergy*, 26(4): 377–388.

Mims, C.A. 1991. *Fundamental Issues in Control of Carbon Gasification Reactivity*. J. Lahaye and P. Ehrburger (Editors). Kluwer Academic Publishers, Dordrecht. Page 383.

Minkova, V., Marinov, S.P., Zanzi, R., Bjornbom, E., Budinova, T., Stefanova, M., and Lakov, L. 2000. Thermochemical Treatment of Biomass in a Flow of Steam or in a Mixture of Steam and Carbon Dioxide. *Fuel Processing Technology*, 62(1): 45–52.

Mokhatab, S., Poe, W.A., and Speight, J.G. 2006. *Handbook of Natural Gas Transmission and Processing*. Elsevier, Amsterdam.

Nordstrand, D., Duong, D.N.B., and Miller, B.G. 2008. *Combustion Engineering Issues for Solid Fuel Systems-Chapter 9 Post-combustion Emissions Control*. B.G. Miller and D. Tillman (Editors). Elsevier, London.

Ochoa, J., Cassanello, M.C., Bonelli, P.R., and Cukierman, A.L. 2001. CO2 Gasification of Argentinian Coal Chars: A Kinetic Characterization. *Fuel Processing Technology*, 74(3): 161–176.

Parkash, S. 2003. *Refining Processes Handbook*. Gulf Professional Publishing, Elsevier, Amsterdam.

Penrose, C.F., Wallace, P.S., Kasbaum, J.L., Anderson, M.K., and Preston, W.E. 1999. Enhancing Refinery Profitability by Gasification, Hydroprocessing and Power Generation. Proceedings. Gasification Technologies Conference. San Francisco, CA. October. https://www.globalsyngas.org/uploads/eventLibrary/GTC99270.pdf.

Pfeifer, C., Puchner, B., and Hofbauer, H. 2009. Comparison of Dual Fluidized Bed Steam Gasification of Biomass with and Without Selective Transport of CO2. *Chemical Engineering Science*, 64(23): 5073–5083.

Probstein, R.F., and Hicks, R.E. 1990. *Synthetic Fuels*. PH Press, Cambridge, MA. Chapter 4.

Rabovitser, I.K., Nester, S., and Bryan, B. 2010. Plasma Assisted Conversion of Carbonaceous Materials into A Gas. United States Patent 7,736,400. June 25.

Roddy, D.J., and Manson-Whitton, C. 2012. Biomass Gasification and Pyrolysis. In *Comprehensive Renewable Energy*. Volume 5: Biomass and Biofuels. D.J. Roddy (Editor). Elsevier, Amsterdam.

Senneca, O. 2007. Kinetics of Pyrolysis, Combustion and Gasification of Three Biomass Fuels. *Fuel Processing Technology*, 88(1): 87–97.

Shen, C.-H., Chen, W.-H., Hsu, H.-W., Sheu, J.-Y., and Hsieh, T.-H. 2012. Co-gasification Performance of Coal and Petroleum Coke Blends in A Pilot-Scale Pressurized Entrained-Flow Gasifier. *International Journal of Energy Research*, 36(4): 499–508.

Sikarwar, V.S., Zhao, M., Clough, P., Yao, J., Zhong, X., Memon, M.Z., Shah, N., Anthony, E.J., and Fennell, P.S. 2016. An Overview of Advances in Biomass Gasification. *Energy and Environmental Science*, 9(10): 2939–2977.

Sikarwar, V.S., Zhao, M., Fennell, P.S., Shah, N., and Anthony, E.J. 2017. Progress in Biofuel Production from Gasification. *Progress in Energy and Combustion Science*, 61: 189–248.

Singh, S.P., Weil, S.A., and Babu, S.P. 1980. Thermodynamic Analysis of Coal Gasification Processes. *Energy*, 5(8–9): 905–914.

Sjöström, K., Chen, G., Yu, Q., Brage, C., and Rosén, C. 1999. Promoted Reactivity of Char in Cogasification of Biomass and Coal: Synergies in the Thermochemical Process. *Fuel*, 78(10): 1189–1194.

Sondreal, E.A., Benson, S.A., Pavlish, J.H., and Ralston, N.V.C. 2004. An Overview of Air Quality III: Mercury, Trace Elements, and Particulate Matter. *Fuel Processing Technology*, 85(6–7): 425–440.

Speight, J.G. 2009. *Enhanced Recovery Methods for Heavy Oil and Tar Sands*. Gulf Publishing Company, Houston, TX.

Speight, J.G. (Editor) 2011a. *The Biofuels Handbook*. Royal Society of Chemistry, London.

Speight, J.G. 2011b. *The Refinery of the Future*. Gulf Professional Publishing, Elsevier, Oxford.

Speight, J.G. 2013a. *The Chemistry and Technology of Coal*, 3rd Edition. CRC Press, Taylor & Francis Group, Boca Raton, FL.

Speight, J.G. 2013b. *Coal-Fired Power Generation Handbook*. Scrivener Publishing, Salem, MA.

Speight, J.G. 2013c. *Oil Sand Production Processes*. Gulf Professional Publishing, Elsevier, Oxford.

Speight, J.G. 2014a. *The Chemistry and Technology of Petroleum*, 5th Edition. CRC Press, Taylor & Francis Group, Boca Raton, FL.

Speight, J.G. 2014b. *Gasification of Unconventional Feedstocks*. Gulf Professional Publishing, Elsevier, Oxford.

Speight, J.G. 2016. Hydrogen in Refineries. In *Hydrogen Science and Engineering – Materials, Processes, Systems, and Technology*. D. Stolten and B. Emonts (Editors). Wiley-VCH Verlag GmbH & Co., Weinheim. Chapter 1, Page 3–18.

Speight, J.G. 2017. *Handbook of Petroleum Refining*. CRC Press, Taylor & Francis Group, Boca Raton, FL.

Speight, J.G. 2019a. *Handbook of Petrochemical Processes*. CRC Press, Taylor & Francis Group, Boca Raton, FL.

Speight, J.G. 2019b. *Natural Gas: A Basic Handbook*, 2nd Edition. Gulf Publishing Company, Elsevier, Cambridge, MA.

Speight, J.G. 2020. *Synthetic Fuels Handbook: Properties, Process, and Performance*, 2nd Edition. McGraw-Hill, New York.

Speight, J.G. 2021. *Coal-Fired Power Generation Handbook*, 2nd Edition. Scrivener Publishing, Beverly, MA.

Speight, J.G. 2024. *Petroleum Refining Technology Series Volume 4: Hydrotreating and Hydrocracking Processes in Petroleum Refining*. CRC Press, Taylor & Francis Group, Boca Raton, FL.

Sundaresan, S., and Amundson, N.R. 1978. Studies in Char Gasification – I: A Lumped Model. *Chemical Engineering Science*, 34(3): 345–354.

Sutikno, T., and Turini, K. 2012. Gasifying Coke to Produce Hydrogen in Refineries. *Petroleum Technology Quarterly*, Q3: 105.

Tijmensen, M.J.A., Faaij, A.P.C., Hamelinck, C.N., and Van Hardeveld, M.R.M. 2002. Exploration of the Possibilities for Production of Fischer Tropsch Liquids and Power via Biomass Gasification. *Biomass and Bioenergy*, 23(2): 129–152.

Tinaut, F.V., Melgar, A., Pérez, J.F., and Horillo, A. 2008. Effect of Biomass Particle Size and Air Superficial Velocity on the Gasification Process in a Downdraft Fixed Bed Gasifier. An Experimental and Modelling Study. *Fuel Processing Technology*, 89(11): 1076–1089.

US Congress. 1976. *Public Law FEA-76-4*. United States Library of Congress, Washington, DC.

Wang, Y., Duan, Y., Yang, L., Jiang, Y., Wu, C., Wang, Q., and Yang, X. 2008. Comparison of Mercury Removal Characteristic between Fabric Filter and Electrostatic Precipitators of Coal-Fired Power Plants. *Journal of Fuel Chemistry and Technology*, 36(1): 23–29.

Yang, H., Xua, Z., Fan, M., Bland, A.E., and Judkins, R.R. 2007. Adsorbents for Capturing Mercury in Coal-Fired Boiler Flue Gas. *Journal of Hazardous Materials*, 146(1–2): 1–11.

Ye, D.P., Agnew, J.B., and Zhang, D.K. 1998. Gasification of a South Australian Low-Rank Coal with Carbon Dioxide and Steam: Kinetics and Reactivity Studies. *Fuel*, 77(11): 1209–1219.

5 Hydrogen Production from Biomass and from Waste

5.1 INTRODUCTION

Although there are several methods by which hydrgoen may be produced for use in the refinery (Chapter 3, Chapter 4) (Conte et al., 2001), the three most important thermal conversion systems for the production of hydrogen that are applicable to biomass resources and to carbonaceous wastes are (1) pyrolysis processes (2) gasification processes, and (3) biochemical processes (Table 5.1) (He et al., 2009; Speight, 2014b). Thermal conversion processes use heat as the dominant mechanism to convert the feedstock into another chemical form. As an example, pyrolysis processes are thermal conversion process in which the feedstock is thermally decomposed in the complete absence of oxidant (Pandey, 2009). Products include synthesis gas (carbon monoxide and hydrogen) and methane as well as other low molecular weight hydrocarbon derivatives), plus liquids and char.

Gasification processes have been the focal point of conversion processes in recent years to estimate efficiency and performance of the gasification process using a variety of carbonaceous feedstocks such as sugarcane residue (Gabra et al., 2001), rice hulls (Boateng et al., 1992), pine sawdust (Lv et al., 2004), almond shells (Rapagnà et al., 2000), almond (Rapagnà and Latif, 1997), and food waste (Ko et al., 2001) as well as woody biomass (Hanaoka et al., 2005). Thus, in the current context, the gasification of various types of biomass is a widely used thermochemical process for obtaining products with more value and potential applications than the raw material itself (Pan et al., 2000; Hu et al., 2012).

In the presence of a gasifying agent (such as oxygen or air), the organic feedstock is converted to a multifunctional gaseous mixture, usually called bio-synthesis gas, which can be used for the production of energy (heat and/or electricity generation), chemicals (ammonia), and biofuels. Furthermore, a solid residue after feedstock conversion (char) is generally found (Brigwater, 2003a, 2003b; Rauch et al., 2014; Molino et al. 2016; Sikarwar et al., 2017). Synthesis gas consists of a mixture of carbon monoxide, hydrogen, methane, carbon dioxide (the primary components) and water, hydrogen sulfide, and ammonia (secondary components), as well as tarry constituents and other trace species with a composition dependent on feedstock type and characteristics, operating conditions (i.e., the gasifying agents, gasifier temperature and pressure, type of bed materials), and gasification technology (Molino et al., 2013; Asadullah, 2014; Rodríguez-Olalde et al., 2015; Ahmad et al., 2016).

Gasification is an appealing process for the utilization of relatively inexpensive and readily available feedstocks, such as biomass and waste, that might in the case of waste be sent to a landfill (where the production of methane – a so-called greenhouse gas – will be produced and often escapes into the atmosphere) or combusted which may not (depending upon the feedstock) be energy efficient. Because of the development of the process using coal as the feedstock (Speight, 2013), biomass is often considered to be an unconventional carbonaceous feedstock which also include heavy crude oil, extra heavy crude oil, tar sand bitumen, and crude oil residua (Speight, 2014b).

The gasification process is optimized to give the highest yield gas rather than to produce bio-oil and/or biochar which can be used to produce synthesis gas through the agency of the gasification process (Demirbaş, 2007). The technology used for the gasification process is a tried-and-true technology for coal (Speight, 2013a) and other carbonaceous feedstocks such as crude oil residua (Speight, 2014b). The majority of gasifiers are partial oxidation reactors, in which just sufficient

DOI: 10.1201/9781003186694-5

TABLE 5.1
Conversion Processes for the Conversion of Biomass to Hydrogen-rich Gas Streams*

Process	Comments
Biochemical	
Gasification	Yield of hydrogen from gasification is higher than yield from pyrolysis
Air-steam gasification	A practical conversion technology
Catalytic gasification	Good yields of bio-oil
Catalytic steam gasification	Suitable for wet biomass
	may be conducted at temperatures as low as 300°C (570 °F)
Hydrothermal gasification	Suitable for the conversion of biomass with high moisture content without drying
Initial moisture process	Oil yield increases with increasing moisture content of the feedstock
Pyrolysis	Simplest process for conversion
Catalytic pyrolysis	High yield of hydrogen is obtained from zinc chloride catalyst
Fast pyrolysis	Conversion of biomass to bio-oil to produce hydrogen
Steam-reforming	Currently, the major process for hydrogen production
Supercritical water gasification	Can produce hydrogen from biomass with high efficiency

*Listed alphabetically and not by any order of preference.

air or oxygen is introduced to burn part of the input feedstock to provide the heat for pyrolysis and gasification (Adhikari et al., 2015). In cases where the oxidant is air, the product gas is diluted by the nitrogen present, and although air is 79% nitrogen, the stoichiometry of partial oxidation is such that the final product gas has approximately 50% v/v nitrogen as a diluent.

As a result the gas heating value of the fuel gas derived by gasification is relatively low although the gaseous product has wide range of applications ranging from power generation as well as a feedstock for the production of chemicals. The power derived from the gasification of carbonaceous feedstocks followed by the combustion of the product gas(es) is considered to be a source of renewable energy of derived gaseous products (Table 5.2) that are generated from a carbonaceous source (such as biomass or waste) (Pakdel and Roy, 1991; Speight, 2014b, 2020; Adhikari et al., 2015).

If the requirement of the process is a high heat content gas, the production of products such as methane, ethane, and propane further increase the calorific value. The use of pure oxygen as the gasification agent eliminates the nitrogen diluent and can produce medium calorific value gaseous products. An alternative strategy is to carry out the gasification process by means of indirect heat input.

However, tar is a significant potential problem in the gasification of biomass as it can lead to equipment blockages, increased maintenance and makes operation difficult (Brage et al., 2000). It is usually a thick, dark-colored liquid with a low condensation temperature and hence can lead to blockages in downstream equipment. In particular, low-boiling hydrocarbon derivatives (C_{2-6}) can actually avoid condensation and instead form tarry aerosols, which in turn degrade the quality of outlet gas and potentially make it unsuitable for use in high-purity applications, *e.g.*, applications other than boilers.

In terms of tar production, during the heating of biomass, the order of mass loss is as follows: first, moisture evaporation takes place from 30 to 120°C (86 to 250°F), followed by hemicellulose breakdown between 150 to 220°C (300 to 430°F), lignin breakdown at temperatures on the order

TABLE 5.2
Gasification Products

Product	Properties
Low-Btu gas	150 to 300 Btu/ft^3
	Approximately 50% v/v nitrogen
	Smaller amounts of carbon monoxide and hydrogen
	Some carbon dioxide
	Trace amounts of methane
Medium Btu gas	300 to 550 Btu/ft^3
	Predominantly carbon monoxide and hydrogen
	Small amounts of methane
	Some carbon dioxide
High-Btu gas	980 to 1,080 Btu/ft^3
	Predominantly methane – typically >85% v/v

of 220 to 400°C (430 to 750°F) and cellulose being the most stable decomposes at temperatures on the order of around 315 to 450°C (600 to 840°F). Lignin degrades over the widest range due to the diverse range of chemical structures and bonds present. Devolatilization and biomass conversion can occur across a range of temperatures for different kinds of biomass; this is commonly investigated using thermogravimetric analysis (TGA). The typical components of tar are (Sikarwar et al., 2016):

- Benzene (38% w/w)
- Toluene (14.5% w/w)
- Single-ring aromatic hydrocarbon derivatives (14% w/w)
- Naphthalene (9.5% w/w)
- Two-ring aromatic hydrocarbon derivatives (8% w/w)
- Heterocyclic compounds (6.5% w/w)
- Phenolic compounds (4.5% w/w)
- Three-ring aromatic hydrocarbon derivatives (3.5% w/w)
- Four-ring aromatic hydrocarbon derivatives (1% w/w)

The composition and amount of tar depends on the gasifier type and gasification temperature. Tar from the fluidized-bed gasifier can be characterized as secondary and tertiary tar. For example, increasing the temperature of the gasification process typically reduces the amount of tar (the primary tar) but increases the share of higher molecular weight (and higher-boiling) constituents (the secondary tar). The effect of pressure on tar composition is variable and process dependent. However, it has been observed that an increase in pressure may drive the composition of the tar to the higher boiling constituents.

The minimization of tar formation can be achieved in two ways: (1) by degradation of the tar inside the gasifier, which is referred to as primary tar reduction and requires an appropriate selection of functional parameters discussed earlier such as the design of the gasifier and the use of catalysts during gasification, (2) by thermal, mechanical or catalytic cracking of tar in a separate step after the biomass gasification, which is referred to as secondary tar reduction where the tar is deposited further downstream.

In terms of the use of biomass feedstocks and, hence, the production of synthesis gas, the process has grown from a predominately coal conversion process used for making *town gas* for industrial lighting to an advanced process for the production of multi-product, carbon-based fuels from a variety of feedstocks such as the biomass alone, biomass mixed with coal, and biomass mixed with the viscous products from crude oil refining and any other carbonaceous feedstocks (Kumar et al., 2009; Speight, 2013a, 2014a, 2014b; Luque and Speight, 2015; Sikarwar et al., 2016, 2017).

A gasifier that accepts biomass as the feedstock needs uniform-sized and dry fuel for smooth and trouble-free operation. Most gasifier systems are designed either for woody biomass (or dense briquettes made from loose biomass) or for loose, pulverized biomass (Chen et al., 1992; Bhattacharya et al., 1999; Hanaoka et al., 2005). The feedstock requirements vary depending upon the biomass fed to the gasifier, for example: (1) for woody biomass: pieces smaller than 5 to 10 cm (2 to 4 inches) in any dimension, depending on design and bulk density of wood or briquettes, and (2) for loose biomass: pulverized biomass, depending on design, moisture content up to 15 to 25%, depending on gasifier design, and a mineral matter content (manifested as product ash) below 5% w/w is preferred; with a maximum limit of 20% w/w (Demirbaş et al., 2013).

Overall, use of gasification technology (Speight, 2013a, 2014b) with the necessary gas cleanup options can have a smaller environmental footprint and lesser effect on the environment than landfill operations or combustion of the biomass or organic waste. In fact, there are strong indications that gasification is a technically viable option for waste conversion, including residual waste from separate collection of municipal solid waste. The process can meet existing emission limits and can have a significant effect on the reduction of landfill disposal using known gasification technologies to produce a synthesis gas product than can be employed as a fuel, a petrochemical feedstock as well as for use as a feedstock to the Fischer-Tropsch process (Arena, 2012; Fabry et al., 2013; Speight, 2014b; Luque and Speight, 2015).

However, compared to a typical fossil fuel (if there is such a fuel), the complex ligno-cellulosic structure of biomass is more difficult to gasify or combust. The nature of the mineral impurities in conjunction with the presence of various inorganic species, as well as sulfur and nitrogen containing compounds, adversely impacts the benign thermal processing of the oxygenated hydrocarbon structure of the biomass. While combustion of biomass feedstocks results in fuel-bound nitrogen and sulfur (being converted to NOx and SOx, respectively) steam gasification involves thermal treatment under a reducing atmosphere resulting in fuel-bound nitrogen release as molecular nitrogen (N_2) and fuel-bound sulfur conversion to hydrogen sulfide (H_2S) that is more easily removed by means of adsorption beds. Unlike combustion, the gasification process is more energy intensive. Careful engineering of the process is necessary if the result is to produce rather than consume a significant amount of energy or power as a result of the thermal treatment (Butterman and Castaldi, 2008).

In general, hydrogen can be generated from carbonaceous feedstocks by (1) thermochemical methods, which include gasification and pyrolysis, (2) an electrochemical method, and (3) biochemical methods and all three methods enable hydrogen production. This chapter presents descriptions of the various types of biomass and carbonaceous wastes that are available for the production of hydrogen which includes the benefits such as (1) the availability of biomass as an inexhaustible resource, (2) a reduction in the total quantity of carbonaceous waste depending upon the waste composition, and (3) the potential for a reduction in environmental pollution.

5.2 BIOMASS

The term 'biomass' encompasses a wide range of materials that produce a variety of useful products by way of the gasification process (Balat, 2011; Demirbaş, 2011; Ramroop Singh, 2011; Speight, 2011a). In the current context, the main basic sources of biomass include: (1) wood, including bark, logs, sawdust, wood chips, wood pellets and briquettes, (2) high yield energy crops, such as wheat, grown specifically for energy applications, (3) agricultural crops and residues, such as straw, and (4) industrial waste, such as wood pulp or paper pulp. For processing, a simple form of biomass such as untreated and unfinished wood may be converted into several physical forms, including pellets and wood chips, for use in biomass boilers and stoves. In addition, the widely varying heat content of the different types of biomass varies widely and must be taken into consideration when designing any conversion process (Jenkins and Ebeling, 1985).

Many forms of biomass contain a high percentage of moisture (along with carbohydrates and sugars) and mineral constituents – both of which can influence the viability of a gasification process – the presence of high levels of moisture in the biomass reduces the temperature inside the gasifier, which then reduces the efficiency of the gasifier. Before it is fed to the gasifier, the biomass feedstock must be processed to a uniform size or shape to ensure that as much of the biomass is gasified as possible. In addition, the widely varying heat content of the different types of biomass varies widely and must be taken into consideration when designing any conversion process (Jenkins and Ebeling, 1985).

5.2.1 Biomass Types

Biomass is biological material that has come from animal, vegetable or plant matter and is considered to be *carbon neutral* – while the plant is growing, it uses the sun's energy to absorb the same amount of carbon from the atmosphere as it releases into the atmosphere. By maintaining this closed carbon cycle it is felt, with some mathematical meandering, that there is no overall increase in carbon dioxide levels through emissions to the atmosphere.

Typical biomass waste materials include wood material (bark, chips, scraps, and saw dust), pulp and paper industry residues, agricultural residues, organic municipal material, sewage, manure, and food processing by-products. Agricultural residues such as straws, nut shells, fruit shells, fruit seeds, plant stalks and stover, green leaves, and molasses are potential renewable energy resources (Balat, 2011; Demirbaş, 2011; Ramroop Singh, 2011; Speight, 2011a).

Biomass can be considered as any renewable feedstock which is in principle be *carbon neutral* (while the plant is growing, it uses the sun's energy to absorb the same amount of carbon from the atmosphere as it releases into the atmosphere) (Speight, 2011a, 2020). It is biological material that has come from animal, vegetable, or plant matter and is considered to be *carbon neutral* – while the plant is growing, it uses the sun's energy to absorb the same amount of carbon from the atmosphere as it releases into the atmosphere. By maintaining this closed carbon cycle it is felt (though a paper exercise), with some mathematical meandering, that there is no overall increase in carbon dioxide levels through emissions to the atmosphere.

More pertinent to the present text, many forms of biomass contain a high percentage of moisture (along with carbohydrates and sugars) and mineral constituents – both of which can influence the viability of a gasifier operation (Chapter 3) – the presence of high levels of moisture in the biomass reduces the temperature inside the gasifier, which then reduces the efficiency of the gasifier. In addition, biomass can come in a range of sizes.

Raw materials that can be used to produce biomass fuels are widely available and arise from a large number of different sources and in numerous forms. Biomass can also be used to produce electricity – either blended with traditional feedstocks, such as coal or by itself. However, each of the biomass materials can be used to produce fuel but not all forms are suitable for all the different types of energy conversion technologies such as biomass gasification (Rajvanshi, 1986; Brigwater, 2003a, 2003b; Dasappa et al., 2004; Speight, 2011a; Basu, 2013). The main basic sources of biomass material are: (1) wood, including bark, logs, sawdust, wood chips, wood pellets and briquettes, (2) high yield energy crops, such as wheat, that are grown specifically for energy applications, (3) agricultural crop and animal residues, like straw or slurry, (4) food waste, both domestic and commercial, and (5) industrial waste, such as wood pulp or paper pulp. For processing, a simple form of biomass such as untreated and unfinished wood may be cut into several physical forms, including pellets and wood chips, for use in biomass boilers and stoves.

Thermal conversion processes use heat as the dominant mechanism to convert biomass into another chemical form. Some of the applications of thermal conversion are combined heat and power (CHP) and co-firing. In a typical dedicated biomass power plant, efficiencies range from 7 to

27% w/w of the energy content of the fuel. In contrast, cofiring biomass with coal, typically occurs at efficiencies near those of the coal combustor (30 to 40% of the energy content of the fuel) (Baxter, 2005; Liu et al., 2011).

Many forms of biomass contain a high percentage of moisture (along with carbohydrates and sugars) and mineral constituents – both of which can influence the viability of a gasification process (Chapter 3) – the presence of high levels of moisture in the biomass reduces the temperature inside the gasifier, which then reduces the efficiency of the gasifier. Also, most small to medium-sized biomass/waste gasifiers are air blown, operate at atmospheric pressure and at temperatures in the range 800 to 100°C (1,470 to 2,190°F).

The efficiency of the gasification process has been examined using various types of biomass such as sugarcane residue (Gabra et al., 2001), rice hulls (Boateng et al., 1992), pine sawdust (Lv et al., 2004), almond shells (Rapagnà and Latif, 10997; Rapagnà et al., 1997, 2000), wheat straw (Ergudenler and Ghali, 1993), food waste (Ko et al., 2001), and wood biomass (Pakdel and Roy, 1991; Bhattacharaya et al., 1999; Chen et al., 1992; Hanaoka et al., 2005). Recently, co-gasification of various biomass and coal mixtures has attracted a great deal of interest from the scientific community. In addition, the inorganic matter present in biomass catalyzes the gasification of coal (Chen et al., 2012). Also, it has been determined that the gasification of biomass involves several steps which are (1) evaporation of moisture at temperatures up to 150°C, 300°F, (2) pyrolysis causing the release of volatile matter in the temperature range 200 to 650°C, 390 to 1,200°F, (3) reaction of volatile constituents in the gas phase between 700 to 1,000°C , 1,290 to 1,830°F, and (4) heterogeneous reaction of the char between 700 to 1,000°C, 1,290 to 1,830°F.

More generally, biomass gasification plants are much smaller than the typical coal or crude oil coke gasification plants used in the power, chemical, fertilizer, and refining industries – the sustainability of the fuel supply is often brought into question. As such, a biomass gasification plant is less expensive to construct and has a smaller environmental footprint. For example, while a large industrial gasification plant may take up 150 acres of land and process 2,500 to 15,000 tons per day of feedstock (such as coal or crude oil coke), the smaller biomass plants typically process 25 to 200 tons of feedstock per day and take up less than 10 acres.

The gasification of biomass offers a great potential of possible product gas applications and there are various gasification technologies available for biomass. The main difference is the used gasification agent and related to this the way of heat supply, which is necessary because of the overall endothermic gasification reactions. This heat can be either added externally, in a so called allothermal gasification process, or generated internally by the full combustion of some biomass, referred to as autothermal gasification process. Another defining feature at different gasification processes is the reactor design, which distinguishes between fixed-bed, fluidized bed, and entrained flow reactors and the used gasification agent (Liu et al., 2010, 2011).

While co-gasification of coal and biomass is advantageous from a chemical viewpoint, some practical problems are present on upstream, gasification, and downstream processes. Also, the decomposition of biomass occurs at a lower temperature than the decomposition of coal and therefore different reactors compatible to the feedstock mixture are required (Speight, 2011; Brar et al., 2012; Speight, 2013a, 2013b). In addition, moisture content and pretreatment (torrefaction) are very important during upstream processing.

Heavy metals (metals with relatively high densities, atomic weights, or atomic numbers) and impurities such as sulfur and mercury present in coal can make synthesis gas difficult to use and unhealthy for the environment. Alkali present in biomass can also cause corrosion problems high temperatures in downstream pipes.

It is recommended that biomass moisture content should be less than 15% w/w prior to gasification. In such a situation, water is removed from the cell wall leading to shrinkage. The long-chain molecules constituents of the cell wall move closer to each other and bind more tightly.

The torrefaction process involves the thermal treatment of biomass in the absence of oxygen at a temperature on the order of 250 to 300°C (480 to 570°F) to drive off moisture, decompose

TABLE 5.3
Advantages of Using Biomass for Hydrogen Production

Advantages
Theoretically inexhaustible fuel source
Minimal environmental impact when direct combustion of plant mass is not used to generate energy (i.e., fermentation, pyrolysis, etc. are used instead), (iii) alcohol derivatives and other fuels produced by biomass are efficient, viable, and relatively clean-burning
Available on a world-wide basis

hemicellulose completely, and partially decompose cellulose (Speight, 2011a). Also, torrefaction of the feedstock can result in a more efficient process with higher yields of hydrogen and carbon monoxide in the gasification process.

Finally, the sorption-enhanced reforming concept (SER, sometimes referred to as adsorption or absorption-enhanced reforming, AER) can result in higher yields of hydrogen by the selective removal of carbon dioxide from the reaction zone in the gasifier which pushes the equilibrium of the water gas shift reaction to the production of more hydrogen. Consequently, a yield of hydrogen on the order of 75% v/v can be reached by application of the sorption-enhanced reforming concept by using calcium oxide in the process which reacts with the carbon dioxide at a temperature on the order of 650 to 700°C (1,200 to 1,290°F):

$$CaO + CO_2 \rightarrow CaCO_3$$

Thus, the formed calcium carbonate ($CaCO_3$) serves as a carbon dioxide carrier which leads to the in situ removal of carbon dioxide from the system which is achieved by the circulating bed material.

Biomass fuel producers, coal producers, and, to a lesser extent, waste companies are enthusiastic in regard to supplying co-gasification power plants and realize the benefits of co-gasification with alternate fuels (Speight, 2011a, 2020; Lee and Shah, 2013; Speight, 2013a, 2013b). The benefits of co-gasification technology involving coal and biomass include the use of a reliable coal supply with gate-fee waste and biomass which allows the economies of scale from a larger plant to be supplied just with waste and biomass. In addition, the technology offers a future option of hydrogen production and fuel development in refineries.

In addition, while biomass may seem to some observers to be the answer to the global climate change issue, the advantages must be considered carefully (Table 5.3). For example, the advantages are (1) biomass is a theoretically inexhaustible fuel source, (2) when direct conversion of combustion of plant mass – such as fermentation and pyrolysis – is not used to generate energy there is minimal environmental impact, (3) alcohols and other fuels produced by biomass are efficient, viable, and relatively clean-burning, and (4) biomass is available on a world-wide basis.

On the other hand, the disadvantages include (1) the highly variable heat content of different biomass feedstocks, (2) the high water content that can affect the process energy balance, and (3) there is a potential net loss of energy when a biomass plant is operated on a small scale – an account of the energy put used to grow and harvest the biomass must be included in the energy balance.

5.2.2 Bio-Oil

Bio-oil is a dark brown viscous liquid composed of highly oxygenated compounds produced from the pyrolysis of biomass and which can be converted into a gaseous mixture (containing hydrogen) by use of a pyrolysis process (Czernik et al., 2000; Demirbaş, 2007; Martino et al., 2021). The pyrolysis process carried out at high temperatures on the order of between 200 to 750°C (390 to 1,380°F). Slow pyrolysis results in the production of gas and biochar while fast pyrolysis produces bio-oil and bio-char.

During the pyrolysis process the biopolymers (i.e., hemicellulose, cellulose, and lignin) decompose to form a solid residue (often referred to as 'charcoal' or 'biochar'), other low-boiling hydrocarbon derivatives, condensable gases, and non-condensable gases (such as carbon dioxide, CO_2, carbon monoxide, CO, and hydrogen, H_2). After cooling at room temperature, the non-permanent gases are condensed to form the bio-oil which is a liquid phase product that typically has a higher energy density than biomass (Demirbaş, 2007; Kumar et al., 2016; Wang et al., 2020; Pafili et al., 2021; Zhang, 2022).

The composition of bio-oil and its properties depend on the raw feedstock composition and the conditions at which pyrolysis was undertaken, such as temperature, heating rate, and residence time (Mohan et al., 2017). The constituents of bio-oil include carboxylic acid derivatives, aldehyde derivatives, alcohol derivatives, ketone derivatives, and substituted furan derivatives, as well as phenol derivatives and cyclic oxygenate derivatives (Pafili et al., 2021).

The addition of water to the crude bio-oil gives rise to two fractions which are (1) a hydrophilic carbohydrate or aqueous phase containing typically 20% w/w organic constituents, which can be catalytically steam reformed and (2) a hydrophobic organic phase, often referred to as 'pyrolytic lignin' which contains furan derivatives and a variety of aromatic-based derivatives. Bio-oil can be readily stored and transported more easily in comparison to biomass and possesses very low amounts of sulfur in comparison to crude oil.

The catalytic steam reforming of bio-oil is another process that is available to produce hydrogen. The process is endothermic in which hydrocarbon derivatives react with steam in the presence of a catalyst resulting in the formation of a mixture of carbon monoxide and hydrogen which, when followed by a water gas shift reaction yields more hydrogen:

$$CxHyOz + (2x-z)H_2O \rightarrow xCO_2 + (2x + y/2-z)H_2$$

The ideal bio-oil reforming catalyst must have a high activity for reforming and should also be highly selective toward hydrogen generation as well as a high resilience toward deactivation by coke deposition.

5.3 WASTE

Waste or 'rubbish' (also called 'garbage' or 'trash' in the United States) is a substance or object, or the collection of substances and objects selected for disposal or that are required to be disposed by the provisions of local, regional, or national laws. In addition, waste is also substances or objects that are not the prime products of a process (or processes) for which the initial user has no further use in terms of his/her own purposes of production, transformation, or consumption, and of which s/he wants to dispose. Wastes may be generated during (1) the extraction of raw materials, (2) the processing of raw materials into intermediate and final products, (3) the consumption of final products, and (4) other human activities.

Whatever the form, waste is an unavoidable by-product of human activity and industrial diversification has added substantial quantities of industrial waste and biomedical waste into the waste. Thus, there is a need for the management and disposal of the growing volume of waste, in addition to waste disposal (landfill) operations that are being stretched to the limit and suitable disposal areas are in short supply. In addition, the potential for rain water (and snow melt) to each chemical constituents from landfills and into the groundwater table is of immediate concern. Other more specific waste sources (excluding hazardous waste) and possibly including crude oil coke may provide niche opportunities for co-utilization (Brigwater, 2003; Arena, 2012; Basu, 2013; Speight, 2013a, 2013b).

Furthermore, the disposal of municipal and industrial waste has become an important problem because the traditional means of disposal, landfill, are much less environmentally acceptable than previously. Much stricter regulation of these disposal methods will make the economics of waste processing for resource recovery much more favorable. One method of processing waste streams is to convert the energy value of the combustible waste into a fuel, such as hydrogen (Richardson et al., 1995). One type of fuel attainable from waste is a low heating value gas, usually 100–150 Btu/scf, which can be used to generate process steam or to generate electricity (Gay et al. 1980). Co-processing such waste with coal is also an option (Speight, 2020).

Gasification is a unique process that transforms any carbon-based material, such as municipal solid waste (MSW), into energy without burning it and converts the carbonaceous waste into gaseous products of which synthesis gas is (in the current context) of prime importance – removal of pollutants and impurities results in clean gas that can be converted into electricity and valuable products (Chapter 2).

Thus, the gasification of waste is a chemical process where carbonaceous waste is heated in a low-oxygen environment to the point that the molecular constituents are decomposed to produce a combustible gas, an inert slag, and/or char.

With gasification, municipal solid waste and other types of wastes are no longer of environmental concern but are feedstocks for a gasifier. Instead of the associated costs of disposal of and landfill management, using waste as a feedstock for a gasification process reduces disposal costs and required landfill space, and converts the carbonaceous wastes to electricity and fuels.

Initially, the gasification process was applied to coal as a means of producing fuel gases and chemicals as well as electricity but has seen considerable evolution in terms of acceptance of feedstocks other tan coal as well as the technologies used for the process and represents *significant* advances over the incineration process (Chapter 2, Chapter 3) (Orr and Maxwell, 2000; Malkow, 2004; Speight, 2011a, 2011b, 2013a, 2013b, 2014a, 2017, 2019b, 2020). In order to understand the advantages of gasification when compared to incineration for waste materials, it is important to understand the differences between the two processes.

Incineration, which does have a place in waste disposal operations, uses municipal solid waste as a fuel. The waste is burned with high volumes of air to form carbon dioxide and heat.

On the other hand, gasification converts municipal solid waste to usable synthesis gas, and it is the production of this synthesis gas which makes gasification different to the incineration process.

In addition, one of the concerns with incineration of municipal solid waste is the formation and reformation of toxic dioxin derivatives and furan derivaitves, especially from PVC plastics (polyvinyl chloride plastics). These toxins end up in exhaust streams by three pathways: (2) by decomposition, as low-molecular weight volatile constituents, (2) by re-forming in which lower molecular weight constituents combine to form new products, and/or (3) by the unusual step of passing through the incinerator without change. Incineration does not always allow adequate control of these processes.

In respect of municipal solid waste, gasification is significantly different from and cleaner than incineration: (1) in the high temperature environment in gasification, higher molecular weight materials such as plastics, are effectively decomposed to synthesis gas, which can be cleaned and processed before any further use, (2) dioxin derivatives and furan derivatives need sufficient oxygen to form and the oxygen-deficient atmosphere in a gasifier does not provide the environment needed for the formation of dioxins and furans, (3) when the synthesis gas is primarily used as a fuel for making heat, it can be cleaned as necessary *before* combustion; this cannot occur in incineration.

Furthermore, the disposal of municipal and industrial waste has become an important problem because the traditional means of disposal, landfill, are much less environmentally acceptable than previously. Much stricter regulation of these disposal methods will make the economics of waste processing for resource recovery much more favorable.

One method of processing waste streams is to convert the energy value of the combustible waste into fuel. One type of fuel attainable from waste is a low heating value gas, usually 100–150

Btu/scf, which can be used to generate process steam or to generate electricity (Gay et al. 1980). Co-processing such waste with coal is also an option (Speight, 2013a, 2013b, 2020). Thus, waste-to-energy plants based on the gasification principle are high-efficiency power plants that utilize municipal solid waste as fuel rather than conventional sources of energy such as coal or crude oil, although the co-gasification of waste with biomass, coal, crude oil residua, and biomass is always an options (Speight, 2011a, 2011b, 2013a, 2013c, 2014a, 2017).

This section presents descriptions of the various types of waste and the recovery of energy from waste by gasification, including the benefits such as: (1) reduction in the total quantity of waste depending upon the waste composition and the gasification technology employed, (2) reduction in environmental pollution, and (3) improved commercial viability of the waste disposal project from the sale of energy/products.

In summary, as hydrogen becomes in greater demand by the refining industry and by the chemical industries, landfill space around major cities is becoming scarce and the disposal and transportation costs of waste are increasing thereby making waste gasification a relatively cheaper source of hydrogen. The flexibility of gasification process enables the process to be tailored to the requirements of (1) hydrogen demands and (2) the needs for synthesis gas as demanded by the local markets.

5.3.1 Solid Waste

Solid waste does not include waste materials that result from activities associated with the exploration, development, or production of oil or gas or geothermal resources, or other substance or material regulated by the local or federal governments.

Such waste includes, but is not limited to bricks, concrete and other masonry materials, soil, rock, wood (including painted, treated, and coated wood and wood products), land clearing debris, wall coverings, plaster, drywall, plumbing fixtures, non-asbestos insulation, roofing shingles and other roof coverings, asphaltic pavement, glass, plastics that are not sealed in a manner that conceals other wastes, empty buckets ten gallons or less in size and having no more than one inch of residue remaining on the bottom, electrical wiring and components containing no hazardous liquids, and pipe and metals that are incidental to any of the above.

Furthermore, the disposal of a wide variety of wastes has become an important problem because the traditional means of disposal to a landfill has become environmentally much less acceptable than previously. Newer and stricter regulation of the conventional disposal method(s) has made the economics of waste processing for resource recovery much more favorable.

However, before moving on to the various aspects of the gasification process, it is worthwhile to describe in more detail the types of waste that arise from human activities and which might be suitable for gasification.

Municipal solid waste (MSW) is solid waste resulting from, or incidental to, municipal, community, commercial, institutional, and recreational activities; it includes garbage, rubbish, ashes, street cleanings, dead animals, medical waste, and all other nonindustrial solid waste.

The composition of municipal solid waste (Table 5.4), which is generated from households, can vary from one community to the next, but the overall differences are not substantial. In fact, there are several types of waste that might also be classified within the municipal solid waste category (Table 5.5). The moisture content of raw municipal solid waste is usually 20% w/w.

Industrial solid waste, which is solid waste resulting from manufacturing, mining, or agricultural operations is classified as either as (1) hazardous waste, which includes any industrial solid waste or combination of industrial solid wastes identified or listed as hazardous waste or (2) non-hazardous waste, which includes any industrial solid waste that is not identified as hazardous waste or listed as hazardous waste.

Thus, industrial solid waste encompasses a wide range of materials of varying environmental toxicity. Typically this range would include paper, packaging materials, waste from food processing, oils, solvents, resins, paints and sludge, glass, ceramics, stones, metals, plastics, rubber, leather,

TABLE 5.4
General Composition of Municipal Solid Waste

Component	% w/w
Paper	33.7
Cardboard	5.5
Plastics	9.1
Textiles	3.6
Rubber, leather, 'other'	2.0
Wood	7.2
Horticultural wastes	14.0
Food wastes	9.0
Glass and metals	13.1

TABLE 5.5
Sources and Types of Waste

Source	Examples	Waste Types
Residential	Single and multifamily dwellings	Food wastes, paper, cardboard
		Plastics, textiles, leather
		Yard wastes, wood, glass, metals
		Ash, special wastes oil, tires
Industrial	Light and heavy manufacturing	Housekeeping wastes,
		Food wastes
	Construction sites	Construction and demolition materials
		Wood steel, concrete
	Power plants	Ash
	Chemicals plants	Chemical waste
Commercial	Stores, hotels, restaurants	Paper, cardboard, plastics
		Food wastes
	Markets, office building	Glass, metals, special wastes,
		Hazardous wastes
Institutional	Schools, hospitals	Paper, cardboard, plastics
	Government centers	Food waste
		Glass, metals, special wastes,
		Hazardous wastes
Municipal	Street cleaning	Street sweepings
		Landscape waste
		Tree trimmings
		Wastes from parks
		Wastes from recreational areas
Agriculture	Crops, orchards, dairies	Agricultural wastes
	Farms, feedlots	Hazardous waste (e.g., pesticides)
Waste class	*Description*	

TABLE 5.6
Types of Non-Hazardous Industrial Solid Waste

Class 1	Any industrial solid waste or mixture of industrial solid wastes that (because of its concentration, or physical or chemical characteristics) is toxic; corrosive; flammable; a strong sensitizer or irritant; or a generator of sudden pressure by decomposition, heat, or other means; or may pose a substantial present or potential danger to human health or the environment when improperly processed, stored, transported, or disposed of or sent for disposal
Class 2	Waste: waste consisting of any individual solid waste or combination of industrial solid wastes that is not described as hazardous waste, Class 1 waste, or Class 3 waste
Class 3	Waste: waste consisting of inert and essentially insoluble industrial solid waste, usually including, but not limited to, materials such as rock, brick, glass, dirt, and certain plastics and rubber, that are not readily decomposable

wood, cloth, straw, abrasives, etc. As with municipal solid waste, the absence of a regularly updated and systematic database on industrial solid waste ensures that the exact rates of generation are largely unknown. Also, a generator of nonhazardous industrial solid waste must further classify the waste (Table 5.6).

5.3.2 Bio-solids

The term 'bio-solids' includes livestock waste, agricultural crop residues and agro-industrial by-products. Similarly, fish farming communities commonly integrate fish rearing with agricultural activities such as livestock husbandry, vegetable and paddy cultivation and fruit farming. Some sewage sludge cannot qualify as bio-solids because of the presence of persistent chemicals, bio-accumulative chemicals, and toxic chemicals, radionuclides, and heavy metals at levels sufficient to contaminate soil and water when applied to land.

Typically, bio-solids are created through the treatment of domestic wastewater generated from sewage treatment facilities and can actually begin before the wastewater reaches the sewage treatment plant. In addition, regulations require that industrial facilities pre-treat their wastewater to remove hazardous contaminants before it is sent to a wastewater treatment plant.

Many countries with agricultural-based economies use agricultural wastes to produce biogas through anaerobic digestion (Speight, 2011a, 2020). The biogas (typically contains methane on the order of 60% v/v) is primarily used directly for cooking, heating, and lighting, whilst the slurry from the anaerobic digesters is used as liquid fertilizer, a feed supplement for cattle and pigs and as a medium for soaking seeds.

However, with the current interest in increasing the use of hydrogen in refineries (and in the petrochemical industry) there is interest in using organic bio-solids (regulations permitting) as gasifier feedstock or as a co-feedstock.

5.3.3 Biomedical Waste

Biomedical waste is the waste materials produced by hospitals and health care institutions, which have been increasing over the past four decades to meet the medical and health care requirements of the growing world population. Until recent years, little attention was paid to the wastes generated from these facilities, which are potentially hazardous to human health and the environment. In fact, serious concern has arisen regarding the potential for spreading pathogens, as well as causing environmental contamination due to the improper handling and management of clinical and biomedical waste.

On the other hand, regulated medical waste (RMW, also referred to as 'regulated medical waste' or 'red bag waste') which may contain items such as blood-soaked bandages. This term emphasizes the role of regulation in defining the actual material and as an alternative to "infectious waste," given the lack of evidence of this type of waste's infectivity. State regulations also address the degree or amount of contamination (such as the amount of blood-soaked gauze) that defines the discarded item as a regulated medical waste.

Biohazardous waste (also called 'infectious waste' or 'regulated medical waste') is a waste stream that contains potentially infectious material (such as such as blood, body fluids, and human cell lines) is waste contaminated with potentially infectious agents or other materials that are deemed a threat to public health or the environment and is regulated state-by-state, but also falls under the Blood-Borne Pathogen Standard as defined by the US Office of Safety and Health Administration (OSHA). Such waste are subject to state and federal regulations and may not be suitable as a general gasification feedstock and require higher temperature to assure complete disposal of the constituents.

5.3.4 Liquid Waste

Liquid waste is a non-solid material that has no further use and must be treated and disposed of according to local, state, and federal regulations. A liquid waste that can seep into the soil and can easily spread, soak into other materials, or contaminate the soil. If the waste should enter a water system it can potentially harm plant, animal life, and human life.

Organic liquid wastes are but not always, such as solutions of the polynuclear hydrocarbon derivatives) often biodegradable materials that can still be toxic if not properly managed. On the other hand, inorganic liquid pollutants include mine waste containing heavy metals such as copper, lead, mercury, zinc, cadmium, sodium, and their oxides and salts.

An example of liquid waste that is relevant to this section is the so-called 'black liquor' which is the spent liquor from the Kraft process in which pulpwood is converted into paper pulp by removing lignin constituents, hemicellulose constituents, and other extractable materials from wood to free the cellulose fibers. Other terms for the spent cooking liquor from the sulfite process include 'brown liquor', 'red liquor', 'thick liquor', and 'sulfite liquor'.

By way of explanation, the sulfite process it typically conducted under acid or neutral conditions, sulfur dioxide and/or bisulfite ions react with lignin to produce water-soluble sulfonated lignin derivatives (referred to as lignosulfonate derivatives) which are degraded by acid hydrolysis reactions in large pressure vessels (referred to as 'digesters'). On the other hand, the kraft puling process for paper manufacture involves digesting wood chips at elevated temperature and pressure in 'white liquor', which is an aqueous solution of sodium sulfide (Na_2S) and sodium hydroxide (NaOH) which chemically dissolves the lignin that binds the cellulose fibers together.

More specifically to the present context, black liquor waste is an aqueous solution of residues from lignin residues and hemicellulose derivatives well as the inorganic chemicals used in the process and 15% w/w solids of which 10% w/w are inorganic and 5% w/w are organic. Typically, the organic constituents in black liquor are 40 to 45% w/w soap derivatives, 35 to 45% w/w lignin derivatives, and 10 to 15% w/w other (miscellaneous, depending upon the source) organic materials.

Typically, the organic constituents in black liquor are made up of water/alkali soluble degradation components from the wood. Lignin is degraded to shorter fragments with sulfur content on the order of 1 to 2% w/w and sodium content at approximately 6% w/w of the dry solids. In the pulping process, cellulose (and hemicellulose) is degraded to aliphatic carboxylic acid derivatives and hemicellulose fragments. The extractable constituents yield 'tall oil soap' and crude turpentine.

Black liquor gasification has the potential to achieve higher overall energy efficiency than the conventional recovery boiler while generating synthesis gas from the liquor.

5.4 MIXED FEEDSTOCKS

In the gasification sense, co-gasification is (as the name implies) a gasification process in which a mixed feedstock (in the current context, such as coal and biomass or coal and waste or biomass and waste) are mixed prior to entering the gasifier and gasified in the same gasification unit. This option allows using both feedstocks to be converted in a more efficient manner and with the potential for reduced environmental pollution as long as the pollutants are efficiently removed before gasification. In fact, any carbonaceous feedstock has the potential to be co-gasified with biomass or with carbonaceous waste or biomass for environmental reasons, technical reasons, or commercial reasons (Speight, 2011a; Brar et al., 2012).

Although the gasification processes used for individual feedstocks are relatively straightforward, process efficiency depends for the most part on the unique characteristics of each feedstock. For example, consideration should be given to the individual constituents of each feedstock to determine if there are any reactions between the constituents that can have an adverse effect on the process. An example is the composition of any mineral matter that may cause the constituents to interact to form a slag that may cause harm to the gasifier.

However caution is advised when using mixed feedstocks for gasification or, for that matter, for any conversion process. It has been the method used in the past (and even continued to be the method in some cases) to assess the properties of the mixed feedstock by calculating an average value for one or more of the properties of the mixed feedstock. This is a dangerous practice because it ignores the potential for interaction of the contents of each feedstock that can cause changes in the chemistry of the process as well as a loss of process efficiency One approach to avoid a significant decrease in conversion efficiency is to develop a formulation of mixed feedstocks in order to produce a more consistent material. Formulation combines various preprocessed resources and/or additives to produce a feedstock that provides process consistency.

Feedstock formulation is not a new concept and has been used for decades by the coal industry. In the current context, biomass blending feedstocks refers to the combination of multiple sources of the same biomass resource to average out compositional and moisture variations, whereas aggregation refers to the combination of different raw or preprocessed biomass resources to produce a single, consistent feedstock with desirable properties. Co-gasification technology varies and is usually site specific with high dependence on the feedstock. At the largest scale, the plant may include the well proven fixed-bed and entrained flow gasification processes. Pyrolysis and other advanced thermal conversion processes are included where power generation is practical using the on-site feedstock produced.

Gasification generally has a non-negative environmental image whereas combustion has a negative public image. The public view is not always based on objective analysis of emissions and other effects but nevertheless it is still a very important factor.

Examples presented here to emphasize the above considerations are presented below and are (1) the gasification of biomass and coal and (2) the gasification of biomass and municipal solid waste.

5.4.1 Biomass and Coal

Recently, co-gasification of various biomass and coal mixtures has attracted a great deal of interest from the scientific community (Usón et al., 2004). Both fixed-bed and fluidized-bed gasifiers have been used in co-gasification of biomass with coal (McKendry, 2002). While the co-gasification of coal and biomass improves the hydrogen/carbon monoxide ratio in the produced gas (Sjöström et al., 1999; Kumabe et al., 2007; Chen et al., 2012), the inorganic matter present in biomass can catalyze the gasification of the coal constituents.

While co-gasification of coal and biomass is advantageous from a chemical viewpoint, some practical problems are present on upstream, gasification, and downstream processes. In addition, moisture content and pretreatment (torrefaction) are very important during upstream processing

(Speight, 2011; Brar et al., 2012; Speight, 2013a, 2013b). Alkali metals (sodium and potassium) present in biomass can also cause corrosion problems high temperatures in downstream pipes. In addition, biomass densification may be conducted to prepare pellets and improve density and material flow in the feeder areas.

It is recommended that biomass moisture content should be less than 15% w/w prior to gasification. In such a situation, water is removed from the cell wall leading to shrinkage. The long-chain molecules constituents of the cell wall move closer to each other and bind more tightly.

The torrefaction process is a thermal treatment of biomass in the absence of oxygen, usually at 250 to 300°C (480 to 570°F) to drive off moisture, decompose hemicellulose completely, and partially decompose cellulose (Speight, 2011a). Torrefaction results in higher yields of hydrogen and carbon monoxide in the gasification process.

The benefits of co-gasification technology involving coal and biomass include the use of a reliable coal supply with gate-fee waste and biomass which allows the economies of scale from a larger plant to be supplied just with waste and biomass. In addition, the technology offers a future option of hydrogen production and fuel development in refineries.

The low melting point of the ash produced from woody biomass leads to agglomeration which causes defluidization of the ash and sintering, deposition as well as corrosion of the gasifier construction metal bed (Vélez et al., 2009). Biomass containing alkali oxides and salts are likely to produce clinkering/slagging problems from ash formation (McKendry, 2002).

5.4.2 Biomass and Waste

Co-utilization of waste and biomass with coal may provide economies of scale that help achieve the above identified policy objectives at an affordable cost. In some countries, governments propose co-gasification processes as being *well suited for community-sized developments* suggesting that waste should be dealt with in smaller plants serving towns and cities, rather than moved to large, central plants (satisfying the so-called *proximity principal*).

Co-gasification technology varies, being usually site specific and high feedstock dependent. At the largest scale, the plant may include the well proven fixed-bed and entrained flow gasification processes. Pyrolysis and other advanced thermal conversion processes are included where power generation is practical using the on-site feedstock produced. However, the needs to be addressed are (1) core fuel handling and gasification/pyrolysis technologies, (2) fuel gas clean-up, and (3) conversion of fuel gas to electric power (Ricketts et al., 2002).

The use of waste as gasification feedstock or as a co-gasification feedstock can mitigate an important problem. The disposal of municipal and industrial wastes has become an important problem because the traditional means of disposal, landfill, has become environmentally much less acceptable than previously. New, much stricter regulation of these disposal methods will make the economics of waste processing for resource recovery much more favorable. One method of processing waste streams is to convert the energy value of the combustible waste into fuel. One type of fuel attainable from wastes is a low heating value gas heat content on the order of 100 to 150 Btu/ft^3), which can be used to generate process steam or to generate electricity.

The ability of a refinery to efficiently accommodate viscous feedstock streams (such as heavy crude oil, extra heavy crude oil, residua, deasphalter bottoms, visbreaker bottoms, and tar sand bitumen) enhances the economic potential of the refinery and the development of heavy crude oil, extra heavy crude oil, and tar sand resources. A refinery with the flexibility to meet the increasing product specifications for fuels through the ability to upgrade high-boiling feedstocks is an increasingly attractive means of extracting maximum value from each barrel of oil produced.

5.5 CHEMISTRY

The gasification process consists of the conversion of the organic carbon in the feedstock into various constituents that make up the synthesis gas constituents by reacting the feedstock at high temperatures (>700°C, 1,290°F), without combustion using a controlled amount of oxygen and/or steam (Marano, 2003; Lee, 2007; Lee et al., 2007; Higman and Van der Burgt, 2008; Sutikno and Turini, 2012; Speight, 2013a, 2014b; Luque and Speight, 2015; Speight, 2020). The products include gaseous hydrocarbon derivatives (methane, ethane, and propane), carbon monoxide, carbon dioxide, and hydrogen. Other products may also be produced depending upon the composition of the feedstock.

$$\text{Feedstock} \rightarrow H_2 + CO + CO_{2(g)} + \text{hydrocarbons}_{(gas)} + \text{tar}_{(liquid)} + \text{char}_{(solid)}$$

Elevated temperatures, high heating rates, and long volatile stage residence times are vital for synthesis gas production or hydrogen production (Chapter 11). Since the liquid yield through fast pyrolysis is high, the gaseous yield is reduced.

It is important to distinguish the gasification process from a pyrolysis process. The main difference between pyrolysis and gasification is the absence of a gasifying agent in a pyrolysis process (Chapter 2, Chapter 4). Pyrolysis is a thermal degradation of organic compounds, at a range of temperatures on the order of 300 to 900°C (570 to 1650°F), under oxygen-deficient circumstances to produce various types of products such as gases (often referred to as synthesis gas), a liquid product (referred to as bio-oil), and a solid product (referred to as biochar).

On the other hand, the gasification process involves thermal cracking of solid or high boiling liquid carbonaceous material (such as bio-oil biomass, biomass char) into a combustible gas mixture, mainly composed of hydrogen, carbon monoxide (CO), carbon dioxide (CO_2), and methane (CH_4) and other gases with some by-products (solid char or slag, oils, and water). The synthesis gas product has a chemical composition and properties that are largely affected by (1) the type and composition of the feedstock, (2) the operational conditions throughout pyrolysis and gasification such as reactor temperature, residence time, pressure, and (3) the reactor geometry.

In the current context, and in addition to the potential reactions that can occur in the gasifier (Table 5.7), the gasification process is often considered to involve two distinct physico-chemical stages: (1) devolatilization of the feedstock to produce volatile matter and char and (2) followed by char gasification, which is complex and specific to the conditions of the reaction – both processes contribute to the complex kinetics of the gasification process (Sundaresan and Amundson, 1978; Luque and Speight, 2015).

TABLE 5.7
Reactions that Occur During the Gasification of a Carbonaceous Feedstock

$2C + O_2 \rightarrow 2CO$

$C + O2 \rightarrow CO_2$

$C + CO_2 \rightarrow 2CO$

$CO + H_2O \rightarrow CO_2 + H_2$ (shift reaction)

$C + H_2O \rightarrow CO + H_2$ (water gas reaction)

$C + 2H_2 \rightarrow CH_4$

$2H_2 + O_2 \rightarrow 2H_2O$

$CO + 2H2 \rightarrow CH_3OH$

$CO + 3H2 \rightarrow CH_4 + H_2O$ (methanation reaction)

$CO_2 + 4H_2 \rightarrow CH_4 + 2H_2O$

$C + 2H_2O \rightarrow 2H_2 + CO_2$

$2C + H_2 \rightarrow C_2H_2$

$CH_4 + 2H_2O \rightarrow CO_2 + 4H_2$

5.5.1 General Aspects

As with other carbonaceous feedstocks, the gasification of biomass includes a series of reaction steps that convert the feedstock into *synthesis gas* (carbon monoxide, CO, plus hydrogen, H_2) or, if biomass is the feedstock, into bio-synthesis gas and other gaseous products. The gaseous products and the proportions of these product gases (such as methane, CH_4, and other hydrocarbon derivatives, water vapor (H_2O), carbon dioxide, (CO_2), carbon monoxide (CO), hydrogen (H_2), hydrogen sulfide (H_2S), and sulfur dioxide, (SO_2) depends on: (1) the type of feedstock, (2) the chemical composition of the feedstock, (3) the gasifying agent or gasifying medium, and (4) the thermodynamics and chemistry of the gasification reactions as controlled by the process operating parameters (Singh et al., 1980; Pepiot et al., 2010; Shabbar and Janajreh, 2013; Speight, 2013a, 2013b, 2014b; Luque and Speight, 2015). In addition, the kinetic rates, the extent of the conversion for any feedstock (irrespective of the composition), as well as the several chemical reactions that are a part of the gasification process are variable and are typically functions of (1) temperature, (2) pressure, and (3) reactor configuration, the (4) gas composition of the product gases, and (5) whether or not the product gases influence the outcome of the reaction (Johnson, 1979; Penner, 1987; Müller et al., 2003; Slavinskaya et al., 2009; Speight, 2013a, 2013b, 2014b; Luque and Speight, 2015).

At a temperature in excess of 500°C (930°F) the conversion of the feedstock to char and ash and char is completed. Furthermore, with an increase in heating rate, feedstock particles are heated more rapidly and are burned in a higher temperature region, but the increase in heating rate may have very little effect on the mechanism (Demirbaş, 2004; Tinaut et al., 2008; Irfan, 2009).

Most notable effects in the physical chemistry of the gasification process are those effects due to the chemical character of the feedstock as well as the physical composition of the feedstock (Speight, 2011a, 2013a, 2014a, 2014b). In more general terms of the character of the feedstock, gasification technologies generally require some initial processing of the feedstock with the type and degree of pretreatment a function of the process and/or the type of feedstock (Table 5.8, Table 5.9). This is especially true for biomass there the feedstock can have several constituents that strongly influence the course and outcome of the conversion process.

Another factor, often presented as very general rule of thumb, is that optimum gas yields and gas quality are obtained at operating temperatures of approximately 595 to 650°C (1,100 to 1,200°F).

With some feedstocks, the higher the amounts of volatile material produced in the early stages of the process the higher the heat content of the product gas. However, in the case of biomass, there is the potential for carbon dioxide to be an initial product of the process which will serve to lower the heat content of the product gas. Also, when the process temperature is not sufficiently high, the

TABLE 5.8
Acidic and Alkaline Methods for Biomass Treatment

Method	Conditions	Outcome
Acid based methods	Low pH using an acid*	Hydrolysis of hemicellulose to sugar monomers
		Minimizes the need for hemicellulase enzymes
Neutral conditions	Steam pretreatment	Solubilizes hydrocarbons by
		Conversion to acetic acid
		Does not usually result in total conversion to monomer sugars
		Requires hemicellulase to act on soluble oligomers
Alkaline methods		Leaves part of the feed in the solid fraction
		Requires hemicellulases to act on hydrocarbon derivatives

*Such as sulfuric acid, H_2SO_4, or phosphoric acid, H_3PO_4.

TABLE 5.9
Summation of the Methods for the Pretreatment of Biomass Feedstocks

Physical Methods	Miscellaneous Methods
Milling:	Explosion*
Ball milling	Steam, NH_3, CO_2, SO_2, acids, alkali
Two-roll milling	NaOH, NH_3, $(NH_4)_2SO_3$
Hammer milling	Acid:
Irradiation:	Sulfuric, hydrochloric, and phosphoric acids
Gamma-ray irradiation	Gas:
Electron-beam irradiation	Chlorine dioxide, nitrous oxide, sulfur dioxide
Microwave irradiation	Oxidation:
Other methods:	Hydrogen peroxide
Hydrothermal	Wet oxidation
High-pressure steaming	Ozone
Extrusion	Solvent extraction of lignin:
Pyrolysis	Ethanol-water extraction
	Benzene-water extraction
	Butanol-water extraction
	Organic solvents
	Ionic liquids

*The feedstock material is subjected to the action of steam and high-pressure carbon dioxide before being discharged through a nozzle.

char oxidation reaction is suppressed, and the overall heat content of the product gas is diminished. All such events serve to complicate the reaction rate and make the attempted derivation of a kinetic relationship that can be applied to all types of feedstock (and gasifier configurations) subject to serious question and doubt.

Pressure also plays a role in the type of product produced and is, as might be anticipated, dependent on the type of feedstock being processed and the gas product that is desired. In fact, some (or all) of the following processing steps will be required: (1) pretreatment of the feedstock, (2) primary gasification of the feedstock, (3) secondary gasification of the carbonaceous residue from the primary gasifier, (4) removal of carbon dioxide, hydrogen sulfide, and other acid gases, (5) shift conversion for adjustment of the carbon monoxide/hydrogen mole ratio to the desired ratio, and (6) catalytic methanation of the carbon monoxide/hydrogen mixture to form methane.

While pretreatment of the biomass feedstock for introduction into the gasifier is often considered to be a physical process in which the feedstock is prepared for gasifier – typically as pellets or finely ground feedstock – there are chemical aspects that must also be considered (Chapter 2, Chapter 4). Co-gasification of other feedstocks, such as coal and with the residuum or coke from crude oil refining (often referred to as petcoke) offers a bridge between the depletion of crude oil stocks when coal, residua, and/or petcoke are used (Speight, 2014b).

The high reactivity of biomass and the accompanying high production of volatile products suggest that some synergetic effects might occur in simultaneous thermochemical treatment of coal, residua, or petcoke, depending on the gasification conditions such as: (1) feedstock type and origin, (2) reactor type, and (3) process parameters (Penrose et al., 1999; Gray and Tomlinson, 2000; McLendon et al., 2004; Lapuerta et al., 2008; Fermoso et al., 2009; Shen et al., 2012; Khosravi and Khadse, 2013; Speight, 2013a, 2014a, 2014b; Luque and Speight, 2015).

TABLE 5.10
Characteristics of the Different Types of Gasifiers

Gasifier Type	Fuel Properties
Fixed/moving bed	Particle size: 1 to 10 cm
	Mechanically stable fuel particles
	Unblocked passage of gas through the bed
	Pellets or briquettes preferred
	Updraft configuration more tolerant to biomass moisture content
Fluidized bed	Drying occurs as biomass moves down the gasifier
	Ash melting temperature of fuel: higher limit for operating temperature
	Fuel particle size relatively small to ensure good contact with bed material; typically <40 mm
	Good fuel flexibility due to high thermal inertia of the bed
Entrained bed	Fuel particle size: <50 micrometers
	Pulverized for high fuel conversion in short residence times
	Low moisture content
	Ash melting behavior can influence reactor/process design

For example, carbonaceous fuels are gasified in reactors, a variety of gasifiers such as the fixed- or moving-bed, fluidized-bed, entrained-flow, and molten bath gasifiers have been developed that have differing feedstock requirements (Table 5.10) (Beenackers, 1999; Shen et al., 2012; Speight, 2014b; Luque and Speight, 2015).

5.5.2 Reactions

The gasification process is not a single step process, but involves a variety of reactions (Table 5.7), and like any gasification process, the gasification of biomass (including bio-oil and biochar) is a complex thermal process that depends on the pyrolysis mechanism to generate gaseous precursors, which in the presence of reactive gases such as oxygen and steam convert the majority of the biomass into a synthesis gas, which is mainly composed of carbon monoxide and hydrogen, and used to produce chemicals and liquid fuels over catalysts (Speight, 2016, 2017; 2019a).

If air is used for combustion, the product gas will have a heat content of approximately 150–300 Btu/ft^3 (depending on process design characteristics) and will contain undesirable constituents such as carbon dioxide, hydrogen sulfide, and nitrogen. The use of pure oxygen, although expensive, results in a product gas having a heat content on the order of 300 to 400 Btu/ft^3 with carbon dioxide and hydrogen sulfide as by-products (both of which can be removed from low or medium heat-content, low-Btu or medium-Btu) gas by any of several available processes (Speight, 2013a, 2014a, 2017.

If a high heat-content (high-Btu) gas (900 to 1,000 Btu/ft^3) is required, efforts must be made to increase the methane content of the gas. The reactions which generate methane are all exothermic and have negative values, but the reaction rates are relatively slow, and catalysts may, therefore, be necessary to accelerate the reactions to acceptable commercial rates. Indeed, the overall reactivity of the feedstock and char may be subject to catalytic effects insofar as the mineral constituents of the feedstock (such as the mineral matter in biomass) may modify the reactivity by a direct catalytic effect (Davidson, 1983; Baker and Rodriguez, 1990; Mims, 1991; Martinez-Alonso and Tascon, 1991).

Biomass begins to rapidly decompose with heat once its temperature rises above 240°C (465°F) and the biomass breaks down into a combination of gases, liquids, and solid (char). In the process,

TABLE 5.11
Biomass Properties that Influence the Gasification Process

Chemical Properties	Elemental Composition	C,H,N,O,S (% w/w)
	Proximate analysis	Moisture
		Fixed carbon
		Volatile matter
		Mineral matter
Physical Properties	Density	
	Porosity	
	Particle size	
	Particle shape	
Thermal Properties	Heating value	
	Elemental composition	
	Types of minerals	Mineral ash
		Ash fusion temperature

the feedstock undergoes three steps in the conversation to gas – the first two steps, pyrolysis and combustion, occur very rapidly – all of which are highly dependent upon the properties of the biomass (Table 5.11). However, there is also a preliminary step in which any moisture in the biomass before it enters the pyrolysis step. All of the moisture needs to be (or will be) removed from the biomass at some point (typically between 100 and 150°C, 212 and 300°F) in the higher temp processes. Moisture removal (often referred to as drying) is one of the major issues that has to be solved for successful gasification of biomass.

In the pyrolysis step, char is produced as the feedstock heats up and volatiles are released. Finally, in the gasification step, the feedstock char reacts with steam to produce hydrogen (H_2) and carbon monoxide (CO).

Drying:

$$\text{Biomass (wet)} \rightarrow \text{biomass (dry)} + H_2O$$

Pyrolysis:

$$\text{Biomass} \rightarrow \text{volatiles} + \text{char}$$

Combustion:

$$2C_{\text{volatiles + char}} + O_2 \rightarrow 2CO + H_2O$$

Gasification:

$$C_{\text{char}} + H_2O \rightarrow H_2 + CO$$

$$CO + H_2O \rightarrow H_2 + CO_2$$

Thus, in the initial stages of gasification, the rising temperature of the feedstock initiates devolatilization and the rupture of the weaker chemical bonds to yield volatile tar, volatile oil, phenol derivatives, and hydrocarbon gases. These products generally react further in the gaseous phase to form hydrogen, carbon monoxide, and carbon dioxide. The char (fixed carbon) that remains after devolatilization reacts with oxygen, steam, carbon dioxide, and hydrogen.

In essence, the direction of the gasification process is subject to the constraints of thermodynamic equilibrium and variable reaction kinetics. The combustion reactions (reaction of the feedstock or char with oxygen) essentially go to completion. The thermodynamic equilibrium of the rest of the gasification reactions are relatively well defined and collectively have a major influence on thermal efficiency of the process as well as on the gas composition.

In fact, very little reliable kinetic information (for general application) on gasification reactions exists, partly because it is highly dependent on (1) the chemical nature of the feed, which varies significantly with respect to composition, mineral impurities, (2) feedstock reactivity, and (3) process parameters, such as temperature, pressure, and residence time. In addition, physical characteristics of the feedstock (or char) also play a role in phenomena such boundary layer diffusion, pore diffusion and ash layer diffusion which also influence the kinetic outcome.

Alkali metal salts are known to catalyze the steam gasification reaction of carbonaceous materials, including biomass – hence the need for pretreatment of biomass to remove such metals (Chapter ?).

Cesium (Cs) > rubidium (Rb) > potassium (K) > sodium (Na) > lithium (Li)

Ruthenium-containing catalysts are used primarily in the production of ammonia. It has been shown that ruthenium catalysts provide five to 10 times higher reactivity rates than other catalysts.

The use of catalysts can enhance or suppress the formation of certain components* in the gaseous product by changing the chemistry of the reaction, the rate of reaction, and the thermodynamic balance of the reaction. For example, in the production of synthesis gas (mixtures of hydrogen and carbon monoxide), methane is also produced in small amounts.

5.6 PRODUCTS

Hydrogen can be produced from biomass and carbonaceous waste by means of thermochemical conversion processes such as pyrolysis, gasification, steam gasification, steam-reforming, supercritical water gasification (SCWG) of the feedstock (Collot et al., 1999; Demirbaş, 2010, 2016). Furthermore, the dramatic increase in the price of crude oil and crude oil products has raised increasing concerns regarding the environmental impact of fossil fuels and fossil fuel products, especially related to greenhouse gas emissions are forcing the search for new energy sources and alternative ways(Balat, 2009; Sgar et al., 2009).

Biomass has the advantage of low environmental impact compared with that for fossil fuels. The price of hydrogen obtained by direct gasification of lignocellulosic biomass, however, is typically higher (but prices do vary) than that for hydrogen produced by steam reforming of natural gas (Spath et al., 2000; Veziroglu and Sahin, 2008).

Hydrogen production from biomass, like hydrogen production from fossil fuel sources, involves conversion technologies such as gasification and pyrolysis which provide a synthesis gas, mainly consisting of hydrogen and carbon monoxide that can be subjected to downstream processes in order to produce pure hydrogen. The processes for biomass conversion are conveniently categorized as (1) thermo-chemical processes and (2) biochemical processes.

The thermo-chemical processes involve either gasification or pyrolysis (heating biomass in the absence of oxygen) to produce a hydrogen-rich stream of gas (synthesis gas which is a blend of hydrogen and carbon monoxide) and a broad range of biomass types can be used as the feedstocks for this type of process. On the other hand, enzyme-based biochemical processes (often referred to as 'digester type processes') use wet, sugar-based feedstocks but could include cellulosic feedstocks.

Also, hydrothermal gasification in supercritical water medium has become a promising technique to produce hydrogen from biomass with high efficiency (Kong et al., 2008; Demirbaş, 2008a; Balat, 2008; Ishida et al., 2009; Demirbaş, 2010, 2016). This process can be applied to the conversion of biomass with high moisture content without drying (Ishida et al., 2009). Main thermochemical processes are pyrolysis and gasification, based on the following pyrolysis reactions (Demirbaş, 2007, 2008a, 2008b):

$$\text{Biomass} \rightarrow H_2 + CO + CO_2 + \text{hydrocarbon gases} + \text{tar} + \text{char}$$

$$\text{Biomass} + H_2O \rightarrow H_2 + CO_2 + CO$$

5.6.1 Gaseous Products

In general, the total hydrogen-rich gaseous products increased with increasing pyrolysis temperature for the biomass sample (Demirbaş, 2001; 2002a). The pyrolysis yields of hydrogen plus paraffin derivatives obtained from biomass samples by using different percentages of additives at different temperatures were investigated (Caglar and Demirbaş, 2002a; 2002b). The chemicals ($ZnCl_2$, Na_2CO_3, and K_2CO_3) can be employed as additives in the process. The yield of hydrogen and the paraffin derivatives (mainly methane, CH_4) from the biomass samples generally increased with increasing the pyrolysis temperature and percent of additive (Demirbaş, 1998, 2002a; Hao et al., 2003, 2005; Byrd et al., 2007; Lu et al., 2007, 2008; Santos et al., 2008; Zhang, 2013a; Ren et al., 2014).

Steam reforming is the major process for the production of hydrogen presently. However, this process has several disadvantages. Hence to meet the increasing demand for this future fuel, alternatives to reforming processes are essential (Wang et al., 1997; Lan et al., 2012; Zhang et al., 2013b; Ren et al., 2014).

Comparison with other biomass thermochemical gasification such as air gasification or steam gasification, the supercritical water gasification (SCWG) can directly deal with the wet biomass without drying and have high gasification efficiency in relatively lower temperature (Demirbaş, 2004; Demirbaş and Arin, 2004; Demirbaş, 2005a, 2005b; Lu et al, 2008; Speight, 2013a).

There are many useful products from the gasification of biomass, which include: (1) synthesis gas, which can be further processed by means of the Fischer-Tropsch process into methanol, dimethyl ether, and other chemical feedstocks (an important aspect in the present context related to the production of hydrogen) (2) heat, (3) power, (4) biofuels, and (5) biochar. Generally, biomass feedstocks are classified into four main groups: woody biomass, herbaceous biomass, marine biomass, and manure (Sikarwar et al., 2016). The gasifier is usually designed to generate a specific product but the feedstock material is an important parameter to specify and optimize where possible (Luque and Speight, 2015; Sikarwar et al., 2016, 2017).

Large scale gasifiers employ one of two types of fluidized-bed configurations: (1) bubbling fluidized bed and circulating fluidized bed or (2) a combination of both indirectly fired to maintain the bed temperature below the ash fusion temperature of the ash produced from the biomass.

A bubbling fluidized-bed gasifier (Chapter 3) consists of fine, inert particles of sand or alumina, which are selected based on their suitability of physical properties such as size, density and thermal characteristics. The dimension of the bed at some height above the distributor plate is increased to reduce the superficial gas velocity below the fluidization velocity to maintain inventory of solids and to act as a disengaging zone. A cyclone is used to trap the smaller size particle that exit the fluidized bed to either return fines to the bed or to remove ash-rich fines from the system.

The feedstock is introduced either through a feed chute to the top of the bed or deep inside the bed. The deeper introduction of the feedstock into the bed of inert solids provides sufficient residence time for fines that would otherwise be entrained in the fluidizing gas. The biomass organic constituents pyrolytically vaporize and are partially combusted in the bed. The exothermic combustion provides the heat to maintain the bed at temperature and to volatilize additional biomass. The bed needs to preheated to the start-up temperature using hydrocarbon resources such as natural gas, fuel oil, either by direct firing or by indirect heating.

After the bed reaches the ignition temperature, the feedstock is slowly introduced into the bed to raise the bed temperature to the desired operating temperature which are normally in the range of 700 to 900°C (1,290 to 1,650°F). The bed temperature is governed by the desire to obtain complete devolatilization versus the need to maintain the bed temperature below the ash fusion temperature of any mineral matter in the feedstock.

The advantages of the fluidized-bed gasifiers are that they (1) yield a uniform product gas, (2) have the ability to accept a wide range of fuel particles sizes, including fines, (3) have a nearly uniform temperature distribution throughout the reactor, and (4) provide a high rate of heat transfer between inert material and biomass, aiding high conversion, with low tar production. The disadvantages are the formation of large bubbles at higher gas velocities, which bypass the bed reducing the high rate of heat/mass transfer significantly.

Autothermal gasifiers provide the necessary heat of reaction by means of partial oxidation within the gasification reactor. Thus, for synthesis gas production either pure oxygen (in entrained flow reactors) or mixtures of oxygen and steam (in fluidized-bed reactors) are used as gasification agent. The advantage of autothermal gasification is the direct heating of the reactants and therefore more efficient energy utilization.

Allothermal (or indirect) gasification is characterized by the separation of the processes of heat production and heat consumption which consists of two reactors, connected by an energy flow (Pfeifer et al., 2009). Allothermal gasifiers generally produce two gas streams: a medium calorific product gas (gasification reactor) with little or no nitrogen and a flue gas (combustion reactor). The production of a nitrogen-free gas without the need of pure oxygen is one of the advantages over autothermal gasification processes. Another advantage is the complete carbon conversion and that there is no problematic waste produced. All carbon-containing streams from the product gas cleaning (such as dust and tar removal) can be recycled to the combustion zone and converted to heat, which is used for the gasification reactions (Tijmensen et al., 2002).

In a fixed-bed–fluidized-bed-entrained flow gasifier (Chapter 3), the differences in the design of the cause differences in composition of the gas which is also influenced other factors. Including feedstock composition, reactor type, and operating parameters (temperature, pressure, oxygen fuel ratio). During oxygen gasification, the combustion products (carbon dioxide and water) are in the product gas and take part in the chemical reactions, mainly in the water-gas shift reaction. On the other hand, higher amount of hydrogen in the product gas can be found during the steam gasification. The hydrogen that occurs in the product gas does not originate only from the fuel as in the case of oxygen gasification but also from the steam. The tar content and composition depend not only on the type of gasifier but also on the composition of the feedstocks as well as on the temperature in the reactor (Table 5.12). In entrained flow gasification, where the temperature is above 1,000°C (1,830°F), there is no tar produced, whereas in fluidized-bed gasification, where the temperature is below 1,000°C (1,830°F), tar is produced and has to be removed from the product gas.

The products of the gasification of biomass are varied insofar as the gas composition varies with the system employed (Speight, 2013a). It cannot be over-emphasized that the gas product must be first freed from any pollutants such as particulate matter and sulfur compounds before further use, particularly when the intended use is a water gas shift or methanation (Cusumano et al., 1978; Probstein and Hicks, 1990).

In all cases, the gas exiting the gasifier contains impurities that need to be removed for the advanced use of biomass gasification gas. For example, if the gas is to be used in a Fischer-Tropsch process, the gas must be ultra-clean. Any particulate matter in the gas exiting the gasifier can be removed by a hot gas filter, whereas the tar constituents and low-boiling hydrocarbon derivatives can be converted by reforming to synthesis gas. For tar removal, another possibility is scrubbing by an organic solvent.

However, the gas stream will also, more than likely, contain other impurities, such as hydrogen sulfide, carbonyl sulfide, ammonia, and alkali metals. The sulfur constituents will further add complications to a gas reforming process by poisoning the catalyst. In addition to hydrocarbon derivatives, ammonia is partly converted to nitrogen in the reformer. The alkali metals, in turn, will

TABLE 5.12
General Classification of Tars

Category	Formation Temperature	Constituents
Primary	400–600°C	Mixed oxygenates
	750--1,110°F	Phenolic ethers
Secondary	600–800°C	Alkyl phenols
	1,110–1,470°F	Heterocyclic ethers
Tertiary	800–1,000°C	Polynuclear aromatic hydrocarbons
	1,470–1,830°F	Phenolic ethers

condense in hot gas filtration taking place at around 500°C (930°F). The final conditioning of gas for synthesis purposes can be realized by conventional commercial technologies for acid gas removal, such as the Rectisol process, and the H2/CO ratio can be adjusted by a water gas shift reactor.

The desirable product from the gasification of carbonaceous feedstocks is synthesis gas is one of the key products of the gasification of biomass and is a vital source of hydrogen (Chadeesingh, 2011). Employing the Fischer-Tropsch process (Chapter 12), naphtha (for the production of gasoline), kerosene, and middle distillates (for the production of diesel fuel and other chemicals) can be manufactured (Chadeesingh, 2011; Speight, 2019a). Hydrogen is produced from synthesis gas by means of the water gas shift reaction (Chapter 11), which has numerous applications including in fuel cells. synthetic natural gas (SNG, CH_4) is also one of the significant uses of synthesis gas.

Contaminants such as particulates, tars, nitrogenous compounds such as ammonia (NH_3) and hydrogen cyanide (HCN), sulfur-containing inorganic compounds such as hydrogen sulfide (H_2S), carbonyl sulfide (COS), and carbon disulfide (CS_2). Halogens such as hydrogen chloride and traces of metals such as sodium (Na) and potassium (K) are present in varying quantities in synthesis gas produced from gasification.

The type of feedstock, operational conditions, and the gasifier type are the variables which determine the amount of tar produced. These contaminants in synthesis gas pose numerous technical and working problems and must be removed (Chapter 9). For example, hydrogen sulfide is responsible for equipment corrosion, tar causes fouling, and catalyst deactivation occurs due to the presence of tar, hydrogen sulfide, ammonia, hydrogen chloride, and trace metals.

Synthesis gas produced from biomass sources is comparable in combustion efficiency to natural gas (Chadeesingh, 2011; Speight, 2020) which reduces the emissions of sulfur, nitrogen oxides, and mercury, resulting in a much cleaner fuel (Nordstrand et al., 2008; Lee et al., 2006; Sondreal et al., 2004, 2006; Yang et al., 2007; Wang et al., 2008). The resulting hydrogen gas can be used for electricity generation or as a transport fuel or in any aspects of refining that requires hydrogen (Speight, 2016). The gasification process also facilitates capture of carbon dioxide emissions from the combustion effluent.

Synthesis gas is mainly used as an intermediary building block for the final production (synthesis) of various fuels such as synthetic natural gas, methanol, and synthetic crude oil (including the production of dimethyl ether, gasoline, and diesel fuel) (Chadeesingh, 2011; Speight, 2013a). At this point, and to dismiss any confusion that may arise, synthesis gas, as generated from biomass, is not the same as synthesis gas. Synthesis gas is a clean and renewable form of energy generated from biomass that could very well substitute for conventional sources of energy. The gas is generally composed of methane (55 to 65%), carbon dioxide (35 to 45%), nitrogen (0 to 3%), hydrogen (0 to 1%), and hydrogen sulfide (0 to 1%).

The use of bio-synthesis gas offers the opportunity to furnish a broad range of environmentally clean fuels and chemicals and there has been steady growth in the traditional uses of synthesis gas. In fact, the major use of synthesis gas is in the manufacture of hydrogen for a growing number of purposes, especially in crude oil refineries (Speight, 2014a, 2016).

The hydroformylation of olefin derivatives (the Oxo reaction), a completely chemical use of synthesis gas, is the fourth largest use of carbon monoxide and hydrogen mixtures. A direct application of synthesis gas as fuel (and eventually also for chemicals) that promises to increase is its use for *integrated gasification combined cycle* (IGCC) units for the generation of electricity (and also chemicals), crude oil coke or viscous feedstocks (Holt, 2001). Finally, synthesis gas is the principal source of carbon monoxide, which is used in an expanding list of carbonylation reactions, which are of major industrial interest.

While synthesis gas is the prime product for the production of hydrogen, other products include (1) Low-Btu gas with a heating value on the order of <300 Btu/ft,[3] (2) medium-Btu gas, with a heating value in the range on the order of 300 to 550 Btu/ft,[3] and (3) high-Btu gas which is essentially pure methane and often referred to as 'synthetic natural gas' or 'substitute natural gas' (SNG) (Speight, 1990, 2013a).

The primary gases of interest in medium heat-content gas are hydrogen and carbon monoxide and the gas is considerably more versatile than low heat-content gas; like low heat-content gas, medium heat-content gas may be used directly as a fuel to raise steam, or used through a combined power cycle to drive a gas turbine, with the hot exhaust gases employed to raise steam, but medium heat-content gas is especially amenable to synthesize methane (by methanation), higher hydrocarbon derivatives (by Fischer-Tropsch synthesis), methanol, and a variety of synthetic chemicals, and to produce hydrogen.

In medium heat-content gas, the H_2/CO ratio varies from 2:3 C to 3:1 and the increased heating value correlates with higher methane and hydrogen contents as well as with lower carbon dioxide contents. Furthermore, the very nature of the gasification process used to produce the medium heat-content gas has a marked effect upon the ease of subsequent processing. For example, the CO_2-acceptor product is amenable to use for methane production because it has (1) the desired H_2/CO ratio just exceeding 3:1, (2) an initially high methane content, and (3) relatively low water and carbon dioxide contents. Other gases may require appreciable shift reaction and removal of large quantities of water and carbon dioxide prior to methanation.

5.6.2 Liquid Products

The production of liquid fuels from a carbonaceous feedstock via gasification is often referred to as the indirect liquefaction of the feedstock (Speight, 2013a, 2014a, 2017.

The synthesis of hydrocarbon derivatives from carbon monoxide and hydrogen (synthesis gas) (the Fischer-Tropsch synthesis) is a procedure for the indirect liquefaction of various carbonaceous feedstocks (Speight, 2011a, 2011b). This process is the only liquefaction scheme currently in use on a relatively large commercial scale (for the production of liquid fuels from coal using the Fischer-Tropsch process (Singh, 1981).

Synthesis gas is produced from the feedstock is converted to gaseous products at temperatures in excess of 800°C (1,470°F), and at moderate pressures:

$$[C]_{feedstock} + H_2O \rightarrow CO + H_2$$

The hydrogen/carbon monoxide ratio is typically on the order of 2/2:1 or 2/5:1, since up to three volumes of hydrogen may be required to achieve the next stage of the liquids production, the synthesis gas must then be converted by means of the water-gas shift reaction) to the desired level of hydrogen:

$$CO + H_2O \rightarrow CO_2 + H_2$$

After this, the gaseous mix is purified and converted to a wide variety of hydrocarbon derivatives:

$$nCO + (2n + 1)H_2 \rightarrow C_nH_{2n+2} + nH_2O$$

Feedstock*	Gasification	Synthesis gas	Water gas shift	Hydrogen
	Pyrolysis	Gas Reforming	Water Gas Shift	Hydrogen

FIGURE 5.1 Representation of the Processes Involved in the Production of Hydrogen.

5.6.3 Solid Products

The solid product (solid waste) of a biomass gasification process is typically ash which is the oxides of metal-containing constituents of the feedstock. The amount and type of solid waste produced is very much feedstock dependent.

At the high temperature of the gasifier, most of the mineral matter of the feedstock is transformed and melted into slag, an inert glass-like material, and, under such conditions, non-volatile metals and mineral compounds are bound together in molten form until the slag is cooled in a water bath at the bottom of the gasifier, or by natural heat loss at the bottom of an entrained-bed gasifier. Slag production is a function of mineral matter content of the feedstock – coal produces much more slag per unit weight than crude oil coke. Furthermore, as long as the operating temperature is above the fusion temperature of the ash, slag will be produced. The physical structure of the slag is sensitive to changes in operating temperature and pressure of the gasifier and a quick physical examination of the appearance of the slag can often be an indication of the efficiency of the conversion of feedstock carbon to gaseous product in the process.

Slag is comprised of black, glassy, silica-based materials and is also known as *frit*, which is a high density, vitreous, and abrasive material low in carbon and formed in various shapes from jagged and irregular pieces to rod and needle-like forms. Depending upon the gasifier process parameters and the feedstock properties, there may also be residual carbon char. Vitreous slag is much preferable to ash, because of its habit of encapsulating toxic constituents (such as heavy metals) into a stable, non-leachable material. Leachability data obtained from different gasifiers unequivocally shows that gasifier slag is highly non-leachable and can be classified as non-hazardous. Because of its particular properties and non-hazardous, non-toxic nature, slag is relatively easily marketed as a by-product for multiple advantageous uses, which may negate the need for its long-term disposal.

The physical and chemical properties of gasification slag are related to (1) the composition of the feedstock, (2) the method of recovering the molten ash from the gasifier, and (3) the proportion of devolatilized carbon particles (char) discharged with the slag. The rapid water-quench method of cooling the molten slag inhibits recrystallization, and results in the formation of a granular, amorphous material. Some of the differences in the properties of the slag may be attributed to the specific design and operating conditions prevailing in the gasifiers.

Char can be recycled back into the gasifier to increase carbon usage and in terms of recycling char to the gasifier, a property that is important to fluidization is the effective particle density.

5.7 THE FUTURE

Currently, in gasification plants designed for the gasification of biomass and waste, clean gas is produced after filtration and scrubbing, limiting its applications. However, if the tar separation is ineffective, the yield and quality of the gas quality is not adequate for the planned use insofar as the gas may be unsuitable for applications where high levels of purity are essential.

An example of a novel hot gas cleanup process is the use of plasma torches to crack tars; this differs from plasma gasification where the plasma is used for energy generation by gasifying biomass, municipal solid waste, and refuse derived fuel.

The future of synthesis gas production by gasification of biomass depends very much on the effect of gasification processes on the surrounding environment. It is these environmental effects and issues that will direct the success of gasification. In fact, there is the distinct possibility that within the foreseeable future the gasification process will increase in popularity in crude oil refineries – some refineries may even be known as gasification refineries (Speight, 2011b). A gasification refinery would have, as the center-piece of the refinery, gasification technology as is the case of the Sasol refinery in South Africa (Couvaras, 1997). The refinery would produce synthesis gas (from the carbonaceous feedstock) from which liquid fuels would be manufactured using the Fischer-Tropsch synthesis technology.

Inorganic components of the feedstock, such as metals and minerals, are trapped in an inert and environmentally safe form as char, which may have use as a fertilizer.

The manufacture of gas mixtures of carbon monoxide and hydrogen has been an important part of chemical technology for more than 100 years. Originally, such mixtures were obtained by the reaction of steam with incandescent coke and were known as 'water gas'.

Partial oxidation processes using oxygen instead of steam also found wide application for synthesis gas manufacture, with the special feature that they could utilize low-value feedstocks such as viscous crude oil residues. In recent years, catalytic partial oxidation employing very short reaction times (milliseconds) at high temperatures (850 to 1,000°C, 1,560 to 1,830°F) is providing still another approach to synthesis gas manufacture (Hickman and Schmidt, 1993).

The conversion of the gaseous products of gasification processes to synthesis gas, a mixture of hydrogen (H_2) and carbon monoxide (CO), in a ratio appropriate to the application, needs additional steps, after purification. The product gases – carbon monoxide, carbon dioxide, hydrogen, methane, and nitrogen – can be used as fuels or as raw materials for chemical or fertilizer manufacture.

Gasification by means other than the conventional methods has also received some attention and has provided rationale for future processes (Rabovitser et al., 2010). In the process, a carbonaceous material and at least one oxygen carrier are introduced into a non-thermal plasma reactor at a temperature in the range of approximately 300°C to approximately 700°C (570 to approximately 1,290°F) and a pressure in a range from atmospheric pressure to approximate 1,030 psi and a non-thermal plasma discharge is generated within the non-thermal plasma reactor. The high operating temperatures (typically above 1,800°C, 3,270°F) allow for the breaking down of all tars, char, and dioxin derivatives.

In the plasma reactor two torches are used together with a gas (such as oxygen, helium, or air) to generate the plasma. The torches which extend into the plasma furnace are fitted with graphite electrodes. An electric current is passed through the electrodes and an electric arc is generated between the tip of the electrodes and the conducting receiver which is the slag at the bottom of the furnace. Because of the electrical resistivity across the system significant heat is required to strip away electrons from the molecules of the gas introduced resulting in an ionized superheated gas stream or plasma. This gas exits the torch at temperatures up to 10,000°C (18,000°F). Such high temperatures are used to break down the waste primarily into elemental gas and solid waste (slag) as well as reduce dioxins, sulfur oxides and carbon dioxide emissions. Unlike waste incineration plasma gasification waste to energy technology has proven to have a benign impact on the environment.

The plasma furnace is the central component of the system where gasification and vitrification take place. Plasma torches are mounted at the bottom of the reactor, they provide high-temperature air (almost three times higher than traditional combustion temperatures) which allow for the gasification of the waste materials.

In the process, the carbonaceous feedstock and the oxygen carrier are exposed to the non-thermal plasma discharge, resulting in the formation of a product gas which comprises substantial amounts of hydrocarbon derivatives, such as methane, hydrogen, and/or carbon monoxide. The products of the process are a fuel or gas known as synthesis gas and an inert vitreous material known as slag.

Furthermore, gasification and conversion of carbonaceous solid fuels to synthesis gas for application of power, liquid fuels, and chemicals is practiced worldwide. Crude oil coke, coal, biomass, and refinery waste are major feedstocks for an on-site refinery gasification unit. The concept of blending of a variety of carbonaceous feedstocks (such as coal, biomass, or refinery waste) with a viscous feedstock of the coke from the thermal processing of the viscous feedstock is advantageous in order to obtain the highest value of products as compared to gasification of crude oil coke alone. Furthermore, based on gasifier type, co-gasification of carbonaceous feedstocks can be an advantageous and efficient process. In addition, the variety of upgrading and delivery options that are available for application to synthesis gas enable the establishment of an integrated energy supply system whereby synthesis gases can be upgraded, integrated, and delivered to a distributed network of energy conversion facilities, including power, combined heat and power, and combined cooling, heating and power (sometime referred to as *tri-generation*) as well as used as fuels for transportation applications.

As a final note, the production of chemicals from biomass is based on thermochemical conversion routes which are, in turn, based on biomass gasification (Roddy and Manson-Whitton, 2012). The products are (1) a gas, which is the desired product and (2) a solid ash residue with a composition depends on the type of biomass. Continuous gasification processes for various feedstocks have been under development since the early 1930s. Ideally, the gas produced would be a mixture of hydrogen and carbon monoxide, but, in practice, it also contains methane, carbon dioxide, and a range of contaminants.

A variety of gasification technologies is available across a range of sizes, from small updraft and downdraft gasifiers through a range of fluidized-bed gasifiers at an intermediate scale and on to larger entrained flow and plasma gasifiers (Brigwater, 2003a, 2003b; Roddy and Manson-Whitton, 2012). In an updraft gasifier, the oxidant is blown up through the fixed gasifier bed with the synthesis gas exiting at the top whereas in a downdraft gasifier, the oxidant is blown through the reactor in a downward direction with the synthesis gas exiting at the bottom.

Gasification processes tend to operate either above the ash melting temperature (typically 1,200°C, 2,190°F) or below the ash melting temperature (typically >1,000°C, >1,830°F). In the higher temperature processes, there is little methane or tar formation. The question of which gasification technology is the most appropriate depends on whether the priority is to (1) produce a very pure bio-synthesis gas, (2) accommodate a wide range of feedstock types, (3) avoid pre-processing of biomass, or (4) operate at a large scale and produce chemical products from a variety of feedstocks.

The introduction of steam as a reactant influent during gasification has been shown to enhance the production of hydrogen during the gasification of a variety of fuels that include coal, biomass, and municipal solid waste (Minkova et al., 2000; Senneca, 2007. Introduction of carbon dioxide will enhance the production of carbon monoxide during high-temperature steam gasification while depressing the production of hydrogen and methane. Both the steam and carbon dioxide have been shown to increase the char reactivity (Ye et al., 1998; Ochoa et al., 2001; Demirbaş, 2006) through a modification of the char pore structure and surface activity. Biomass gasification using steam can result in an increased concentration of hydrogen in the synthesis gas with higher concentrations corresponding to higher gasification temperatures in the 700 to 1,100°C (1,290 to 2,110°F) range. If a methane fuel stream is the preferred product, careful monitoring of the process to operate below 600°C (1,110°F) would permit optimizing the production of methane while minimizing the production of hydrogen. A general characteristic of the fuel stream produced through gasification is a clean stream with a minimum of tar, soot and particulates. Biomass gasification results in a producer gas whose main components are carbon monoxide, hydrogen, and methane. Other volatile components include carbon dioxide, acetaldehyde (CH_3CHO), acetic acid (CH_3CO_2H), phenol (C_6H_5OH), formaldehyde ($HCHO$), formic acid (HCO_2H), and acetone (CH_3COCH_3). If the gasification process is performed under high levels of nitrogen dilution, many of these species may not be detectable since they are present in the parts per million range (Butterman and Castaldi, 2008).

REFERENCES

Adhikari, U., Eikeland, M.S., and Halvorsen, B.M. 2015. Gasification of Biomass for Production of Syngas for Biofuel. Proceedings. 56th SIMS. Linköping, Sweden. October 7–9. Page 255–260. http://www.ep.liu.se/ecp/119/025/ecp15119025.pdf

Ahmad, A.A., Zawawi, N.A., Kasim, F.H., Inayat, A., and Khasri, A. 2016. Assessing the Gasification Performance of Biomass: A Review on Biomass Gasification Process Conditions, Optimization and Economic Evaluation. *Renewable and Sustainable Energy Reviews*, 53: 1333–1347.

Arena, U. 2012. Process and Technological Aspects of Municipal Solid Waste Gasification. A Review. *Waste Management*, 32(4): 625–639.

Asadullah, M. 2014. Barriers of Commercial Power Generation Using Biomass Gasification Gas: A Review. *Renewable and Sustainable Energy Reviews*, 29: 201–215.

Baker, R.T.K., and Rodriguez, N.M. 1990. *Fuel Science and Technology Handbook*. Marcel Dekker Inc., New York. Chapter 22.

Balat, M. 2008. Hydrogen-Rich Gas Production from Biomass via Pyrolysis and Gasification Processes and Effects of Catalyst on Hydrogen Yield. *Energy Sources Part A*, 30(6): 552–564.

Balat, M. 2009. Possible Methods for Hydrogen Production. *Energy Sources Part A*, 31(1): 39–50.

Balat, M. 2011. Fuels from Biomass – An Overview. In *The Biofuels Handbook*. J.G. Speight (Editor). Royal Society of Chemistry, London. Part 1, Chapter 3.

Basu, P. 2013. *Biomass Gasification, Pyrolysis and Torrefaction*, 2nd Edition. Practical Design and Theory. Academic Press, Inc., New York.

Baxter, L. 2005. Biomass-Coal Co-combustion: Opportunity for Affordable Renewable Energy. *Fuel*, 84(10): 1295–1302.

Beenackers, A.A.C.M. 1999. Biomass Gasification in Moving Beds. A Review of European Technologies. *Renewable Energy*, 16(1–4): 1180–1186.

Bhattacharya, S., Md. Mizanur Rahman Siddique, A.H., and Pham, H.-L. 1999. A Study in Wood Gasification on Low Tar Production. *Energy*, 24: 285–296.

Biermann, C.J. 1993. *Essentials of Pulping and Papermaking*. Academic Press Inc., New York.

Boateng, A.A., Walawender, W.P., Fan, L.T., and Chee, C.S. 1992. Fluidized-Bed Steam Gasification of Rice Hull. *Bioresource Technology*, 40(3): 235–239.

Brage, C., Yu, Q., Chen, G., and Sjöström, K. 2000. Tar Evolution Profiles Obtained from Gasification of Biomass and Coal. *Biomass and Bioenergy*, 18(1): 87–91.

Brar, J.S., Singh, K., Wang, J., and Kumar, S. 2012. Cogasification of Coal and Biomass: A Review. *International Journal of Forestry Research*: 1–10.

Brigwater, A.V. 2003a. Renewable Fuels and Chemicals by Thermal Processing of Biomass. *Chemical Engineering Journal*, 91(2–3): 87–102.

Brigwater, A.V. (Editor) 2003b. *Pyrolysis and Gasification of Biomass and Waste*. CPL Press, Newbury, Berkshire.

Butterman, H.C., and Castaldi, M.J. 2008. CO2 Enhanced Steam Gasification of Biomass Fuels. Proceedings. NAWTEC16. 16th Annual North American Waste-to-Energy Conference. Philadelphia, Pennsylvania. May 19–21. https://asmedigitalcollection.asme.org/NAWTEC/proceedings-abstract/NAWTEC16/42932/157/328514.

Byrd, A.J., Pant, K.K., and Gupta, R.B. 2007. Hydrogen Production from Glucose Using Ru/A12 O3 Catalyst in Supercritical Water. *Industrial and Engineering Chemistry Research*, 46(11): 3574–3579.

Caglar, A., and Demirbaş, A. 2002a. Hydrogen-Rich Gas Mixture from Olive Husk via Pyrolysis. *Energy Conversion and Management*, 43(1): 109–117.

Caglar, A., and Demirbaş, A. 2002b. Conversion Cotton Cocoon Shell to Hydrogen-Rich Gaseous Products by Pyrolysis. *Energy Conversion and Management*, 43(4): 489–497.

Chadeesingh, R. 2011. The Fischer-Tropsch Process. In *The Biofuels Handbook*. J.G. Speight (Editor). The Royal Society of Chemistry, London. Part 3, Chapter 5, Page 476–517.

Chen, G., Sjöström, K., and Bjornbom, E. 1992. Pyrolysis/Gasification of Wood in a Pressurized Fluidized Bed Reactor. *Industrial and Engineering Chemistry Research*, 31(12): 2764–2768.

Collot, A.G., Zhuo, Y., Dugwell, D.R., and Kandiyoti, R. 1999. Co-pyrolysis and Cogasification of Coal and Biomass in Bench-Scale Fixed-Bed and Fluidized Bed Reactors. *Fuel*, 78(6): 667–679.

Conte, M., Iacobazzi, A., Ronchetti, M., and Vellone, R. Different Hydrogen Production Methods. *Journal of Power Sources*, 100: 171–187.

Couvaras, G. 1997. Sasol's Slurry Phase Distillate Process and Future Applications. Proceedings. Monetizing Stranded Gas Reserves Conference, Houston. December.

Cusumano, J.A., Dalla Betta, R.A., and Levy, R.B. 1978. *Catalysis in Coal Conversion*. Academic Press Inc., New York.

Czernik, S., French, R., Feik, C., and Chornet, E. 2000. Production of Hydrogen from Biomass-Derived Liquids. Proceedings. DOE Hydrogen Program Review. NREL/CP-570-28890. https://www1.eere.energy.gov/hydrogenandfuelcells/pdfs/28890a.pdf.

Dasappa, S., Paul, P.J., Mukunda, H.S., Rajan, N.K.S., Sridhar, G., and Sridhar, H.V. 2004. Biomass Gasification Technology – A Route to Meet Energy Needs. *Current Science*, 87(7): 908–916.

Davidson, R.M. 1983. Mineral Effects in Coal Conversion. Report No. ICTIS/TR22. International Energy Agency, London.

Demirbaş, A. 1998. Yields of Oil Products from Thermochemical Biomass Conversion Processes. *Energy Conversion and Management*, 39(7): 685–690.

Demirbaş, A. 2001. Yields of Hydrogen and Gaseous Products via Pyrolysis from Selected Biomass Samples. *Fuel*, 80(13): 1885–1891.

Demirbaş, A. 2002a. Hydrogen Production from Biomass by Gasification Process. *Energy Sources*, 24(1): 59–68.

Demirbaş, A. 2002b. Gaseous Products from Biomass by Pyrolysis and Gasification: Effects of Catalyst on Hydrogen Yield. *Energy Conversion and Management*, 43(7): 897–909.

Demirbaş, A. 2004. Effects of Temperature and Particle Size on Bio-char Yield from Pyrolysis of Agricultural Residues. *Journal of Analytical and Applied Pyrolysis*, 72(2): 243–248.

Demirbaş, A., and Arın, G. 2004. Hydrogen from Biomass via Pyrolysis: Relationships Between Yield of Hydrogen and Temperature. *Energy Sources*, 26(11): 1061–1069.

Demirbaş, A. 2005a. Hydrogen Production via Pyrolytic Degradation of Agricultural Residues. *Energy Sources*, 27(8): 769–775.

Demirbaş, A. 2005b. Thermochemical Conversion of Hazelnut Shell to Gaseous Products for Production of Hydrogen. *Energy Sources*, 27(4): 339–347.

Demirbaş, A. 2006. Production and Characterization of Bio-chars from Biomass via Pyrolysis. *Energy Sources, Part A: Recovery, Utilization and Environmental Effects*, 28(5): 413–422.

Demirbaş, A. 2007. The Influence of Temperature on the Yields of Compounds Existing in Bio-oils Obtained from Biomass Samples via Pyrolysis. *Fuel Processing Technology*, 88(6): 591–597.

Demirbaş, A. 2008a. Hydrogen Production from Carbonaceous Solid Wastes by Steam Reforming. *Energy Sources Part A*, 30(10): 924–931.

Demirbaş, A. 2008b. Biohydrogen Generation from Organic Wastes. *Energy Sources Part A*, 30(5): 475–482.

Demirbaş, A. 2010. Hydrogen Production from Biomass via Supercritical Water Gasification. *Energy Sources Part A*, 32(14): 1342–1354.

Demirbaş, A. 2011. Production of Fuels from Crops. In *The Biofuels Handbook*. J.G. Speight (Editor). Royal Society of Chemistry, London. Part 2, Chapter 1.

Demirbaş, A. N., Acar, S., Tasesen, E., and Dumrul, H. 2013. Effect of Initial Moisture Content on the Yield of Hydrogen from Pyrolysis of Beech Wood. *Energy Education Science and Technology*, C 5: 75–81.

Demirbaş, A. 2016. Comparison of Thermochemical Conversion Processes of Biomass to Hydrogen-Rich Gas Mixtures. *Energy Sources, Part A: Recovery, Utilization, and Environmetnal Effects*, 38(2): 2971–2976.

Ergudenler, A., and Ghaly, A.E. 1993. Agglomeration of Alumina Sand in a Fluidized Bed Straw Gasifier at Elevated Temperatures. *Bioresource Technology*, 43(3): 259–268.

Fabry, F., Rehmet, C., Rohani, V.-J., and Fulcheri, L. 2013. Waste Gasification by Thermal Plasma: A Review. *Waste and Biomass Valorization*, 4(3): 421–439.

Fermoso, J., Plaza, M.G., Arias, B., Pevida, C., Rubiera, F., and Pis, J.J. 2009. Co-gasification of Coal with Biomass and Petcoke in a High-Pressure Gasifier for Syngas Production. Proceedings. 1st Spanish National Conference on Advances in Materials Recycling and Eco-Energy. Madrid, Spain. November 12–13.

Gabra, M., Pettersson, E., Backman, R., and Kjellström, B. 2001. Evaluation of Cyclone Gasifier Performance for Gasification of Sugar Cane Residue – Part 1: Gasification of Bagasse. *Biomass and Bioenergy*, 21(5): 351–369.

Gay, R.L., Barclay, K.M., Grantham, L.F., and Yosim, S.J. 1980. Fuel Production from Solid Waste. Symposium on Thermal Conversion of Solid Waste and Biomass. Symposium Series No. 130. American Chemical Society, Washington, DC. Chapter 17. Page 227–236.

Gray, D., and Tomlinson, G. 2000. Opportunities for Petroleum Coke Gasification Under Tighter Sulfur Limits for Transportation Fuels. Proceedings. Gasification Technologies Conference, San Francisco, CA. October 8–11.

Hanaoka, T., Inoue, S., Uno, S., Ogi, T., and Minowa, T. 2005. Effect of Woody Biomass Components on Air-Steam Gasification. *Biomass and Bioenergy*, 28(1): 69–76.

Hao, X.H., Guo, L.J., Mao, X., Zhang, X.M., and Chen, X.J. 2003. Hydrogen Production from Glucose Used as a Model Compound of Biomass Gasified in Supercritical Water. *International Journal of Hydrogen Energy*, 28(1): 55–64.

Hao, X.H., Guo, L.J., Zhang, X.M., and Guan, Y. 2005. Hydrogen Production from Catalytic Gasification of Cellulose in Supercritical Water. *Chemical Engineering Journal*, 110(1–3): 57–65.

He, M., Hu, Z., Xiao, B., Li, J., Guo, A., Luo, S., Yang, F., Feng, Y., and Yang, G. 2009. Hydrogen-Rich Gas from Catalytic Steam Gasification of Municipal Solid Waste (MSW): Influence of Catalyst and Temperature on Yield and Product Composition. *International Journal of Hydrogen Energy*, 34(1): 195–203.

Hickman, D.A., and Schmidt, L.D. 1993. Syngas Formation by Direct Catalytic Oxidation of Methane. *Science*, 259(5093): 343–346.

Higman, C., and Van der Burgt, M. 2008. *Gasification*, 2nd Edition. Gulf Professional Publishing, Elsevier, Amsterdam.

Holt, N.A.H. 2001. *Integrated Gasification Combined Cycle Power Plants. Encyclopedia of Physical Science and Technology*, 3rd Edition. Academic Press Inc., New York.

Hu, J., Fei Yu, F., and Lu, Y. 2012. Application of Fischer–Tropsch Synthesis in Biomass to Liquid Conversion. *Catalysts*, 2(2): 303–326.

Irfan, M.F. 2009. Research Report. Pulverized Coal Pyrolysis & Gasification in N2/O2/CO2 Mixtures by Thermo-Gravimetric Analysis. Novel Carbon Resource Sciences Newsletter. Kyushu University, Fukuoka. Volume 2. Page 27–33.

Ishida, Y., Kumabe, K., Hata, K., Tanifuji, K., Hasegawa, T., Kitagawa, K., Isu, N., Funahashi, Y., and Asai, T. 2009. Selective Hydrogen Generation from Real Biomass Through Hydrothermal Reaction at Relatively Low Temperatures. *Biomass and Bioenergy*, 33(1): 8–13.

Jenkins, B.M., and Ebeling, J.M. 1985. Thermochemical Properties of Biomass Fuels. *California Agriculture* (May–June): 14–18.

Johnson, J.L. 1979. *Kinetics of Coal Gasification*. John Wiley and Sons Inc., Hoboken, NJ.

Khosravi, M., and Khadse, A. 2013. Gasification of Petcoke and Coal/Biomass Blend: A Review. *International Journal of Emerging Technology and Advanced Engineering*, 3(12): 167–173.

Ko, M.K., Lee, W.Y., Kim, S.B., Lee, K.W., and Chun, H.S. 2001. Gasification of Food Waste with Steam in Fluidized Bed. *Korean Journal of Chemical Engineering*, 18(6): 961–964.

Kong, L., Li, G., Zhang, B., He, W., and Wang, H. 2008. Hydrogen Production from Biomass Wastes by Hydrothermal Gasification. *Energy Sources Part A*, 30(13): 1166–1178.

Kumabe, K., Hanaoka, T., Fujimoto, S., Minowa, T., and Sakanishi, K. 2007. Cogasification of Woody Biomass and Coal with Air and Steam. *Fuel*, 86(5–6): 684–689.

Kumar, A., Jones, D.D., and Hanna, M.A. 2009. Thermochemical Biomass Gasification: A Review of the Current Status of the Technology. *Energies*, 2(3): 556–581.

Kumar, A., Chakraborty, J.P., and Singh, R. 2016. Bio-oil: The Future of Hydrogen Generation. *Biofuels*, 8(6): 663–674. https://www.researchgate.net/publication/299569308_Bio-oil_the_future_of_hydrogen_generation.

Lan, P., Xu, L.-H., Lan, Xie, D., Zhang, S.-H., and Yan, Y.-J. 2012. Steam Reforming of Model Compounds and Fast Pyrolysis Bio-oil on Supported Nickel Metal Catalysts for Hydrogen Production. *Energy Sources Part A*, 34(21): 2004–2014.

Lapuerta, M., Hernández, J.J., Pazo, A., and López, J. 2008. Gasification and Co-gasification of Biomass Wastes: Effect of the Biomass Origin and the Gasifier Operating Conditions. *Fuel Processing Technology*, 89(9): 828–837.

Lee, S. 2007. Gasification of Coal. In *Handbook of Alternative Fuel Technologies*. S. Lee, J.G. Speight, and S. Loyalka (Editors). CRC Press, Taylor & Francis Group, Boca Raton, FL.

Lee, S., Speight, J.G., and Loyalka, S. 2007. *Handbook of Alternative Fuel Technologies*. CRC-Taylor & Francis Group, Boca Raton, FL.

Lee, S., and Shah, Y.T. 2013. *Biofuels and Bioenergy*. CRC Press, Taylor & Francis Group, Boca Raton, FL.

Li, S., Xu, S., Liu, S., Yang, C., and Lu, Q. 2004. Fast Pyrolysis of Biomass in Free-Fall Reactor for Hydrogen-Rich Gas. *Fuel Processing Technology*, 85(8–10): 1201–1211.

Liu, G., Larson, E.D., Williams, R.H., Kreutz, T.G., and Guo, X. 2011. Making Fischer-Tropsch Fuels and Electricity from Coal and Biomass: Performance and Cost Analysis. *Energy and Fuels*, 25(1): 415–437.

Liu, K., Song, C., and Subramani, V. 2010. *Hydrogen and Syngas Production and Purification Technologies*. John Wiley & Sons Inc., Hoboken, NJ.

Lu, Y., Guo, L., Zhang, X., and Yan, Q. 2007. Thermodynamic Modeling and Analysis of Biomass Gasification for Hydrogen Production in Supercritical Water. *Chemical Engineering Journal*, 131(1–3): 233–244.

Lu, Y.J., Jin, H., Guo, L.J., Zhang, X.M., Cao, C.Q., and Guo, X. 2008. Hydrogen Production by Biomass Gasification in Supercritical Water with a Fluidized Bed Reactor. *International Journal of Hydrogen Energy*, 33(21): 6066–6075.

Luque, R., and Speight, J.G. (Editors) 2015. *Gasification for Synthetic Fuel Production: Fundamentals, Processes, and Applications.* Woodhead Publishing, Elsevier, Cambridge.

Lv, P.M., Xiong, Z.H., Chang, J., Wu, C.Z., Chen, Y., and Zhu, J.X. 2004. An Experimental Study on Biomass Air-Steam Gasification in a Fluidized Bed. *Bioresource Technology*, 95(1): 95–101.

Malkow, T. 2004. Novel and Innovative Pyrolysis and Gasification Technologies for Energy Efficient and Environmentally Sound MSW Disposal. *Waste Management*, 24(1): 53–79.

Marano, J.J. 2003. *Refinery Technology Profiles: Gasification and Supporting Technologies.* Report prepared for the United States Department of Energy. National Energy Technology Laboratory. United States Energy Information Administration, Washington, DC. June.

Martinez-Alonso, A., and Tascon, J.M.D. 1991. In *Fundamental Issues in Control of Carbon Gasification Reactivity.* J. Lahaye and P. Ehrburger (Editors). Kluwer Academic Publishers, Dordrecht.

McKendry, P. 2002. Energy Production from Biomass Part 3: Gasification Technologies. *Bioresource Technology*, 83(1): 55–63.

McLendon, T.R., Lui, A.P., Pineault, R.L., Beer, S.K., and Richardson, S.W. 2004. High-Pressure Co-gasification of Coal and Biomass in a Fluidized Bed. *Biomass and Bioenergy*, 26(4): 377–388.

Mohan, D., Pittman, C.U., and Philip, S. 2017. Pyrolysis of Wood/Biomass for Bio-oil: A Critical Review. *Progress in Energy and Combustion Science*, 62: 848–889.

Mims, C.A. 1991. *Fundamental Issues in Control of Carbon Gasification Reactivity.* J. Lahaye and P. Ehrburger (Editors). Kluwer Academic Publishers, Dordrecht. Page 383.

Minkova, V., Marinov, S.P., Zanzi, R., Bjornbom, E., Budinova, T., Stefanova, M., and Lakov, L. 2000. Thermochemical Treatment of Biomass in a Flow of Steam or in a Mixture of Steam and Carbon Dioxide. *Fuel Processing Technology*, 62(1): 45–52.

Mokhatab, S., Poe, W.A., and Speight, J.G. 2006. *Handbook of Natural Gas Transmission and Processing.* Elsevier, Amsterdam.

Molino, A., Iovane, P., Donatelli, A., Braccio, G., Chianese, S., and Musmarra, D. 2013. Steam Gasification of Refuse-Derived Fuel in a Rotary Kiln Pilot Plant: Experimental Tests. *Chemical Engineering Transactions*, 32: 337–342.

Molino, A., Chianese, S., and Musmarra, D. 2016. Biomass Gasification Technology: The State of the Art Overview. *Journal of Energy Chemistry*, 25(1): 10–25.

Nordstrand, D., Duong, D.N.B., and Miller, B.G. 2008. *Combustion Engineering Issues for Solid Fuel Systems-Chapter 9 Post-combustion Emissions Control.* B.G. Miller and D. Tillman (Editors). Elsevier, London.

Ochoa, J., Cassanello, M.C., Bonelli, P.R., and Cukierman, A.L. 2001. CO2 Gasification of Argentinian Coal Chars: A Kinetic Characterization. *Fuel Processing Technology*, 74(3): 161–176.

Orr, D., and Maxwell, D. 2000. A Comparison of Gasification and Incineration of Hazardous Wastes. Report No. DCN 99.803931.02. United States Department of Energy, Morgantown, West, VA. March 30.

Pafili, A., Charisiou, N.D., Douvartzides, S.L., Siakavelas, G.I., Wang,W., Liu, G., Papadakis, V.G., and Goula, M.A. 2021. Recent Progress in the Steam Reforming of Bio-oil for Hydrogen Production: A Review of Operating Parameters, Catalytic Systems and Technological Innovations. *Catalysts*, 11(12): 1526. https://www.mdpi.com/2073-4344/11/12/1526.

Pakdel, H., and Roy, C. 1991. Hydrocarbon Content of Liquid Products and Tar from Pyrolysis and Gasification of Wood. *Energy and Fuels*, 5(3): 427–436.

Pan, Y.G., Velo, E., Roca, X., Manyà, J.J., and Puigjaner, L. 2000. Fluidized-Bed Cogasification of Residual Biomass/Poor Coal Blends for Fuel Gas Production. *Fuel*, 79(11): 1317–1326.

Pandey, A. 2009. *Handbook of Plant-Based Biofuels.* CRC Press, Taylor and Francis Group, Boca Raton, FL.

Penrose, C.F., Wallace, P.S., Kasbaum, J.L., Anderson, M.K., and Preston, W.E. 1999. Enhancing Refinery Profitability by Gasification, Hydroprocessing and Power Generation. Proceedings. Gasification Technologies Conference, San Francisco, CA. October. https://www.globalsyngas.org/uploads/eventLibrary/GTC99270.pdf.

Pepiot, P., Dibble, and Foust, C.G. 2010. Computational Fluid Dynamics Modeling of Biomass Gasification and Pyrolysis. In *Computational Modeling in Lignocellulosic Biofuel Production.* M.R. Nimlos and M.F. Crowley (Editors). ACS Symposium Series. American Chemical Society, Washington, DC.

Pfeifer, C., Puchner, B., and Hofbauer, H. 2009. Comparison of Dual Fluidized Bed Steam Gasification of Biomass with and Without Selective Transport of CO2. *Chemical Engineering Science*, 64(23): 5073–5083.

Probstein, R.F., and Hicks, R.E. 1990. *Synthetic Fuels*. PH Press, Cambridge, MA. Chapter 4.

Rabovitser, I.K., Nester, S., and Bryan, B. 2010. Plasma Assisted Conversion of Carbonaceous Materials into a Gas. United States Patent 7,736,400. June 25.

Rajvanshi, A.K. 1986. Biomass Gasification. In *Alternative Energy in Agriculture*. D.Y. Goswami (Editor). CRC Press, Boca Raton, FL. Page 83–102.

Ramroop Singh, N. 2011. Biofuel. In *The Biofuels Handbook*. J.G. Speight (Editor). Royal Society of Chemistry, London. Part 1, Chapter 5.

Rapagnà, S., and Latif, A. 1997. Steam Gasification of Almond Shells in a Fluidized Bed Reactor: The Influence of Temperature and Particle Size on Product Yield and Distribution. *Biomass and Bioenergy*, 12(4): 281–288.

Rapagnà, S., Jand, N., Kiennemann, A., and Foscolo, P.U. 2000. Steam Gasification of Biomass in a Fluidized-Bed of Olivine Particles. *Biomass and Bioenergy*, 19(3): 187–197.

Rauch, R., Hrbek, J., and Hofbauer, H. 2014. Biomass Gasification for Synthesis Gas Production and Applications of the Syngas. *Wiley Interdisciplinary Reviews: Energy and Environment*, 3(4): 343–362.

Ren, Z.-Z., Lan, P., Ma, H.-R., Wang, T., Shi, X.-H., Zhang, S.-P., Xu, Q.-L., and Yan, Y.-J. 2014. Hydrogen Production via Catalytic Steam Reforming of Bio-Oil Model Compounds in A Two-Stage Reaction System. *Energy Sources Part A*, 36(17): 1921–1930.

Richardson, J.H., Rogers, R.S., Thorsness, C.B., Wallman, P.H., Leininger, T.F., Richter, G.N., Robin, A.M., Wiese, H.C., and Wolfenbarger, J.K. 1995. Conversion of Municipal Solid Waste to Hydrogen. Report No. UCRL-JC-120142. Proceedings. DOE Hydrogen Program Review, in Coral Gables, FL. April 19–21. Available from United States Department of Energy, Washington, DC.

Ricketts, B., Hotchkiss, R., Livingston, W., and Hall, M. 2002. Technology Status Review of Waste/Biomass Co-gasification with Coal. Proceedings. *Institution of Chemical Engineers*. Fifth European Gasification Conference. Noordwijk. April 8–10.

Roddy, D.J., and Manson-Whitton, C. 2012. Biomass Gasification and Pyrolysis. In *Comprehensive Renewable Energy*. D.J. Roddy (Editor). Elsevier, Amsterdam. Volume 5: Biomass and Biofuels.

Rodríguez-Olalde, N.E., Mendoza-Chávez, E., Castro-Montoya, A.J., Saucedo-Luna, J., Maya-Yescas, R., Rutiaga-Quiñones, J.G., and Ponce Ortega, J.M. 2015. Simulation of Syngas Production from Lignin Using Guaiacol as a Model Compound. *Energies*, 8(7): 6705–6714.

Santos, J., da Guarda Souza, M.O., and do Carmo Rangel, M. 2008. Effect of Magnesium on the Properties of Nickel and Lanthanum-Based Catalysts in Steam Reforming. *Fuel*, 87(17–18): 3627–3630.

Senneca, O. 2007. Kinetics of Pyrolysis, Combustion and Gasification of Three Biomass Fuels. *Fuel Processing Technology*, 88(1): 87–97.

Shabbar, S., and Janajreh, I. 2013. Thermodynamic Equilibrium Analysis of Coal Gasification Using Gibbs Energy Minimization Method. *Energy Conversion and Management*, 65: 755–763.

Shen, C.-H., Chen, W.-H., Hsu, H.-W., Sheu, J.-Y., and Hsieh, T.-H. 2012. Co-gasification Performance of Coal and Petroleum Coke Blends in A Pilot-Scale Pressurized Entrained-Flow Gasifier. *International Journal of Energy Research*, 36(4): 499–508.

Sikarwar, V.S., Zhao, M., Clough, P., Yao, J., Zhong, X., Memon, M.Z., Shah, N., Anthony, E.J., and Fennell, P.S. 2016. An Overview of Advances in Biomass Gasification. *Energy and Environmental Science*, 9(10): 2939–2977.

Sikarwar, V.S., Zhao, M., Fennell, P.S., Shah, N., and Anthony, E.J. 2017. Progress in Biofuel Production from Gasification. *Progress in Energy and Combustion Science*, 61: 189–248.

Singh, S.P., Weil, S.A., and Babu, S.P. 1980. Thermodynamic Analysis of Coal Gasification Processes. *Energy*, 5(8–9): 905–914.

Sjöström, K., Chen, G., Yu, Q., Brage, C., and Rosén, C. 1999. Promoted Reactivity of Char in Cogasification of Biomass and Coal: Synergies in the Thermochemical Process. *Fuel*, 78(10): 1189–1194.

Sondreal, E.A., Benson, S.A., Pavlish, J.H., and Ralston, N.V.C. 2004. An Overview of Air Quality III: Mercury, Trace Elements, and Particulate Matter. *Fuel Processing Technology*, 85(6–7): 425–440.

Spath, P., Lane, J., Mann, M., and Amos, W. 2000. *Update of Hydrogen from Biomass: Determination of the Delivered Cost of Hydrogen*. NREL Milestone Report, April.

Speight, J.G. 1990. *Fuel Science and Technology Handbook*. J.G. Speight (Editor). Marcel Dekker Inc., New York. Chapter 33.

Speight, J.G. 2020. *Synthetic Fuels Handbook: Properties, Process, and Performance*, 2nd Edition. McGraw-Hill, New York.

Speight, J.G. (Editor) 2011a. *The Biofuels Handbook.* Royal Society of Chemistry, London.

Speight, J.G. 2011b. *The Refinery of the Future.* Gulf Professional Publishing, Elsevier, Oxford.

Speight, J.G. 2013a. *The Chemistry and Technology of Coal*, 3rd Edition. CRC Press, Taylor & Francis Group, Boca Raton, FL.

Speight, J.G. 2013b. *Coal-Fired Power Generation Handbook.* Scrivener Publishing, Salem, MA.

Speight, J.G. 2014a. *The Chemistry and Technology of Petroleum*, 5th Edition. CRC Press, Taylor & Francis Group, Boca Raton, FL.

Speight, J.G. 2014b. *Gasification of Unconventional Feedstocks.* Gulf Professional Publishing, Elsevier, Oxford.

Speight, J.G. 2016. Hydrogen in Refineries. In *Hydrogen Science and Engineering – Materials, Processes, Systems, and Technology.* D. Stolten and B. Emonts (Editors). Wiley-VCH Verlag GmbH & Co., Weinheim. Chapter 1, Page 3–18.

Speight, J.G. 2017. *Handbook of Petroleum Refining.* CRC Press, Taylor & Francis Group, Boca Raton, FL.

Speight, J.G. 2019a. *Handbook of Petrochemical Processes.* CRC Press, Taylor & Francis Group, Boca Raton, FL.

Speight, J.G. 2019b. *Natural Gas: A Basic Handbook*, 2nd Edition. Gulf Publishing Company, Elsevier, Cambridge, MA.

Sundaresan, S., and Amundson, N.R. 1978. Studies in Char Gasification – I: A Lumped Model. *Chemical Engineering Science*, 34(3): 345–354.

Sutikno, T., and Turini, K. 2012. Gasifying Coke to Produce Hydrogen in Refineries. *Petroleum Technology Quarterly*, Q3: 105.

Tijmensen, M.J.A., Faaij, A.P.C., Hamelinck, C.N., and Van Hardeveld, M.R.M. 2002. Exploration of the Possibilities for Production of Fischer Tropsch Liquids and Power via Biomass Gasification. *Biomass and Bioenergy*, 23(2): 129–152.

Tinaut, F.V., Melgar, A., Pérez, J.F., and Horillo, A. 2008. Effect of Biomass Particle Size and Air Superficial Velocity on the Gasification Process in a Downdraft Fixed Bed Gasifier. An Experimental and Modelling Study. *Fuel Processing Technology*, 89(11): 1076–1089.

Usón, S., Valero, A., Correas, L., and Martínez, Á. 2004. Co-gasification of Coal and Biomass in an IGCC Power Plant: Gasifier Modeling. *International Journal of Thermodynamics*, 7(4): 165–172.

Vélez, F.F., Chejne, F., Valdés, C.F., Emery, E.J., and Londoño, C.A. 2009. Cogasification of Colombian Coal and Biomass in a Fluidized Bed: An Experimental Study. *Fuel*, 88(3): 424–430.

Veziroglu, T.N., and Sahin, S. 2008. 21st Century's Energy: Hydrogen Energy System. *Energy Conversion and Management*, 49(7): 1820–1831.

Wang, D., Czernik, S., Montane, D., Mann, M., and Chornet, E. 1997. Biomass to Hydrogen via Fast Pyrolysis and Catalytic Steam Reforming of the Pyrolysis Oil or Its Fractions. *Industrial and Engineering Chemistry Research*, 36(5): 1507–1518.

Wang, Y., Duan, Y., Yang, L., Jiang, Y., Wu, C., Wang, Q., and Yang, X. 2008. Comparison of Mercury Removal Characteristic between Fabric Filter and Electrostatic Precipitators of Coal-Fired Power Plants. *Journal of Fuel Chemistry and Technology*, 36(1): 23–29.

Wang, G., Dai, Y., Yang, H., Xiong, Q., Wang, K., Zhou, J., Li, Y., and Wang, S. 2020. A Review of Recent Advances in Biomass Pyrolysis. *Energy and Fuels*, 34(12): 15557–15578.

Yang, H., Xua, Z., Fan, M., Bland, A.E., and Judkins, R.R. 2007. Adsorbents for Capturing Mercury in Coal-Fired Boiler Flue Gas. *Journal of Hazardous Materials*, 146(1–2): 1–11.

Ye, D.P., Agnew, J.B., and Zhang, D.K. 1998. Gasification of a South Australian Low-Rank Coal with Carbon Dioxide and Steam: Kinetics and Reactivity Studies. *Fuel*, 77(11): 1209–1219.

Zhang, Y.H., Xin, S.Z., Li, T.C., Xu, Q.L., Zhang, S.P., Ren, Z.W., Z.W., and Yan, Y.J. 2013a. A Study on the Integrated Process for Hydrogen-Rich Gas Production from Biomass. *Energy Sources Part A*, 35(20): 1905–1913.

Zhang, Y., Li, W., Zhang, S., Xu, Q., and Yan, Y. 2013b. Hydrogen Production by the Catalytic Reforming of Volatile from Biomass Pyrolysis Over a Bimetallic Catalyst. *Energy Sources Part A*, 35(21): 1975–1982.

Zhang, C. 2022. Review of Catalytic Reforming of Biomass Pyrolysis Oil for Hydrogen Production. *Frontiers in Chemistry*, 10: 962587. https://www.frontiersin.org/articles/10.3389/fchem.2022.962587/full.

6 Hydrogen Purification and Properties

6.1 INTRODUCTION

Hydrogen is the most abundant element in the universe; it is the lightest element with the highest energy content among all existing fuels. Over the past several decades hydrogen has become important not only for product improvement but also as an agent for the upgrading of viscous feedstocks (Chapter 1, Chapter 2). Also, hydrogen is a very efficient and clean energy carrier (with only water as a by-product of the chemical reaction to produce energy), and with the move away from fossil fuels as a source of energy, the importance of hydrogen as a source of energy will increase. Thus, hydrogen shows great potential regarding future energy prospects, especially in the potential low-carbon energy system known as 'the hydrogen economy'. Some of the benefits that promote hydrogen as an energy carrier include the following. Hydrogen is the most abundant element in the universe; it is the lightest element with the highest energy content among all the existing fuels (Balat, 2008; Abe et al., 2019). It is a highly efficient and clean energy carrier (El-Shafie et al., 2019), with only water as a by-product at conversion to energy (Abe et al., 2019; Abe et al., 2019; El-Shafie et al., 2019) and, furthermore, hydrogen lends itself to a variety of storage methods (e.g., gaseous, liquid) (Rusman and Dahari, 2016; Abe et al., 2019). Furthermore, hydrogen can be stored by a variety of storage methods (e.g., gaseous hydrogen or liquid hydrogen) (Holladay et al., 2004; Balat, 2008; Rusman and Dahari, 2016; Abe et al., 2019; Chapman et al., 2019; El-Shafie et al., 2019).

Gas processing (also referred to as 'gas refining') usually involves the use of several integrated unit processes to remove: (1) gaseous hydrocarbon derivatives, (2) water, (3) constituents that contain elements such as sulfur and nitrogen, (4) other gases such as helium and carbon dioxide, and (5) the hydrocarbon mixture known as 'natural gas liquids'. In addition, it is often necessary to install scrubbers and heaters at or near the wellhead that serve primarily to remove sand and other large-particle impurities. Thus, hydrogen purification is an extension of conventional gas processing insofar as there are several physical processes and chemical processes that are available for the purification of hydrogen streams. The purification or separation of hydrogen has traditionally been based on solid-state diffusion technology utilizing noble metal membranes.

Although the ability of metals and particularly palladium to selectively permeate hydrogen has been known for over a century, the commercial utilization of this property for the purification of hydrogen is relatively recent. The development of silver-palladium alloys as diffusion membrane materials in the late 1950s overcame the principal technical difficulties associated with the technique and enabled the successful development of commercial equipment, a technology with which Johnson Matthey has been closely associated from the beginning.

Since that time several other techniques for hydrogen purification have been developed and are at various stages of commercial exploitation. As hydrogen is becoming increasingly important in many major areas, including the electronics manufacturing industry, it is an appropriate time to review the status of membrane diffusion for hydrogen purification and to highlight the current developments in membrane materials, especially palladium alloys, in the context of their potential influence on hydrogen purification technology and the application within the refinery system.

However, as with all such hydrogen purification processes (Table 6.1), there are many variables in the choice of process or the choice of the processing sequence that dictate the choice of process or processes to be employed.

DOI: 10.1201/9781003186694-6

TABLE 6.1
Common Methods for the Removal of Carbon Dioxide and Hydrogen Sulfide from Gas Streams

Method	Description
Carbon Dioxide Removal	
Water scrubbing	Uses principle of the higher solubility of carbon dioxide in water to separate the carbon dioxide from biogas
	The process uses high pressure and removes hydrogen sulfide as well as carbon dioxide
	The main disadvantage of this process is that it requires a large volume of water that must be purified and recycled
Glycol scrubbing	This process is more efficient than water scrubbing
	It also requires the regeneration of a large volume of polyethylene glycol
Chemical absorption	Chemical reaction between carbon dioxide and amine-based solvents (olamines)
Carbon molecular sieves	Use differential adsorption characteristics to separate methane and carbon dioxide, which is carried out at high pressure (pressure swing adsorption)
	Hydrogen sulfide should be removed before the adsorption process
Membrane separation	Selectively separates hydrogen sulfide from carbon dioxide from methane
	The carbon dioxide and hydrogen sulfide dissolve while the methane is collected for use
Cryogenic separation	Cooling until condensation or sublimation of carbon dioxide
Hydrogen Sulfide Removal	
Biological desulfurization	Natural bacteria can convert hydrogen sulfide into elemental sulfur in the presence of oxygen and iron. This can be done by introducing a small amount (2–5%) of air into the headspace of the digester. As a result, deposits of elemental sulfur will be formed in the digester
Hydrogen Sulfide Removal	
Iron/iron oxide reaction	Hydrogen sulfide reacts readily with either iron oxide or iron chloride to form insoluble iron sulfide
	The reaction can be exploited by adding the iron chloride to the digester feed material or passing the biogas through a bed of iron oxide-containing material
	The iron oxide media needs to be replaced periodically
	The regeneration process is highly exothermic and must be controlled to avoid problems
Activated carbon	Activated carbon impregnated with potassium iodide can catalytically react with oxygen and hydrogen sulfide to form water and sulfur
	The activated carbon beds need regeneration or replacement when saturated

In this choice, several factors must be considered: (1) the types and concentrations of contaminants in the gas stream, (2) the degree of contaminant removal desired, (3) the temperature, pressure, volume, and composition of the gas to be processed, and (4) the carbon dioxide-hydrogen sulfide ratio in the gas (Table 6.2).

Process selectivity indicates the preference with which the process removes one acid gas component relative to (or in preference to) another. It is very important to consider the process selectivity for, say, hydrogen sulfide removal compared to carbon dioxide removal that ensures minimal concentrations of these components in the product, thus the need for consideration of the carbon dioxide to hydrogen sulfide ratio in the gas stream.

Although there has been a brief description of hydrogen purification in a previous chapter (Chapter 3), it is appropriate at this point to present more detail on the needs for hydrogen purification and the processes by which this can be accomplished.

Therefore, the focus of this chapter is a selection of the processes that are integral part of the concept of production of a usable product (hydrogen) for use in the refinery and for sale to the consumer.

TABLE 6.2
Examples of the Amount of Hydrogen in Various Gas Streams

Gas Stream*	Hydrogen, % w/w
Biomass gasification	25–35
Coal gasification	25–35
Methanol reforming	75–80
Natural gas reforming*	70–75

*Listed alphabetically.

6.2 HYDROGEN PURIFICATION

The purification of hydrogen is the process of removing contaminants from hydrogen gas to produce a product that will satisfy the future needs of the refiner. Hydrogen gas can be contaminated with a variety of impurities including water vapor, carbon monoxide, and other gases that depend upon the source of the gas. For example, hydrogen gas produced from a natural gas source may contain higher levels of impurities than hydrogen gas produced from water by electrolysis. In the refinery, as in most other cases, the quality of the hydrogen gas will have a significant impact on the designated use of the gas. For example, impurities such as water vapor and carbon monoxide can interfere with the behavior of the hydrogenation process and operation of the fuel cell, while impurities such as oxygen and nitrogen can reduce the efficiency of the system.

Hydrogen that is produced by the electrolysis of water (also known as 'green hydrogen') is typically of higher purity than the hydrogen produced from fossil fuel sources, also known as 'gray hydrogen'. Each of the production processes requires different types of hydrogen purifiers because the production of green hydrogen involves the electrolysis of water, which produces hydrogen gas that is relatively free of impurities. In contrast, the production of 'gray hydrogen' involves the steam reforming of fossil fuels, which can produce hydrogen gas that contains higher levels of impurities, such as carbon monoxide and water vapor.

In many cases, a refinery hydrogen network (Chapter 5) can only operate efficiently if high-quality hydrogen is provided. Thus, purification of the produced hydrogen technology is a crucial link in the hydrogen production to hydrogen utilization chain in the refinery. Stable, reliable, and low-cost hydrogen sources represent a base for large-scale applications of hydroprocessing technology. Thus, the purification of the hydrogen used in a refinery (or, for that matter, in any industrial process that employs hydrogen as a reactant) is a key aspect of any process in which hydrogen is employed (Table 3.4). Although covered in more detail elsewhere in this text, the processes that are used to purify the hydrogen are worthy of a brief note here.

The purification methods used to produce high-quality hydrogen can be classified as (1) physical methods and (2) chemical methods (Aasadnia, et al., 2021). The former methods (i.e., the physical methods) include adsorption methods such as pressure swing adsorption, PSA, temperature swing adsorption, TSA, vacuum adsorption, low-temperature separation methods such as cryogenic distillation and low-temperature adsorption), and membrane separation methods, such as the use of inorganic membranes and organic membranes, while the latter involve a metal hydride separation and catalysis method.

The selection of an appropriate purification method is closely related to the hydrogen supply mode and gas source. For hydrogen production by centralized large-scale coal gasification and natural gas reforming, purification by pressure swing adsorption is primarily adopted after transformation, desulfurization, and decarbonization.

For an organic membrane combined with a pressure swing adsorption process is used for obtaining methanol purge gas, while a two-stage or multi-stage pressure swing adsorption process is adopted for obtaining coke oven gas and by-product gas from the refinery.

Low-temperature adsorption can effectively eliminate multiple impurities, such as sulfide, formaldehyde (HCHO), and formic acid (HCOOH). However, it requires high energy consumption, and it is a complex process suitable for special small-scale and cold source applications (Schorer et al., 2019).

At the same time, the inherent disadvantage is that purified materials react with impure gas during hydrogen recovery, reducing the purification efficiency (Dunikov et al., 2012; Dunikov and Blinov, 2020). New membrane technologies, such as carbon molecular sieve membranes (CMSMs) (Hamm et al., 2017), ionic liquid membranes (Sasikumar et al., 2018), and electrochemical hydrogen pump membranes (Bernardo et al., 2020) are currently in development.

When the hydrogen content of the refinery gas is greater than 50% v/v, the gas should first be considered for hydrogen recovery, using a membrane or pressure swing adsorption unit (Brüschke, 1995, 2003; Mokhatab et al., 2006; Speight, 2019). The tail gas or reject gas that will still contain a substantial amount of hydrogen can then be used as steam reformer feedstock. Generally, the feedstock purification process uses three different refinery gas streams to produce hydrogen. First, high-pressure hydrocracker purge gas is purified in a membrane unit that produces hydrogen at medium pressure and is combined with medium pressure off-gas that is first purified in a pressure swing adsorption unit. Finally, low-pressure off-gas is compressed, mixed with reject gases from the membrane and pressure swing adsorption units, and used as steam reformer feed.

Various processes (some of which are presented alphabetically in the following sub-sections) are available to purify the hydrogen stream, but since the product streams are available in a wide variety of composition, flows, and pressures, the best method of purification will vary. Additionally, several factors that must be taken into consideration in the selection of a purification method. These factors include: (1) hydrogen recovery, (2) product purity, (3) pressure profile, (4) reliability, and (5) cost. Another equally important parameter is not considered here since the emphasis is on the technical aspects of the purification process.

Hydrogen extracted from gas streams or storage systems is prone to be contaminated by a large variety of pollutants, and numerous purification techniques are often required to produce hydrogen, which can be consumed by refinery conversion systems or other refinery processes.

Currently, large-scale hydrogen production is commonly carried out by means of steam methane reforming, coal gasification, or partial oxidation of hydrocarbons (Thomassen et al., 2010; Acer and Dincer, 2014; Dincer and Acer, 2015). Although not widely implemented, there are several small-scale hydrogen production methods, including partial oxidation of biomass, electrolysis and autothermal processes, and small-scale steam methanol reformers (Thomassen et al., 2010; Baykara, 2018). However, both large- and small-scale hydrogen production methods produce hydrogen-rich streams commonly accompanied by impurities such as carbon oxides (i.e., carbon dioxide, carbon monoxide), hydrogen sulfide, nitrogen, and methane (Lee et al., 2004; Ockwig and Nenoff, 2007; Grande et al., 2008; Granite and O'Brien, 2005; Perry et al., 2008; David, 2012; Chen et al., 2014). There are several methods for hydrogen recovery and purification which can be applied to the purification of hydrogen and can be categorized as (1) physical methods and chemical methods (Table 6.3) (Mokhatab et al., 2006; Speight, 2019).

The former methods (i.e., the physical methods for hydrogen purification) include adsorption methods such as pressure swing adsorption (more commonly referred to as PSA), temperature swing adsorption (more commonly referred to as TSA), vacuum adsorption, low-temperature separation methods (such as cryogenic distillation and low-temperature adsorption), and membrane separation methods (such as the use of an inorganic membrane and an organic membrane), while the latter (i.e., the chemical methods for hydrogen purification) involve the use of a metal hydride separation and methods that use a variety of catalysts.

Hydrogen gas streams can contain a varied number and types of contaminants. For example, it is possible to find entrained particles of dirt and rust in a hydrogen electrolyzer product stream. Some electrolyzers have steel or iron alloy containers, valves, and piping. This metal may produce low

TABLE 6.3
General Categorization of Methods for Hydrogen Separation and Purification

Physical Methods	
Adsorption	Pressure swing adsorption
	Temperature swing adsorption
	Vacuum adsorption
	Desiccant separation
Low-temperature separation	Cryogenic distillation
	Low-temperature adsorption
Membrane separation	
	Carbon molecular sieve
	Inorganic membrane
	Metal membrane
	Organic membrane
	Polymer membrane
Wet scrubbing	
Chemical Methods	
	Catalysis
	Metal hydride separation

*Listed alphabetically by category and within each category and not by any preference.

levels of rust particles over time. Electrolyzers frequently use high surface area electrode materials such as nickel that may lose small particles into the electrolyte.

Particles of either plastic or sealant will enter the electrolyte system and can occur due to poor housekeeping during manufacturing or repair. These particulates can be removed from your hydrogen gas stream with an appropriate scrubber or filter.

The electrolyte is the typical liquid contaminant in an alkaline electrolysis system. In this case, it is water plus potassium hydroxide (KOH). A small amount of potassium hydroxide and water escape from the electrolyzer with the flow of hydrogen. An aerosol of fine electrolyte droplets is produced as bubbles of hydrogen gas that rise to the surface of the liquid electrolyte. Another source of KOH/water aerosol hydrogen contamination is the electrolyte-wetted float valves that control the discharge of gas from the unit. Any other liquid contaminant in the hydrogen gas would be unexpected. A coalescing filter can remove solids and liquid KOH/water aerosol droplets.

Under certain conditions, gaseous contaminants such as oxygen, nitrogen, argon, and water vapor occur in hydrogen gas streams.

The concentration of water saturation is governed by the temperature and pressure in that environment. The objective of a purifier system is to remove excess water vapor. This prevents condensation from occurring downstream, with subsequent flooding of other components. The presence of some water vapor in hydrogen that has been purified for storage in pressure tanks is not harmful, if a compressor is not used.

However, practical and economic considerations impose restraints so that only two techniques are actually utilized over this whole range, namely catalytic purification and polymer membrane diffusion. The two physical techniques, cryogenic separation and pressure swing adsorption, are best suited to large-scale applications and the remaining techniques, including palladium alloy membrane diffusion, are utilized for small to medium outputs.

There are two other aspects of importance to be considered in selecting a purification technique besides the scale of operation. These include both the selectivity of the technique against gases other than hydrogen and its resistance to 'poisoning' by constituents present as impurities in the feed gas.

All techniques can operate well with hydrogen-rich gas feedstocks. Catalytic purification removes oxygen by reacting with hydrogen to form water, and carbon monoxide by oxidation or methanation; this technique is used to upgrade relatively pure hydrogen produced, for example, by electrolysis. Cryogenic separation can tolerate a wider range of hydrogen content in the feed gas, typically 30 to 80%, but is limited in gas composition to those constituents that will selectively condense at cryogenic temperatures.

Metal hydride separation and the two diffusion techniques (i.e., use of a palladium membrane and a polymer membrane) have the ability to deal with gas streams that are lean in hydrogen.

A palladium membrane purifier operates on the basis of pressure-driven diffusion across the membranes, and only hydrogen can diffuse through the membrane. The palladium diffuser can take a number of forms, including an array of tubes, a coiled tube, or membrane foil and employs a palladium and silver alloy material which allows monatomic hydrogen to pass through the crystal lattice at temperatures on the order of >300°C (>570°F). The hydrogen molecule (H_2) that comes into contact with the palladium membrane surface dissociates into monatomic hydrogen, which passes through the membrane. On the other surface of the palladium membrane, the monatomic hydrogen is recombined into diatomic hydrogen.

On the other hand, the polymer membrane diffusion method is based on a differential diffusion rate principle, and the purity of the output hydrogen is affected by the hydrogen concentration as well as by the nature of the other constituents that are present in the input gas stream.

The level of recovery of hydrogen from gas streams varies considerably from one technique to another. Metal hydride separation, pressure swing adsorption, and polymer membrane diffusion have relatively poor recovery levels, typically in the range of 70 to approximately 85% v/v, while cryogenic separation and solid polymer electrolyte techniques can attain recovery levels on the order of about 95% v/v. On the other hand, palladium membrane diffusion and catalytic purification techniques can achieve high recovery levels of up to 99% from hydrogen-rich gases.

At the lowest levels, typically low to mid 90% v/v are the cryogenic separation and polymer membrane diffusion methods. Metal hydride separation can achieve a 99% v/v purity level with the solid polymer electrolyte technique attaining almost a factor of 10 better purity. Two methods are capable of producing moderately high purity levels of 99.999% v/v, namely pressure swing adsorption and catalytic purification, but where very high purities of 99.9999% v/v or better are required, only palladium membrane diffusion comes into consideration.

Thus, the selection of an appropriate method for the purification of hydrogen is related to the mode of generation of the hydrogen as well as the source of hydrogen. For example, when hydrogen is produced by coal gasification and natural gas reforming, the pressure swing adsorption method is typically applied, but only after the desulfurization and decarbonization of the gas stream. The cryogenic distillation method of purification is also applicable to large-scale production, but the typical purity of the hydrogen stream may be on the order of 85 to 99% v/v, which may be insufficient for the application requirements.

Thus, for the production of high-purity hydrogen, it may require the application of more than one process based on the types of impurities in the hydrogen stream. For example, an organic membrane combined with the pressure swing adsorption process may be used to purge methanol from the stream, while a two-stage or multi-stage pressure swing adsorption process may be adopted for obtaining coke oven gas and by-product gas from the refinery. Also, the low-temperature adsorption method, the metal hydride method, and the metal membrane method are available processes according to the types and amounts of impurities.

Low-temperature adsorption can effectively eliminate multiple impurities, such as sulfide derivatives, formaldehyde (HCHO), and formic acid (HCOOH). Also, the metal hydride separation method and the palladium (Pd) membrane separation method are effective for separating gas sources with a high content of inert components. At the same time, their inherent disadvantage is that purified materials react with impure gas during the hydrogen recovery, thereby reducing the efficiency of the purification method. New membrane technologies, such as carbon molecular sieve membranes (often referred to by the acronym CMSMs), ionic liquid membranes, and electrochemical hydrogen pump membranes are becoming available.

Typically, hydrogen gas purification is often required for applications where high-purity hydrogen gas is critical, such as industrial processes and power generation, since impurities in the hydrogen can affect the performance of the system. For example, impurities such as water vapor, carbon, oxygen, and nitrogen can reduce the efficiency of the system.

To maintain a certain level of hydrogen purity throughout the hydrogen supply chain, it is important to implement appropriate purification processes at each stage of the chain. This may involve the use of specialized equipment and techniques to remove impurities and contaminants from the hydrogen gas. In addition, strict quality control measures must be implemented to ensure that the hydrogen gas meets the required purity standards throughout the supply chain. This may involve the use of tools and methods such as gas chromatography, mass spectrometry, and gas analyzers to assess the purity of the gas at various stages of the supply chain.

Finally, in terms of determining the purity of the hydrogen, one of the most common methods used to analyze the composition of gases, including hydrogen, is gas chromatography. This method involves the use of a specialized instrument that separates the gases based on their chemical properties and then measures the concentrations of each gas. Mass spectrometry is another method used to analyze the composition of hydrogen gas and involves the use of a specialized instrument that ionizes the gases and then measures the masses of the ions to determine the composition of the gas. Gas analyzers can also be used for measuring the purity of hydrogen. These are sophisticated instruments that use a variety of techniques, such as infrared absorption or electrochemical sensors, to measure the concentrations of impurities in the gas.

6.3 PROCESS OPTIONS

The processes that have been developed for hydrogen purification vary from a simple once-through wash operation to complex multi-step systems with options for recycling of the gases (Mokhatab et al., 2006; Speight, 2019, 2020). In some cases, process complexities arise because of the need for recovery of the materials used to remove the contaminants or even recovery of the contaminants in their original, or altered, form.

In more general terms, hydrogen purification is divided into the removal of particulate impurities and the removal of gaseous impurities. For the purposes of this chapter, the latter operation includes the removal of hydrogen sulfide, carbon dioxide, sulfur dioxide, and products that are not related to synthesis gas and hydrogen production. However, there is also a need for a subdivision of these two categories as dictated by needs and process capabilities: (1) coarse purification, whereby substantial amounts of unwanted impurities are removed in the simplest, most convenient manner; (2) fine purification for the removal of residual impurities to a degree sufficient for the majority of normal chemical plant operations, such as catalysis or the preparation of normal commercial products; or cleaning to a degree sufficient to discharge an effluent gas to the atmosphere through a chimney; and (3) ultra-fine purification where the extra step (as well as the extra expense) is justified by the nature of the subsequent operations or the need to produce a particularly pure product.

Contrary to the general belief of some scientists and engineers, all hydrogen purification systems are *not* alike, and having a good understanding of the type of gaseous effluents from coal-based processes is necessary for implementing the appropriate solution. The design of a hydrogen purification system must always take into account the operation of the upstream installations since every process will have specific set of requirements. In some cases, the application of a dry dusting removal unit may not be possible and thus requires a special process design for the wet hydrogen purification plant. Thus, the hydrogen purification process must always be of optimal design for both the upstream and downstream processes.

Hydrogen purification, although generally simple in chemical and/or physical principles, is often confusing because of the frequent changes in terminology and, often, lack of cross-referencing

TABLE 6.4
Summation of the Methods Used for Purification of Hydrogen and Approximate Purity of the Hydrogen*

Method	Purity
Catalytic purification	ca. 99% v/v
Cryogenic separation	90–98% v/v
Metal hydride separation	75–95% v/v
Palladium membrane diffusion	ca. 99% v/v
Polymer membrane diffusion	92–98% v/v
Pressure swing adsorption	70–85% v/v
Solid polymer electrolyte cell	95% v/v

*Listed alphabetically and not by preference or performance.

(Mokhatab et al., 2006; Speight, 2013a, 2013b, 2014, 2019, 2020). Although gas processing employs different process types, there is always overlap between the various concepts. And, with the variety of possible constituents and process operating conditions, a universal purification system cannot be specified for economic application in all cases.

Nevertheless, the first step in hydrogen purification is usually a device to remove large particles of carryover (entrained) material, coal, and other solid materials (Mokhatab et al., 2006; Speight, 2019, 2020). This is followed by cooling, quenching, or washing, to condense tars and oils and remove dust and water-soluble materials from the gas stream. Water washing is desirable for simplicity in hydrogen purification; however, the purification of this water is not simple.

Cleanup steps and their sequence can be affected by the type of gas produced and its end use (Mokhatab et al., 2006; Speight, 2019, 2020). The minimum requirement in this respect would be the application of low heat-value (low-Btu) gas produced from low-sulfur anthracite coal as a fuel gas. The gas may pass directly from the gasifier to the burners and, in this case, the burners are the cleanup system. Many variations on this theme are possible and, in addition, the order of the cleanup stages may be varied.

Each technique (Table 6.4) has limitations in terms of the purity of the hydrogen and it is appropriate to consider these in a comparative way, prior to discussing the development and status of noble metal membrane diffusion in more detail.

However, practical considerations often impose restraints so that only two techniques are actually utilized over this whole range, namely (1) catalytic purification and (2) polymer membrane diffusion. The two physical techniques, cryogenic separation and pressure swing adsorption, are best suited to large-scale applications and the remaining techniques, including Palladium alloy membrane diffusion, are utilized for small to medium outputs.

The selection of a particular process type for hydrogen purification is not simple, and several factors have to be considered, not the least of which is the constitution of the gas stream that requires treatment. Indeed, process selectivity indicates the preference with which the process will remove one acid gas component relative to (or in preference to) another. For example, some processes remove both hydrogen sulfide and carbon dioxide, while other processes are designed to remove hydrogen sulfide only (Mokhatab et al., 2006; Speight, 2019, 2014).

The purification or separation of hydrogen has traditionally been based on solid-state diffusion technology utilizing noble metal membranes (Grashoff et al., 1983). In fact, despite the emergence of alternative techniques, diffusion technology based on silver-palladium membranes remains the most suitable technique for the production of high-purity hydrogen.

Although the ability of metals and particularly palladium to selectively permeate hydrogen has been known for over a century, the commercial utilization of this property for the purification of hydrogen is relatively recent. The development of silver-palladium alloys as diffusion membrane materials in the late 1950s overcame the principal technical difficulties associated with the technique and enabled the successful development of commercial equipment, a technology with which Johnson Matthey has been closely associated from the beginning.

Since that time, several other techniques for hydrogen purification have been developed and are at various stages of commercial exploitation. As hydrogen is becoming increasingly important in many major areas, including the electronics manufacturing industry, it is an appropriate time to review the status of membrane diffusion for hydrogen purification and to highlight the current developments in membrane materials, especially palladium alloys, in the context of their potential influence on hydrogen purification technology and its industrial applications.

Since the introduction of the technique for hydrogen purification by selective diffusion through palladium alloy membranes, several other purification techniques have been developed. These fall broadly into the following categories:

Chemical:
- Catalytic Purification

Physical:
- Metal Hydride Separation
- Pressure Swing Adsorption
- Cryogenic Separation.

Selective Diffusion –
- Noble Metal Membrane Diffusion
- Polymer Membrane Diffusion
- Solid Polymer Electrolyte Cells.

There are two other aspects of importance to be considered in selecting a purification technique besides the scale of operation, which are (1) the ability of the technique to cope with a range of gas feedstocks in terms of hydrogen content and (2) the limitations imposed by the technique that are typically based on the chemical composition of the feedstock. These two aspects include the selectivity of the technique against gases other than hydrogen and its resistance to 'poisoning' by constituents present as impurities in the feed gas.

All techniques can operate well with hydrogen-rich gas feedstocks. Catalytic purification removes oxygen by reacting with hydrogen to form water, and carbon monoxide by oxidation or methanation; this technique is used to upgrade relatively pure hydrogen produced, for example, by electrolysis. In fact, the pressure swing adsorption method also requires hydrogen-rich gas streams since it functions by selectively adsorbing impurities. On the other hand, cryogenic separation can tolerate a wider range of hydrogen content in the feed gas, typically 30 to 80%, but is limited in gas composition to those constituents that will selectively condense at cryogenic temperatures.

Metal hydride separation can achieve a 99% purity level with the solid polymer electrolyte technique attaining a near 100% purity. Two methods are capable of producing moderately high purity levels of 99.999%, namely pressure swing adsorption and catalytic purification, but where very high purities of 99.9999% or better are required, only palladium membrane diffusion comes into consideration.

Metal hydride separation and the two diffusion techniques, palladium membrane and polymer membrane, have the ability to deal with feed gases lean in hydrogen. The hydride separation and palladium diffusion techniques are based on the very selective adsorption and diffusion of hydrogen, respectively, and the purity of the output hydrogen is not affected by the leanness of the feed gas. However, polymer membrane diffusion is based on a differential diffusion rate principle, and purity of the output hydrogen will be affected by the hydrogen concentration as well as by the nature of the other constituents that are present in the input gas stream.

The effect of membrane thickness on the diffusion of hydrogen through palladium and silver-palladium alloys is well established and conforms closely to a relationship where the rate of diffusion is inversely proportional to membrane thickness.

The selection of the optimum purification technique for specific industrial applications must be based on both technical and economic considerations. The degree of purification of hydrogen obtained from the different methods varies from around 90% for the cryogenic and polymer membrane techniques to 99.9999% for palladium alloy membrane diffusion.

Traditionally, analytical and metallurgical processing requirements utilize palladium alloy diffusion technology coupled with an impure hydrogen feed from electrolytic plants or commercial bottled gas. Here, diffusion technology is used profitably to upgrade the hydrogen, removing impurities and moisture down to 0.5 vpm. A major and growing application is in semiconductor manufacturing, particularly in the epitaxial growth stage. Recent experience suggests that the use of high-purity hydrogen, produced by the application of palladium alloy diffusion technology, leads to significantly improved yields in manufacturing and hence decreased production costs.

The techniques currently available for the purification or separation of hydrogen have been described in terms of their flexibility with respect to feed gas and the quality of hydrogen produced.

Hydrogen purification technology is a crucial link from hydrogen production to hydrogen utilization. Stable, reliable, and low-cost hydrogen sources represent a base for large-scale applications in refineries. Thus, high-efficient and low-power hydrogen purification technologies for refinery hydroprocesses play a major role in the development of the refinery.

A refinery can operate efficiently only if high-quality hydrogen is available from one or more of several processes. Typically, the hydrogen produced by any of these methods cannot be used directly without purification (1) by physical methods or (2) by chemical methods.

The physical methods include former include adsorption methods such as pressure swing adsorption (PSA) or temperature swing adsorption (TSA), low-temperature separation methods (such as cryogenic distillation and low-temperature adsorption), and membrane separation methods (the use of an inorganic membrane or n organic membrane), while the latter involve a metal hydride separation and catalysis method (Table 6.3). The selection of an appropriate H_2 purification method is closely related to the hydrogen supply mode and gas source. For H_2 production by centralized large-scale coal gasification and natural gas reforming with an H_2 supply amount ≥10,000 Nm^3/h, pressure swing adsorption purification is primarily adopted after transformation, desulfurization, and decarbonization. The pressure swing adsorption technology is a mature technology and is characterized by low operation costs and a long service life. However, the production of very high-purity hydrogen by the traditional pressure swing adsorption method with a standard impurity content results in a decreased recovery rate and yield. It is also not cost-efficient due to low requirements for specific impurity removal (e.g., $CO \leq 0.2$ ppm).

Cryogenic distillation (Aasadnia et al., 2021) is also applicable to large-scale production, but standard hydrogen purity may be on the order of 85 to 99% v/v, which may not satisfy all of the process requirements. To produce high-purity hydrogen, versatile processes should be applied that are selected based on the contaminants to be removed. For example, an organic membrane combined with a pressure swing adsorption process can be used for obtaining methanol purge gas, while a two-stage or multi-stage pressure swing adsorption process is adopted for obtaining coke oven gas and by-product gas from the refinery. In addition, low-temperature adsorption, metal hydride separation, and metal membrane separation processes are available processes according to the types and amounts of impurities. The low-temperature adsorption process can effectively eliminate multiple impurities, such as sulfide derivatives, formaldehyde (HCHO), and formic acid (HCOOH) (Schorer et al., 2019). At the same time, the inherent disadvantage is that purified materials react with impure gas during hydrogen recovery, reducing the purification efficiency (Dunikov et al., 2012; Dunikov and Blinov, 2020). New membrane technologies, such as carbon molecular sieve membranes (CMSMs) (Hamm et al., 2017) and ionic liquid membranes (Sasikumar et al., 2018) have been developed for hydrogen purification.

Thus, hydrogen purification is an essential part of the overall hydrogen technology in order to meet the end specification for the gas, dependent upon whether the gas is to be used for refinery operations (such as hydrotreating processes and hydrocracking processes), chemical manufacturing, or for the pure hydrogen that is required for use in fuel cells. The hydrogen purification system may include one or more filter (or reactor) vessels coupled in series for removing a variety of contaminants. The system may be operated by receiving a hydrogen gas stream at a first temperature and pressure and dropping the temperature of the stream as the gas flows through the system.

The process options for the purification of the various hydrogen streams are presented below in alphabetical order rather than in any order or preference and are (1) cryogenic separation, (2) membrane separation processes, (3) metal hydride separation, (4) pressure swing adsorption units, and (5) wet scrubbing.

The purification of hydrogen actually commences during the production stage when the water gas shift reaction is used not only to increase the yield of hydrogen from, for example, the steam methane reforming process, but also to lower the content of carbon monoxide in the water gas shift process product gas. Thus:

$$CO + H_2O \rightarrow H_2 + CO_2$$

The water gas shift process is an older and well=established process in many industrial plants that produce hydrogen not only for petrochemical processes but also to adjust the carbon monoxide-hydrogen ratio for the Fischer-Tropsch process (Chadeesingh, 2011; Speight, 2013a).

However, in the sense of 'purification proper', the following types of processes are the more usual processes that are termed 'purification process' which include processes (listed alphabetically rather than in any order of use or importance): (1) cryogenic separation, (2) desiccant separation systems, (3) membrane separation, (4) metal hydride separation, (5) pressure swing adsorption, and (6) wet scrubbing, as well as (7) other purification processes such as processes for moisture removal and oxygen removal.

6.3.1 Cryogenic Separation

Distillation is another method used to purify hydrogen gas and the process involves the use of a distillation column that separates the impurities from the hydrogen gas based on the respective boiling points.

The cryogenic separation processes (sometime referred to as the 'cryogenic distillation process') is a process that uses low temperatures to liquefy and separate hydrogen from other gases and has the ability to produce pure hydrogen of a very high purity (>99.999% v/v). The principle of cryogenic distillation is to separate and purify hydrogen by utilizing the difference in the relative volatility of different components in feed gases. Compared with methane and other low-boiling hydrocarbon derivatives, hydrogen has relatively high volatility, such that hydrocarbon derivatives, carbon monoxide, and nitrogen condense before hydrogen with temperature reductions (Song et al., 2019).

The low-temperature separation method assures a high hydrogen recovery rate, but it is challenging to adapt the method for treating different feed gases. As such, it is necessary to remove carbon dioxide, water, and other impurities from the feed gases before the separation so as to avoid equipment blockage at a low temperature. Besides this, high costs and energy consumption accompany the requirements for gas compressors and cooling equipment in the actual operation. Although most impurities are liquefied at a low temperature, some remain in the gas phase as saturated steam, so it is difficult to directly obtain hydrogen that meets the purity standards of fuel cell vehicles.

Cryogenic separation units operate by cooling the gas and condensing some or all, of the constituents for the gas stream. Depending on the product purity required, separation may involve flashing or distillation. Cryogenic units offer the advantage of being able to separate a variety of products from a single feed stream. One specific example is the separation of low-boiling olefin derivatives from a hydrogen stream. Hydrogen recovery is in the range of 95% v/v, with a purity above 98% v/v.

6.3.2 Desiccant Separation Systems

The use of silica desiccant columns (such as silica gel beads) is another common purification method and is popular due to its simplicity. The hydrogen stream then flows through a stainless-steel desiccant cartridge for moisture removal.

A molecular sieve is an effective purification process that is required. For example, a 5A molecular sieve is a strong adsorbent which is available in beads and pellet form to absorb different liquids and gases. It is the calcium form of the Type A crystal structure and is widely used for drying and purifying hydrogen due to its high adsorbing quality, speed, and high crushing strength.

The 5A molecular sieve is generally an alkali aluminosilicate with an effective pore size of 5 angstroms and is suitable for PSA hydrogen purification by pressure swing adsorption, where unwanted moisture and gases are adsorbed to obtain pure hydrogen. The 5A zeolite adsorbent can be regenerated either through the thermal swing process or by decreasing the pressure in the pressure swing process. A temperature on the order of 250 to 300°C (480 to 570°F) is needed to remove the moisture from the molecular sieve

General membrane separation depends upon a pressure gradient to drive one component of the mixture (Liu and Zhang, 2004). Pressure differences of several hundred pounds per square inch are typical for this separation. The process does not produce a high-purity product, nor does it remove one component with perfect selectivity. One attempt to solve these problems has been the electrolytic transport of hydrogen in the form of protons across a solid polymer electrolyte in an electrolysis cell.

Although pressure swing adsorption and cryogenic distillation technologies are both commercial processes, multiple units are required and, in some instances, supplementary wash columns are required to remove carbon monoxide and carbon dioxide. Pressure-driven membrane processes are considered to be better candidates for hydrogen production because they are not as energy intensive, and they yield high-purity hydrogen (Adhikari and Fernando, 2006).

6.3.3 Membrane Separation

In membrane separation processes, membranes are used for the separation and purification of hydrogen gas. The membrane is chosen to be selectively permeable (in this current context) to hydrogen gas.

As an alternative to pressure-driven membranes, electrochemical membrane technology has been utilized for hydrogen separation and it has been used to effectively recover hydrogen from a variety of hydrogen-rich gas mixtures (Sakai et al., 2008). Electrochemical hydrogen separation (EHS) offers several advantages over other processes, including high-purity hydrogen from one-step operation (Grigoriev et al., 2011). An additional benefit of this electrochemical method is that carbon dioxide, if present in the gas stream to be treated, can be captured and stored (Gardner and Ternan, 2007).

Membrane separation methods have the advantages of flexible and simple operation, compact structure, low energy consumption, and environmental friendliness (Adhikari and Fernando, 2006; Schorer et al., 2019; Vermaak et al., 2021).

Metal alloys or composite metal membranes have been used for hydrogen purification. Therefore, ceramic membranes, inert to poisonous gases, are desirable and, in microporous membranes, the flux is directly proportional to the pressure, whereas in palladium membranes, the flux is proportional to the square root of the pressure.

In the membrane separation technology, with a permeable-selective membrane as a separation medium, the raw material components can selectively permeate the membrane under the action of driving forces (pressure difference, concentration difference, and potential difference), thereby achieving separation and purification.

The performance of membrane materials is the most critical factor determining the hydrogen separation and purification effects of the membrane. Commonly used membrane materials primarily include metal and polymer membranes, and novel membrane materials, such as nanomaterial membrane (CMSM), and a metal-organic framework membrane (MOF membrane), represent the largest known class of porous crystalline materials ever synthesized.

In the water gas shift reactor, carbon monoxide is first converted to carbon dioxide, and then the hydrogen and carbon dioxide are separated using an absorption process. A compressor is used to pressurize the purified hydrogen for immediate use or storage. Also, the use of high-temperature proton-selective polymer electrolyte membranes (PEMs) is required to separate hydrogen from other gas molecules such as carbon dioxide and carbon monoxide.

As an increasingly important gas separation technology, membrane separation has the advantages of flexible and simple operation, compact structure, low energy consumption, and environmental friendliness. In the membrane separation technology, with a perm-selective membrane as a separation medium, the raw material components can selectively permeate the membrane under the action of driving forces (pressure difference, concentration difference, and potential difference), thereby achieving separation and purification. The performance of membrane materials is the most critical factor determining the hydrogen separation and purification effects of the membrane.

There are various types of membranes that can be utilized to purify hydrogen H2 from a gas mixture that mainly consists of hydrogen and carbon dioxide which are (1) the hydrogen-selective membranes and (2) the membranes that allow passage of carbon dioxide.

In the first case (i.e., hydrogen-selective membranes), hydrogen passes through the membrane while carbon dioxide and other impurities (such as methane, CH4, and carbon monoxide, CO) are retained by the membrane to produce low-pressure hydrogen and high-pressure carbon dioxide. In the second case (membranes that allow passage of carbon dioxide), membranes retain hydrogen and other impurities (such as methane, CH4, and carbon monoxide, CO), to produce high-pressure hydrogen and low-pressure carbon dioxide.

Most of the membranes used for the separation of hydrogen and carbon dioxide are hydrogen-selective membranes. However, carbon dioxide-selective membranes may be favored in certain applications, including applications where the purified hydrogen remains at high pressure.

Thus, membrane separation processes are an alternative compared to mature technologies such as pressure swing adsorption and cryogenic distillation. Metal alloys or composite metal membranes have been used for hydrogen purification but metallic membranes are sensitive to some gases such as carbon monoxide and hydrogen sulfide and therefore, ceramic membranes which are inert to some gases, may be desirable for hydrogen purification. Inorganic microporous membranes offer advantages over thin-film palladium membranes.

In summary, membrane-based processes have a high potential for hydrogen purification, both at high pressure and low pressure but, as is the case, with many purification processes, must be selected based on the composition of the gas stream (Adhikari and Fernando, 2006; Speight, 2019; Sazali et al., 2020).

In general, carbon-based membranes are essentially formed by the pyrolysis of polymeric starting materials and the structure of polymer as well as the pyrolysis process parameters is among the key factors that decide transport properties of resultant carbon membranes (Sazali et al., 2020). Also, carbon molecular sieve (CMS) membranes with rigid and uniform pore structures are ideal candidates for high-temperature- and pressure-demanded separations, such as hydrogen purification from the steam methane reforming process.

Carbon molecular sieve membranes (CMSMs) with an amorphous microporous structure are the most common carbon-based membranes, usually obtained by the carbonization or pyrolysis of polymer precursors in the inert gas or vacuum environment. Common polymer precursors include polyimide and its derivatives, polyfurfuryl alcohols, and phenolic resins (Li et al., 2015; Xu et al., 2020; Lei et al., 2021).

6.3.3.1 Carbon-Based Membranes

Graphene and GO exhibit a single-atom thickness, high mechanical strength, and good chemical stability. Keeping in mind that the membrane thickness is inversely proportional to its permeability, graphene-based membranes have become ideal membranes with minimum transmission resistance and maximum permeation flux because of their ultra-low thickness. Therefore, to improve the gas separation performance of graphene-based membranes, the emphasis of the research is on imparting uniformly distributed nanopores with an appropriate size and shape and high porosity in graphene sheets.

Most of the membranes for the separation of hydrogen and carbon dioxide are hydrogen-selective membranes, due to their more advanced development compared to carbon dioxide-selective membranes. However, carbon dioxide-selective membranes may be favored in certain applications, including applications where the purified hydrogen remains at high pressure.

6.3.3.2 Metal Membranes

In the process, hydrogen is catalyzed to protons (H^+) and electrons (e^-)on the compact structure of metal membranes in which the protons pass through the metal membrane and bind electrons on the other side to form hydrogen. The metal membrane blocks carbon dioxide, methane, nitrogen, and oxygen, thereby achieving the selective permeation of hydrogen.

The palladium (Pd) membrane is related to high manufacturing costs, and it is prone to hydrogen embrittlement at a low temperature (Rahimpour et al., 2017).

The palladium membrane hydrogen purification units operate via pressure driven diffusion across palladium membranes through which only hydrogen can diffuse. The palladium diffuser can take a number of forms but typically comprises a palladium and silver alloy material possessing the unique property of only allowing monatomic hydrogen to pass through its crystal lattice when it is heated above nominally 300°C (570°F).

A palladium alloy membrane can be formed by adding other metal elements (such as silver, Ag, gold, Au, copper, and Cu, nickel, Ni) into the palladium membrane to (1) solve the problem of hydrogen embrittlement, (2) enlarge the palladium lattice, and (3) increase the hydrogen permeation rate (Nayebossadri et al., 2019; Zhao et al., 2020).

In fact, ultra-pure hydrogen can be obtained by diffusion through palladium alloys and hydrogen purity in excess of 99.99999% v/v can be obtained. The mechanism of hydrogen diffusion through palladium alloy has six steps which are (1) adsorption, (2) dissociation, (3) ionization, (4) diffusion, (5) reassociation, and (6) desorption. However, little-to-no hydrogen will flow through the membrane at room temperature and the operating conditions required for palladium alloy membranes include temperature on the order of 200 to 800°C (392 to 1,472°F) and a differential pressure across the membrane on the order of 44 to 180 psi. Very little routine maintenance is required and the typical life expectancy of a palladium diffuser in the purifier being around 5 years dependent on the application it is being used for and the level of usage.

Both pure palladium membranes and palladium alloy membranes are self-supporting membranes. The thickness of these membranes is limited to several tens of micrometers to several hundreds of micrometers to ensure sufficient mechanical strength. When the membrane thickness is too high, the hydrogen permeation rate may be limited. As such, it is possible to deposit a palladium membrane or a palladium alloy membrane on the surface of a porous material to prepare a supported palladium composite membrane, and the support increases the mechanical strength of the membrane while decreasing the amount of palladium and the thickness of the membrane.

6.3.3.3 Metal-Organic Framework Membranes

Metal-organic frameworks (MOFs) represent the largest known class of porous crystalline materials ever synthesized, and the metal–organic frameworks generally represent a novel class of organic–inorganic hybrid porous solid materials with regular geometric and crystal structures. They are composed of metal ions or metal ion clusters connected by organic connectors. Compared with other porous materials, metal-organic frameworks have the advantages of structural variability, ultra-high porosity, uniform and adjustable apertures, and adjustable properties of the inner surface (Wang et al., 2020).

6.3.3.4 Polymer Membranes

The working principle of polymer membrane separation is based on the different permeation rates of the constituents of gas streams through the polymer membrane – polysulfone (PSF), polyimide (PI), and polyamide are commonly used as polymer membrane materials. Some membrane separators use bundles of tiny hollow fibers inside a containment vessel or pipe. Each gas has a characteristic permeation rate which allows 'fast gases' (such as hydrogen) to be separated from 'slow gases' (such as oxygen).

During the process, a differential pressure develops across the fibers, and it is this pressure that drives the flow of the faster gas (in this case, hydrogen) through the wall of the fiber. This results in the production of a purified hydrogen stream called the 'permeate'. To accomplish an efficient separation, the differential pressure must be on the order of 100 psi, or greater.

Since the trade-off between selectivity and permeability limits the use of polymer membranes, researchers attempted to prepare mixed matrix membranes (MMMs) by incorporating chemical such as zeolite and silicon dioxide into the polymer to improve the overall performance (Strugova et al., 2018). Typically, polymer membranes have high hydrogen.

6.3.4 Metal Hydride Separation

Metals that form hydride derivatives can reversibly absorb hydrogen and release hydrogen when appropriate temperatures and pressures are applied to the hydride.

A metal hydride (often represented simply as MH_x, where $\times = 1$ or greater, typically 2, 3, or 4) is synthesized through the reaction of a metal with hydrogen which is formed on industrial scale either by the electrolysis of water, by heating coke with steam in the water gas shift reaction or using hydrocarbons with steam. The hydride derivative is a hydrogen atom which has an extra electron and, thus, is a negatively charged ion, or anion (Chen et al., 2013; Dunikov and Blinov, 2020).

Metal hydrides (MH_x) are the most technologically relevant class of hydrogen storage materials because they can be used in a range of applications including neutron moderation, electrochemical cycling, thermal storage, heat pumps, and purification/separation. The thermodynamics of the adsorption/desorption of hydrogen govern the temperature range and the pressure range of applicability.

In the case of the metal hydride separation process, hydrogen is chemically bonded or absorbed by the metal without the need for compression. Instead, the solid metal hydride functions in a similar manner to a sponge that absorbs and releases the gas. Recovery of the hydrogen from the metal hydride is through the application of heat.

6.3.5 Pressure Swing Adsorption

Pressure swing adsorption (PSA) is one of the most common methods used for purification of hydrogen gas. Hydrogen purification by pressure swing adsorption involves the use of a vessel (sometimes referred to as the reactor) filled with a suitable adsorbent material that is selectively adsorptive of certain impurities, such as water vapor or carbon monoxide. Pressure swing adsorption hydrogen purification is an industrial process only suited for stationary installations.

Gas separation and purification by pressure swing adsorption is implemented by periodical pressure changing based on the difference in the adsorbent capacity for different gases. The pressure swing adsorption separation effect primarily depends on the type of adsorbent and the technical process used. Hydrogen is significantly different from the majority of gas molecules, such as carbon monoxide, carbon dioxide, and methane, in terms of static capacity, so it is very suitable for PSA separation and purification.

Thus, gas separation and purification by the pressure swing adsorption method is implemented by periodical pressure changing based on the difference in the adsorbent capacity for different gases.

Modifications and innovations of these adsorbents were reported in regard to different impurities and have mainly focused on the removal of carbon dioxide. The purity of the hydrogen from the process can exceed 90% v/v, and purity of the hydrogen stream in excess of 98% v/v has been reported. The improvement and optimization of the PSA process are two crucial factors that are necessary to increase the efficiency of the hydrogen purification process (Sircar and Golden, 2000, 2010; Lively et al., 2012; Agueda et al., 2015; Kuroda et al., 2018; Shamsudin et al., 2019; He et al., 2020).

The method involves regenerative technology, and there are no high temperatures or associated high electrical currents. A continuous flow of hydrogen with no pressure fluctuations or pulsing effects and maintenance requirements limited to replacement of the de-ionizer cartridge. Also, there is no need for the replacement of desiccants or hazardous caustics required. A typical flow scheme uses a two-bed or a four-bed pressure swing adsorption system – the latter system is often employed to recover hydrogen from coal gas streams that have nitrogen as a major impurity. As might be expected, the four-bed system is superior in performance to the performance of the two-bed system.

The use of silica desiccant columns is another common purification method and is popular due to its simplicity.

The improvement and optimization of the pressure swing adsorption process is a crucial ways to increase the efficiency of the purification process. The flow scheme of the typical pressure swing adsorption system uses a two-bed system and a four-bed system to recover hydrogen from coal gas with nitrogen as a major impurity. The performance of the 4-bed process is (as might be expected) superior to that of the 2-bed process, insofar as the 4-bed process can yield a hydrogen purity on the order of 96 to 99.5% v/v and a recovery rate of 71 to 85% (Abdeljaoued et al., 2018).

6.3.6 Sorption-Enhanced Reforming

The sorption-enhanced reforming process (often referred to as the SER process or absorption-enhanced reforming, AER) can produce streams with a high content of hydrogen by selectively removing carbon dioxide from the reactive zone in the gasifier. This moves the equilibrium of the water gas shift reaction to the side of hydrogen, allowing a hydrogen content on the order of 75% v/v to be achieved:

$$CO + H_2O \rightarrow H_2 + CO_2$$

The downside (if there is one) is the production of more carbon dioxide but this can be counteracted by the use of calcium oxide as an adsorbent for the carbon dioxide.

In the process, limestone (calcium carbonate, $CaCO_3$) is used as bed material which leads to an in situ removal of carbon dioxide to produce calcium oxide which then reacts with carbon dioxide in the gas stream materials. Thus, the calcium carbonate acts as a carbon dioxide carrier which leads to an in-situ carbon dioxide removal from the product gas. This is accomplished by the circulating bed material. CaO reacts with CO_2 in the gasification reactor at a temperature on the order of 650 to 700°C (1,200 to 1,290°F):

$$CaO + CO_2 \leftrightarrow CaCO_3$$

The calcium carbonate that is produced is transported to the combustion reactor were it is regenerated again to calcium oxide thereby releasing the carbon dioxide (at a temperature on the order of 900°C, 1,650).

6.3.7 Wet Scrubbing

Wet scrubbing systems, particularly amine systems or potassium carbonate systems, are used for removal of acid gases such as hydrogen sulfide or carbon dioxide (Speight, 1993; Dalrymple et al., 1994; Mokhatab et al., 2006; Speight, 2019). Most systems depend on chemical reactions and can be designed for a wide range of pressures and capacities. They were once widely used to remove carbon dioxide in steam reforming plants but have generally been replaced by pressure swing adsorption units except where carbon monoxide is to be recovered. Wet scrubbing is still used to remove hydrogen sulfide and carbon dioxide in partial oxidation plants.

Wet scrubbing systems remove only acid gases or high molecular weight hydrocarbon derivatives, but they do not remove methane or other hydrocarbon gases, hence they have little influence on product purity. Therefore, wet scrubbing systems are most often used as a pretreatment step, or where a hydrogen-rich stream is to be desulfurized for use as fuel gas.

A wet scrubbing system is, as the name implies, the use of an aqueous solution which has the ability (or at least the potential) to remove contaminants from hydrogen streams.

Wet scrubbers use a liquid to remove pollutants from an exhaust stream. In gaseous emission control applications, wet scrubbers remove pollutants by absorption. For this reason, wet scrubbers used for gaseous pollutant control are often are referred to as absorbers. Wet scrubbers rely on the creation of large surface areas of scrubbing liquid that allow intimate contact between the liquid and gas.

For example, Amine scrubbing has been used to separate carbon dioxide from gaseous streams since 1930 (Mokhatab et al., 2006; Rochelle, 2009; MacDowell et al., 2010; Speight, 2019. In the process, the gas stream rich on carbon dioxide is contacted with an aqueous amine solution which reacts with the carbon dioxide to produce water-soluble salts.

The loading capacity for chemical solvents, such as the like amine-type solvents, is primarily dependent on the concentration of the active components and the achievable loading according to the thermodynamic equilibrium. For the range of alkanolamines, the primary amines (monoethanolamine, MEA, $HOCH_2CH_2NH_2$) will be more favorable in terms of reaction rates compared to secondary amines [such as diethanolamine, DEA, $HN(CH_2CH_2OH)_2$], or tertiary amines [such as methyl diethanolamine, MDEA, $CH_3N(C_2H_4OH)_2$.]. However, achievable loadings and heat requirement for regeneration will be also higher for primary amines. (Bailey and Feron, 2005)

In the process, the gas stream that has a high content of carbon dioxide passes through several stages which are (1) the gas stream is fed into the absorber where the carbon dioxide reacts with the amine solution resulting in the lean gas stream after which (2) the liquid amine solution, which is often referred to as a lean amine solution with a low content of carbon dioxide (3) enters from the top of the absorber and leaves at the bottom of the absorber as a rich amine stream, (4) which is then preheated and sent to the stripper. In the stripper, the rich amine solution flows in counter-current flow down to the bottom. At the bottom, the aqueous amine solution is heated in order to desorb the carbon dioxide and to generate steam, which heats the rich amine solution falling from the top of the column. The steam and the desorbed carbon dioxide leave the stripper at the top after which the steam and a small part of evaporated amine is condensed and fed back to the stripper and the desorbed carbon dioxide leaves the stripper. To complete the cycle, the regenerated amine solution is cooled and recycled to the absorber (Bailey and Feron, 2005; Ryckebosch et al., 2011; Bauer et al., 2013).

In general, there are several types of wet scrubber designs, including (1) the spray tower-type, (2) the tray-type, and (3) the packed-bed type which are generally referred to as low-energy scrubbers. Packed-bed scrubbers provide excellent gas-liquid contact and efficient mass transfer; they can generally be smaller in size than spray scrubbers and so are an effective option when space is limited.

Plugging may occur in packed-bed scrubbers if there is a high loading of particulate matter, but the packing can be removed for cleaning. Some packed-bed scrubbers employ mobile spherical packing; the movement of the packing increases turbulence and helps keep the packing clean. Tray-type (or plate-type) scrubbers provide a film of liquid for the gas to pass through. Contact between gas and liquid is obtained by forcing the gas to pass upward through small orifices and bubbling through a liquid layer flowing across the plates. However, the use of a wet scrubbing system also has the potential to increase the proportion of water in the treated gas stream.

Wet scrubbers can also be used for heat recovery from hot gases by flue-gas condensation. In this option, water from the scrubber drain is circulated through a cooler to the nozzles at the top of the scrubber. The hot gas enters the scrubber at the bottom and, if the gas temperature is above the dew point of the water, it is initially cooled by evaporation of the water drops. Further cooling causes water vapor to condense, thereby adding to the amount of circulating water.

In summary, the selection of the optimum purification technique for specific applications in the refinery must be based on both technical and economic considerations. The degree of purification of hydrogen obtained from the different methods varies from a purity on the order of 90% v/v for the Cryogenic method and the Polymer Membrane method to a purity on the order of 99.9999% v/v for the palladium alloy membrane diffusion method.

Traditionally, analytical and metallurgical processing requirements utilize palladium alloy diffusion technology coupled with an impure hydrogen feed from electrolytic plants or commercial bottled gas. Here, diffusion technology is used profitably to upgrade the hydrogen, removing impurities and moisture down to 0.5 ppm.

The ready availability and nature of the fuel and the ability to produce from it low-cost hydrogen of over one thousand times its own volume combine to produce a useful portable supply of hydrogen.

6.3.8 Other Purification Processes

The level of recovery (and purity) of hydrogen from various sources varies considerably from one technique to another. Only palladium membrane diffusion and catalytic purification techniques can achieve high recovery levels of up to 99% v/v from hydrogen-rich gases.

In addition to the process described in the above sections, there are two other processes that must be used to complete the purification processes, and they are (1) moisture removal and (2) oxygen removal.

6.3.8.1 Moisture Removal

Water contamination of hydrogen can be reduced by coalescers, refrigeration dryers, membrane dryers, molecular sieve dryers, and desiccant dryers. Coalescers are designed to cause the combination of smaller aerosols into larger droplets, susceptible to the effects of gravity. Coalescers remove sub-micron solids and aerosols by three different mechanisms.

First, particles in the range from 0.001 to 0.2 microns collide with the filter media and are subject to diffusion coalescing. Second, particles in the range of 0.2 to 2 microns are intercepted by 0.5-micron glass fibers. The efficiency of the interception mechanism increases as the pore size decreases. Third, particles 2 microns and larger are removed by direct inertial impacting, because of their larger mass and momentum. A coalescer is one of the most cost-effective water aerosol removers.

Refrigeration dryers cool the gas and dry the gas by condensation to a low dew point by using electric motor compressors and refrigeration fluids in a conventional Joule-Thompson expansion scheme.

Membrane dryers utilize a hygroscopic ion exchange membrane to selectively remove water vapor from mixed gas streams. They can be thought of as desiccants in tubular form. Tubes are bundled together inside a pipe shell so that the wet feed gas stream flows through the tubes and wets the inside walls. A counter-current dry gas stream flows on the outside of the tubes and purges water from the shell. A disadvantage of this type of dryer is that the dry purge-gas flow rate is greater than the wet product feed rate.

Another substance used for drying gas is zeolite. Zeolites are alumino-silicate mineral particles that absorb and desorb large quantities of water reversibly. Water absorption on these so-called molecular-sieve dryers is by way of a physical rather than chemical reaction. Zeolite dryers are regenerated thermally in their piping containers by heating them up to 550°C (1,000°F) while purging them with a dry gas. For continuous operation, two units are used. One is drying gas while the other is regenerating.

6.3.8.2 Oxygen Removal

The catalytic recombination of hydrogen and contaminant oxygen is usually accomplished at room temperature or higher and results in the formation of water vapor, which is easily removed if necessary. Complete removal of the undesirable oxygen impurity can be done in the presence of a stoichiometric or greater quantity of hydrogen.

Usually, the catalysts that are used for removal of the oxygen impurity from hydrogen (or vice versa) are based on platinum group metals (PGM). Thin films of PGM are supported on the surface of alumina pellets. The catalyst pellets are contained by screens in a section of the pipe. The hydrogen gas flows through the pipe as it is being treated.

The catalyst will not work while wet. Care should be taken to avoid water condensation on the catalyst. If it becomes wet, simply dry it out before using. Removal of oxygen from hydrogen is called deoxygenation. A catalytic recombiner (D) can be designed to be capable of removing up to 3% oxygen from a hydrogen production stream, reducing the oxygen content to less than 1 ppm.

6.4 HYDROGEN PROPERTIES

Hydrogen, the simplest element in the periodic table of the elements (Figure 6.1) is composed of only one proton and one electron, makes up a major part of the composition of the universe and exhibits a variety of chemical and physical properties (McCarty et al., 1981). Physically, hydrogen is lighter than air and insoluble in water (Table 6.5).

Hydrogen has three known isotopes, and the mass numbers of these isotopes are 1, 2, and 3, with the most abundant being the mass 1 isotope generally referred to as hydrogen (symbol H, or ^{1}H) but also known as 'protium'. On the other hand, the ^{2}H isotope has a nucleus of one proton and one neutron and is generally referred to as 'deuterium' or 'heavy hydrogen' (symbol D, or ^{2}H). The third isotope, generally referred to as 'tritium' (symbol T, or ^{3}H) has one proton and two neutrons in each nucleus.

Hydrogen has many actual and potential uses and is used in: (1) Industrial processes, such as the crude oil refining industry. (2) Fuel cells for producing electricity. (3) Combustion systems for electricity generation and heating. (4) The production of rocket fuel. Hydrogen also has the potential for effectively storing energy for electric power generation, and biofuel producers use hydrogen to produce hydrotreated vegetable oil (HVO) for use as 'renewable diesel fuel'.

The most important chemical properties of hydrogen as an energy medium are the ones related to combustion (Table 6.5) shows the basic and chemical properties of hydrogen.

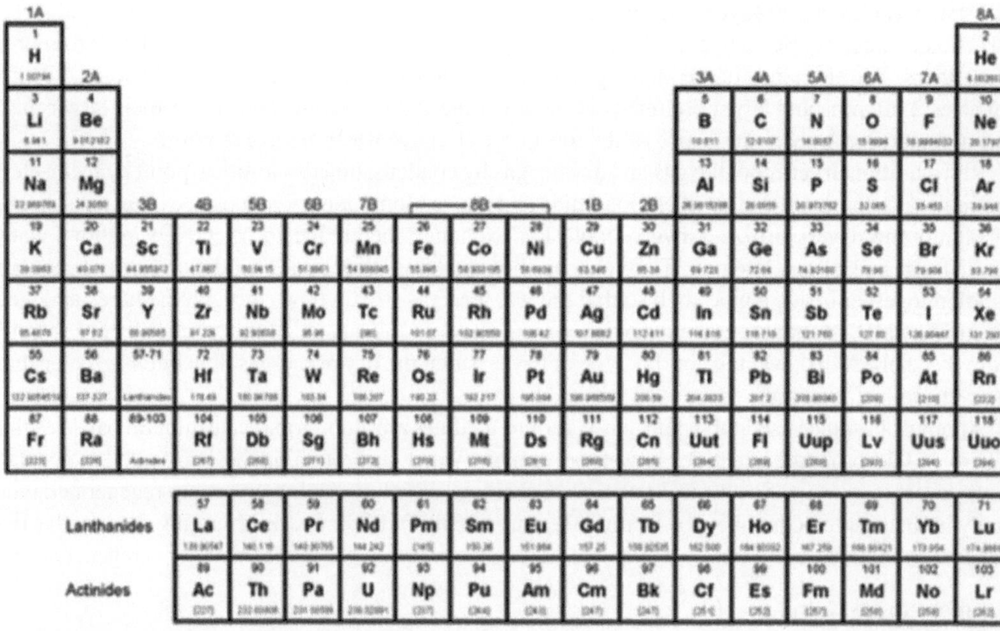

FIGURE 6.1 The Periodic Table of the Elements.

TABLE 6.5
Selected Properties of Hydrogen ($^1H_2$)

Property*	Value
Appearance	Colorless and odorless gas at room temperature
Auto-ignition temperature	500–571°C (930–1,060°F)
Atomic number	1
Atomic weight	1.00782519
Boiling point	−252.88°C/(−422.9°F) (at 1 atm/14.7 psi)
Bond distance	0.7416 angstrom
Chemical structure	H–H
Critical density	0.0310 gram per cubic centimeter
Critical pressure	13.0 atmospheres
Critical temperature	−240.0°C (−400°F)
Density (gas)	0.08988 g L^{-1} (0°C/32°F, 1 atm/14.7 psi)
Density (liquid)	70.8 g L^{-1} (at −253°C/−423°F)
Density (solid)	0.08671 gram per cubic centimeter
Dissociation energy (25°C)	104.19 kilocalories per mole
*Explosive limits	4–76% v/v (in air)
Formula	H2
Heat of combustion to water (g)	−57.796 kilocalories per mole
Heat of fusion	28 calories per mole
Heat of vaporization	216 calories per mole
Melting point	−259.35°C (−434.6°F)
Molecular weight	2.0159
Relative vapor density (air = 1)	0.07
Solubility in water	0.0214 cm^3 g^{-1} (0°C/32°F, 1 atm)

*Listed alphabetically rather than in any order of preferences.

TABLE 6.6
Examples of the Reactions of Hydrogen with Other Elements

1. **Reactions with the halogen family:**
   Hydrogen reacts with halogen elements to produce hydrogen halides
   $H_2 + X_2 \rightarrow 2HX$
   In this equation, X represents a halogen
2. **Reaction with the oxygen molecule:**
   Hydrogen reacts with the oxygen molecule to form water. This reaction is highly exothermic in nature
   $2H_2 + O_2 \rightarrow 2H_2O$
3. **Reaction with metals:**
   Hydrogen reacts with different metals to form their respective hydrides
   $H_2 + 2M \rightarrow 2MH$
   In this equation, M represents an alkali metal

In addition to a variety of chemical reactions (Table 6.6), hydrogen finds a variety of applications due to its dual nature.

The following are some common uses of hydrogen. (1) Used in the synthesis of ammonia and the manufacture of nitrogenous fertilizers. (2) Used in the hydrogenation of unsaturated vegetable oils for manufacturing vanaspati fat. (3) Used in the manufacture of many organic compounds, for example, methanol (CH_3OH). (4) Used in the reduction of metal oxides to metals by metallurgical processes. (5) Used as rocket a fuel. (6) Used for experiments within the automotive industry with hydrogen fuel cells. (7) Used as hydrogen chloride, which is manufactured by the reaction of hydrogen with chlorine.

6.5 STORAGE

Typically, large amounts of hydrogen are produced by the refining industry and this hydrogen is consumed at the refinery. Thus, there is the need to store hydrogen at the site and, to accomplish this, there are several methods exist for hydrogen storage which include (1) physical methods such as the use of high pressure and low temperature or (2) chemical methods which rely on the formation of chemical compounds that release hydrogen under certain treatment conditions (Sankir and Sankir, 2018).

Hydrogen can be stored using a variety of methods which include: (1) high-pressure gas cylinders up to 12,000 psi, (2) liquid hydrogen in cryogenic tanks, (3) adsorbed hydrogen on materials with a large specific surface area (4) absorbed on interstitial sites in a host metal at ambient pressure and temperature, (5) chemically bonded in covalent and ionic compounds at ambient pressure, or (6) through oxidation of reactive metals, e.g. Li, Na, Mg, Al, Zn with water.

However, there are issues that need attention when a hydrgoen storage system is planned and these are (1) hydrogen storage facilities should be provided with adequate ventilation for both normal operating conditions and emergency situations, (2) vent lines for hydrogen (including pressure relief lines and boil-off from cryogenic systems) should be vented to an appropriate exhaust system or a safe outside location, (3) the vent should be designed to prevent moisture or ice from accumulating in the line, and there is also the recommendation that (4) unused hydrogen should be disposed of by venting or possibly flaring. Whether or not the last item (i.e. item 4) is the safest way to dispose of any unused hydrogen is under debate and caution is needed because of the tendency for hydrogen to burn (flare) explosively.

6.5.1 Physical Storage

In terms of the physical methods, hydrogen is stored as a compressed gas or as a cryogenic liquid, and transported as such in cylinders, tubes, and cryogenic tanks for use at site where hydrogen is not produced in sufficient to match the needs at that site. However, the continuing challenge for hydrogen storage is the low boiling point of hydrogen (-252.9°C or -423.2°F) which requires expending a significant amount of energy. On the other hand, the storage of hydrogen as a gas typically requires high-pressure tanks (on the order of 5,000 to 10,000 psi tank pressure).

The storage of hydrogen as a liquid requires cryogenic temperatures because the boiling point of hydrogen at one atmosphere pressure is –252.8°C. Thus, hydrogen storage can be categorized into five approaches (1) adsorbent, (2) liquid organic, (3) interstitial, (4) complex hydride, and (5) chemical hydrogen.

Several methods exist for storing hydrogen and include high pressure and low temperature methods or by employing chemical compounds that release hydrogen upon demand. The methods for the storage of hydrogen can be generally classified in three ways which are (1) as a compressed gas in high-pressure tanks, (2) as a in thermally-insulated containers – often referred to as dewars – at –253°C (–423°F), or (3) as a solid by either absorbing the hydrgoen or by reacting the hydrgoen with metals.

While large amounts of hydrogen are produced by various industries, it is mostly consumed at the site of production, notably for the synthesis of ammonia. However, the ever-present challenge is the very low boiling point of hydrogen (–252.882°C, –423.2°F).

6.5.1.1 Compressed Hydrogen

The most common storage systems are high-pressure gas cylinders with a maximum pressure on the order of 20 3,000 psi.

Compressed hydrogen in hydrogen tanks at 5,000 to 10,000 psi is used for hydrogen tank systems in vehicles, based on type IV carbon-composite technology. Hydrogen can also be stored in the liquid form by reducing the temperature to −253°C (−423°F) similar to liquefied natural gas (LNG) which is stored at −162°C (−269°F).

Furthermore, compressed hydrogen storage is a well-established hydrogen storage technology that involves the physical storage of compressed hydrogen gas in high-pressure vessels, which can be classified into four standard types: Type I, Type II, Type III, and Type IV.

The Type I vessel is an all-metal vessel (usually steel) and hence the heaviest, typically employed in industry for stationary use. Type I vessels store only about 1 wt% hydrogen at 200–300 bar [1]. Type II is a metal liner hoop-wrapped composite cylinder, weighing less than Type I cylinder. However, both Types I and II vessels are unsuitable for vehicle applications due to (1) the low hydrogen storage density arising from the heavyweights of the vessels as well as (2) from challenges related to hydrogen embrittlement. The Type III vessels comprise a fully wrapped composite cylinder with a metal liner that serves as the hydrogen permeation barrier and is typically made of aluminum reduced the potential for hydrogen embrittlement. Type IV vessels comprise a fully wrapped composite cylinder with a plastic liner (typically high density polyethylene), which acts solely as the hydrogen permeation barrier.

6.5.1.2 Liquid Hydrogen

Hydrogen is liquefied by reducing the temperature to −253°C (−423°F), similar to liquefied natural gas (LNG) which is stored at −162°C (−260°F). However, hydrogen has a low volumetric energy density compared to other gases and requires significantly more space. Liquid hydrogen is stored in cryogenic tanks at −252°C (−412°F) and ambient pressure. But because of the low critical temperature of hydrogen (−240°C, −464°F), liquid hydrogen can only be stored in open systems.

Cryogenic liquid storage tanks, also referred to as dewars, are the most common way to store large quantities of hydrogen. Super-insulated low-pressure vessels are needed to store liquid hydrogen at −253°C (−423°F) and the pressure of the liquid hydrogen is no more than 73 psi. Regardless of the quality of the insulation, however, some heat will reach the tank over time and cause the liquid hydrogen to boil. The result is that hydrogen gas accumulates at the top of the liquid tank and causes the pressure inside the tank to increase. To keep the pressure from rising above the limits of the tank, the gaseous hydrogen must be vented from the liquid tank and either released or recompressed by a boil-off compressor to be stored as gaseous hydrogen.

Furthermore, the storage of cryogenic liquids (such as hydrogen) is less efficient than storing compressed hydrogen. Also, a refrigeration system is required to keep the liquid cooled, and if the system temperature rises above hydrogen's critical point (−241°C), the liquid will become a gas. To prevent the possibility of a temperature rise, there must be vacuum insulation between the inner and outer walls of the liquid hydrogen tank system to prevent heat from passing to the liquid.

Moreover, a liquid storage system requires excessively large footprints – on the order of three times the size of the crude oil industry-standard crude oil tanks or naphtha-gasoline tanks. Also, with a high-pressure gas storage system, access to the hydrogen is much more convenient than accessing the hydrogen from liquid storage.

Typically, on-site hydrogen storage is used at central hydrogen production facilities, transport terminals, and end-use locations, and the options for storage of the hydrogen include insulated liquid tanks and gaseous storage tanks. The four types of common high-pressure gaseous storage vessels are: (1) Type I, which is an all-metal cylinder and is the most common storage cylinder, (2) Type II, which is a load-bearing metal liner hoop wrapped with resin-impregnated continuous filament, (3) Type III, which is a non-load-bearing metal liner axial and hoop wrapped with resin-impregnated continuous filament, and (4) Type IV, which is a non-load-bearing, non-metal liner axial and hoop wrapped with resin-impregnated continuous filament.

6.5.1.3 Use of Adsorbents

Adsorption involves a surface process that leads to the transfer of a molecule from a fluid bulk to a solid surface, and adsorption occurs when a particle adheres to the surface of a solid substance without entering it and is held there by intermolecular forces. Adsorption of a gas on a metal surface (the metals which adsorb hydrogen include iron, Fe, gold, Au, platinum, Pt, and palladium, Pd).

Zeolites (which are microporous and highly crystalline aluminosilicate materials) exhibit cage structures and tunnel structures that offer the potential for the encapsulation of non-polar gases such as hydrogen. Hydrogen is physically adsorbed (physiosorbed) on the surface of the zeolite pores through a mechanism of adsorption that involves hydrogen being forced into the pores under pressure and low temperature.

Therefore, similar to other porous materials, hydrogen storage capacity of a zeolite depends on (1) the surface area, (2) the pore volume, (3) the interaction of molecular hydrogen with the internal surfaces of the micropores, and (4) the parameters such as pressure and temperature (Dong et al., 2007; Ren et al., 2017).

Activated carbons are highly porous amorphous carbon materials that have a high apparent surface area, and the physisorption of hydrogen can be increased in these materials by increasing the apparent surface area and optimizing the pore diameter to a size on the order of 7 Ångstrom units (7Å).

Metal-organic frameworks (MOFs) represent another class of synthetic porous materials that store hydrogen and energy at the molecular level.

Also, the enhancement of sorption kinetics and the storage capacity can be improved through nanomaterial-based catalyst doping. They also have a favorable surface-area-to-volume ratio.

6.5.2 Chemical Storage

This method of hydrogen storage offers a much wider range of potential storage systems and is possible because of the high storage density of hydrogen. For example, supercritical hydrogen at 30°C (86°F) and 7,350 psi only has a density of 15.0 mol/L while methanol has a hydrogen density of 49.5 mol hydrogen per liter. Methanol and saturated dimethyl ether at 30°C (86°F) and 103 psi have a density of 42.1 mol hydrogen per liter dimethyl ether.

The dehydrogenation of chemically stored hydrogen can be accomplished either hydrolytically by reaction with water or thermally by heating the compound, as with, for example, ammonia borane (NH_3BH_3) and aluminum trihydride (also known as 'alane', AlH_3) (Kojima et al., 2004; Graetz and Reilly, 2007). However, the reaction pathway for dehydrogenation of chemical hydrogen is not always straightforward and may be governed by the extent to which the dehydrogenation product remains as a liquid phase or as a slurry phase that can be handled easily.

More specifically, hydrogen release can be induced by hydrolysis reactions or by catalyzed dehydrogenation reactions. Illustrative storage compounds are hydrocarbon derivatives, boron hydride derivatives, and ammonia. Nanomaterials also offer advantages for hydrogen storage systems, and the nanomaterials offer an alternative that overcomes the two major barriers of bulk materials which are (1) the rate of sorption and (2) the release temperature.

Examples of chemical storage systems are (1) metal hydride storage and (2) organic hydrogen carriers.

6.5.2.1 Metal Hydride Storage

The highest volumetric densities of hydrogen are found in metal hydrides such as the group one, two, and three light metals (e.g., Li, Mg, B, Al) (Figure 6.1) which can be combined with hydrogen to form a large variety of metal-hydrogen complexes.

Hydrogen can also be stored indirectly in reactive metals such as lithium (Li), sodium (Na), aluminum (Al), or zinc (Zn) (Figure 6.1), and these metal hydrides react with water to the corresponding hydroxide as well as hydrogen from the water. For example, metal hydride derivatives (such as magnesium hydride, MgH_2, and sodium aluminum hydride, $NaAlH_4$) represent sources of stored hydrogen.

Again, the persistent problems are the amount of hydrogen (measured as % w/w that the hydride derivative can carry) and the reversibility of the storage process. However, the hydride derivatives that are chosen for storage applications should provide low reactivity (high safety) and high hydrogen storage densities.

Metal hydrides, such as magnesium hydride (MgH2), sodium aluminum hydride (NaAlH4), lithium aluminum hydride (LiAlH4), and lithium hydride (LiH), represent sources of stored hydrogen.

An interesting chemical hydride is ammonium borohydride (NH_3BH_3) which is a waxy solid at room temperature with a melting point on the order of 90°C (194°F) at which temperature hydrogen is released because of thermal decomposition.

The release of hydrogen from this hydride occurs stepwise, where the onset temperature for the first equivalent is 90°C °C (194°F), and the temperature for the second equivalent is 150°C °C (302°F). The remaining hydrogen will be released at a temperature higher than 150°C (302°F).

6.5.2.2 Organic Hydrogen Carriers

The use of liquid organic hydrogen carriers (often referred to as LOHCs) allows for the transportation and long-cycle hydrogen storage. In particular, traditional organic hydrogen storage liquids are derived from nonrenewable fossil fuels through costly refining procedures, resulting in unavoidable environmental contamination. The technology is based on reversible hydrogen storage and release reactions using unsaturated liquid organics (such as toluene, naphthalene, and N-ethylcarbazole) as hydrogen storage agents and the corresponding saturates (e.g., methylcyclohexane, decalin, and dodecahydro-N-ethylcarbazole) as hydrogen carriers.

Unsaturated organic compounds can store large amounts of hydrogen. These liquid organic hydrogen carriers are hydrogenated for storage and dehydrogenated again at the point when (and where) the hydrogen is needed. Using liquid organic hydrogen carriers, relatively high gravimetric storage densities can be reached and the overall energy efficiency is higher than for other chemical storage options such as producing methane from the hydrogen. Both hydrogenation and dehydrogenation of the liquid organic hydrogen carriers requires catalysts. Also, biomass holds great promise for the preparation of liquid organic hydrogen carriers due to its unique carbon-balance properties and feasibility to manufacture aromatic and nitrogen-doped compounds (Chu et al., 2023; Modisha and Bessarabov, 2023).

In contrast with traditional hydrogen storage technologies, the liquid organic hydrogen carrier technology has following advantages: (1) most of the carriers have a high boiling point and low melting point, allow-ing them to maintain a stable liquid phase at room temperature while remaining non-volatile, (2) this 'carrier' system has stable catalytic hydrogenation and dehydrogenation processes, (3) the reactants and catalysts can be recycled, and (4) the storage, transportation, and maintenance of hydrogen-storage materials are safer and more convenient, allowing for large-scale and long-distance distribution. However, there are also several issues that need attention and are (1) the reaction requires professional hydrogenation and dehydrogenation equipment, (2) the dehydrogenation reaction must be carried out at high temperature, which can result in catalyst coking and deactivation, and (3) the hydrogen produced by the dehydrogenation reaction may not be of the highest purity, and inappropriate conditions and reactants are more likely to induce side reactions (Chu et al., 2023).

REFERENCES

Abe, J., Popoola, A., Ajenifuja, E., and Popoola, O. 2019. Hydrogen Energy, Economy and Storage: Review and Recommendation. *International Journal of Hydrogen Energy*, 44(29): 15072–15086.

Abdeljaoued, A., Relvas, F., Mendes, A., and Chahbani, M.H. 2018. Simulation and Experimental Results of a PSA Process for Production of Hydrogen Used in Fuel Cells. *Journal of Environmental Chemical Engineering*, 6(1): 338–355.

Abe, J., Popoola, A., Ajenifuja, E., and Popoola, O. 2019. Hydrogen Energy, Economy, and Storage: Review and Recommendation. *International Journal of Hydrogen Energy*, 44(29): 15072–15086.

Acar, C., and Dincer, I. 2014. Comparative Assessment of Hydrogen Production Methods from Renewable and Non-renewable Sources. *International Journal of Hydrogen Energy*, 39(1): 1–12.

Adhikari, S., and Fernando, S. 2006. Hydrogen Membrane Separation Techniques. *Industrial & Engineering Chemistry Research*, 45(3): 875–881.

Agueda, V.I., Delgado, J.A., Uguina, M.A., Brea, P., Spjelkavik, A.I., Blom, R., and Grande, C. 2015. Adsorption and Diffusion of H2, N2, CO, CH4 and CO2 in UTSA-16 Metal-Organic Framework Extrudates. *Chemical Engineering Science*, 124: 159–169.

Aasadnia, M., Mehrpooya, M., and Ghorbani, B. 2021. A Novel Integrated Structure for Hydrogen Purification Using the Cryogenic Method. *Journal of Cleaner Production*, 278: 123872. https://www.sciencedirect.com/science/article/abs/pii/S0959652620339172.

Bailey, D.W., and Feron, P.H.M. 2005. Post-combustion Decarbonization Processes. *Oil & Gas Science and Technology*, 60(3): 461–474.

Balat, M. 2008. Potential Importance of Hydrogen as a Future Solution to Environmental and Transportation Problems. *International Journal of Hydrogen Energy*, 33(15): 4013–4029.

Bauer, F., Persson, T., Hulteberg, C., and Tamm, D. 2013. Biogas Upgrading - Technology Overview, Comparison and Perspectives for the Future. *Biofuels, Bioproducts and Biorefining*, 7(5): 499–511. https://onlinelibrary.wiley.com/doi/10.1002/bbb.1423.

Baykara, S.Z. 2018. Hydrogen: A Brief Overview on Its Sources, Production and Environmental Impact. *International Journal of Hydrogen Energy*, 43(23): 10605–10614.

Brüschke, H. 1995. Industrial Application of Membrane Separation Processes. *Pure and Applied Chemistry*, 345, 67(6): 993–1002.

Brüschke, H. 2003. Separation of Hydrogen from Dilute Streams (e.g. Using Membranes). Proceedings. DGMK Conference on Innovation in the Manufacture and Use of Hydrogen. Dresden. October 15–17. Page 47.

Chadeesingh, R. 2011. The Fischer-Tropsch Process. In *The Biofuels Handbook*. J.G. Speight (Editor). Royal Society of Chemistry, London. Part 3, Chapter 5.

Chapman, A., Itaoka, K., Hirose, K., Davidson, F.T., Nagasawa, K., Lloyd, A.C., Webber, M.E., Kurban, Z., Managi, S., and Tamaki, T. 2019. A Review of Four Case Studies Assessing the Potential for Hydrogen Penetration of the Future Energy System. *International Journal of Hydrogen Energy,* 44(13): 6371–6382.

Chen, X., Wei, L., Deng, L., Yang, F., and Zhang, Z. 2013. A Review of the Metal Hydride Based Hydrogen Purification and Separation Technology. *Applied Mechanics and Materials*, 448453: 3027–3036. https://www.researchgate.net/publication/269382823_A_Review_on_the_Metal_Hydride_Based_Hydrogen_Purification_and_Separation_Technology.

Chen, C.-Y., Lai, W.-H., Chen, Y.-K., and Su, S.S. 2014. Characteristic Studies of a PBI/H3PO4 High Temperature Membrane PEMFC Under Simulated Reformate Gases. *International Journal of Hydrogen Energy*, 39(25): 13757–13762.

Chu, C., Wu, K., Luo, B., Cao, Q., and Zha, H. 2023. Hydrogen Storage by Liquid Organic Hydrogen Carriers: Catalyst, Renewable Carrier, and Technology – A Review. *Carbon Resources Conversion*, 6(4): 334–351. https://www.sciencedirect.com/science/article/pii/S2588913323000248#:~:text=LOHCs%20technology%20is%20based%20on,hydrogen%20carriers%20%5B7%5D%2C%20%5B.

Dincer, I., and Acar, C. 2015. Review and Evaluation of Hydrogen Production Methods for Better Sustainability. *International Journal of Hydrogen Energy,* 40(34): 11094–11111.

Dong, J., Wang, X., Xu, H., Zhao, Q., and Li, J. 2007. Hydrogen Storage in Several Microporous Zeolites. *International. Journal of Hydrogen Energy*, 32(18): 4998–5004.

Dunikov, D., Borzenko, V., and Malyshenko, S. 2012. Influence of Impurities on Hydrogen Absorption in a Metal Hydride Reactor. *International Journal of Hydrogen Energy*, 37(18): 13843–13848. https://www.sciencedirect.com/science/article/abs/pii/S0360319912009329.

Dunikov, D., and Blinov, D. 2020. Extraction of Hydrogen from a Lean Mixture with Methane by Metal Hydride. *International Journal of Hydrogen Energy*, 45(16): 9914–9926. https://www.sciencedirect.com/science/article/abs/pii/S0360319920303797.

Edlund, D. 2010. Hydrogen Membrane Technologies and Application in Fuel Processing. In *Hydrogen and Syngas Production and Purification Technologies*. K. Liu, C. Song, and V. Subramani (Editors). John Wiley & Sons, Inc., Hoboken, NJ. Page 357–384.

El-Shafie, M., Kambara, S., and Hayakawa, Y. 2019. Hydrogen Production Technologies Overview. *Journal of Power Energy Engineering*, 7(1): 107–154.

Gardner, C., and Ternan, M. 2007. Electrochemical Separation of Hydrogen from Reformate Using PEM Fuel Cell Technology. *Journal of Power Sources*, 171(2): 835–841.

Grande, C.A., Lopes, F.V.S., Ribeiro, A.M., Loureiro, J.M., and Rodrigues, A.E. 2008. Adsorption of Off-Gases from Steam Methane Reforming (H2, CO2, CH4, CO, and N2) on Activated Carbon. *Separation Science and Technology*, 43(6): 1338–1364.

Granite, E.J., and O'Brien, T. 2005. Review of Novel Methods for Carbon Dioxide Separation from Flue and Fuel Gases. *Fuel Processing Technology,* 86(14–15): 1423–1434.

Grashoff, G.J., Pilkington, C.E., and Corti, C.W. 1983. The Purification of Hydrogen. *Platinum Metals Review*, 27(4): 157–169 https://technology.matthey.com/article/27/4/157-169/.

Graetz, J., and Reilly, J.J. 2007. Kinetically Stabilized Hydrogen Storage Materials. *Scripta Materialia*, 56(10): 835–839. https://www.sciencedirect.com/science/article/abs/pii/S1359646207000449.

Grigoriev, S., Shtatniy, I., Millet, P., Porembsky, V., and Fateev, V. 2011. Description and Characterization of an Electrochemical Hydrogen Compressor/Concentrator Based on Solid Polymer Electrolyte Technology. *International Journal of Hydrogen Energy*, 36(6): 4148–4155.

Hamm, J.B.S., Ambrosi, A., Griebeler, J.G., Marcilio, N.R., Tessaro, I.C., and Pollo, L.D. 2017. Recent Advances in the Development of Supported Carbon Membranes for Gas Separation. *International Journal of Hydrogen Energy*, 42(39): 24830–24845. https://www.sciencedirect.com/science/article/abs/pii/S0360319917332974.

He, B., Liu, J., Zhang, Y., Zhang, S., Wang, P., and Xu, H. 2020. Comparison of Structured Activated Carbon and Traditional Adsorbents for Purification of H2. *Separation and Purification Technology*, 239: 116529. https://www.sciencedirect.com/science/article/abs/pii/S1383586619334240.

Holladay, J.D., Wang, Y., and Jones, E. 2004. Review of Developments in Portable Hydrogen Production Using Microreactor Technology. *Chemical Reviews*, 104(10): 4767–4790.

Kojima, Y., Kawai, Y., Kimbara, M., Nakanishi, H., and Matsumoto, S. 2004. Hydrogen Generation by Hydrolysis Reaction of Lithium Borohydride. *International Journal of Hydrogen Energy*, 29(12): 1213–1217. https://www.sciencedirect.com/science/article/abs/pii/S0360319904000102.

Kuroda, S., Nagaishi, T., Kameyama, M., Koido, K., Seo, Y., and Dowaki, K. 2018. Hydroxyl Aluminum Silicate Clay for Biohydrogen Purification by Pressure Swing Adsorption: Physical Properties, Adsorption Isotherm, Multicomponent Breakthrough Curve Modelling, and Cycle Simulation. *International Journal of Hydrogen Energy*, 43(34): 16573–16588. https://www.sciencedirect.com/science/article/abs/pii/S0360319918321864.

Lee, H., Choi, H., Choi, K., Park, J., and Lee, T. 2004. Hydrogen Separation Using Electrochemical Method. *Journal of Power Sources*, 132(1–2): 92–98.

Lei., L., Lindbråthen, A., Hillestad, M., and He, X. 2021. Carbon Molecular Sieve Membranes for Hydrogen Purification from a Steam Methane Reforming Process. *Journal of Membrane Science*, 627: 119241. https://www.sciencedirect.com/science/article/pii/S0376738821001915.

Li, P., Wang, Z., Qiao, Z., Liu, Y., Cao, X., Li, W., Wang, J., and Wang, S. 2015. Recent Developments in Membranes for Efficient Hydrogen Purification. *Journal of Membrane Science*, 495: 130–168.

Liu, F., and Zhang, N. 2004. Strategy of Purifier Selection and Integration in Hydrogen Networks. *Chemical Engineering Research and Design*, 82(10): 1315–1330.

Lively, R.P., Bessho, N., Bhandari, D.A., Kawajiri, Y., and Koros, W.J. 2012. Thermally Moderated Hollow Fiber Sorbent Modules in Rapidly Cycled Pressure Swing Adsorption Mode for Hydrogen Purification. *International Journal of Hydrogen Energy*, 37(20): 15227–15240. https://www.sciencedirect.com/science/article/abs/pii/S0360319912017405.

MacDowell, N., Florin, N., Buchard, A., Hallett, J., Galindo, A., Jackson, G., Adjiman, C.S., Williams, C.K., Shah, N., and Fennell, P. 2010. An Overview of CO_2 Capture Technologies. *Energy & Environmental Science*, 3(11): 1645–1669. https://pubs.rsc.org/en/content/articlelanding/2010/ee/c004106h.

McCarty, R.D., Hord, J., and Roder, H.M. 1981. *Selected Properties of Hydrogen (Engineering Design Data)*. United States Department of Commerce, National Bureau of Standards, Washington, DC.

Modisha, P., and Bessarabov, D. 2023. Aromatic Liquid Organic Hydrogen Carriers for Hydrogen Storage and Release. *Current Opinion in Green and Sustainable Chemistry*, 42: 100820. https://www.sciencedirect.com/science/article/pii/S245222362300069X.

Mokhatab, S., Poe, W.A., and Speight, J.G. 2006. *Handbook of Natural Gas Transmission and Processing*. Elsevier, Amsterdam.

Muller, B. 2011. Energiespeicherung mittels Methan und energietragenden Staffen – ein thermodynamischer Vergleich (Energy Storage by Carbon Dioxide Methanization and Energy Carrying Compounds: A Thermodynamic Comparison). *Chemie lngenieur Technik (in German)*, 83(11): 2002–2013.

Nayebossadri, S., Speight, J.D., and Book, D. 2019. Hydrogen Separation from Blended Natural Gas and Hydrogen by Pd-Based Membranes. *International Journal of Hydrogen Energy*, 44(55): 29092–29099.

Ockwig, N.W., and Nenoff, T.M. 2007. Membranes for Hydrogen Separation. *Chemical Reviews*, 107(10): 4078–4110.

Perry, K.A., Eisman, G.A., and Bieniewicz, B.C. 2008. Electrochemical Hydrogen Pumping Using a High-Temperature Polybenzimidazole (PBI) Membrane. *Journal of Power Sources*, 177(2): 478–484.

Rahimpour, M., Samimi, F., Babapoor, A., Tohidian, T., and Mohebi, S. 2017. Palladium Membranes Applications in Reaction Systems for Hydrogen Separation and Purification: A Review. *Chemical Engineering and Processing: Process Intensification*, 121: 24–49.

Ren, J., Musyoka, N.M., Langmi, H.W., Mathe, M., and Liao, S. 2017. Current Research Trends and Perspectives on Materials-Based Hydrogen Storage Solutions: A Critical Review. *International Journal of Hydrogen Energy*, 42(1): 289–311.

Rochelle, G.T. 2009. Amine Scrubbing for CO2 Capture. *Science*, 325(5948): 1652–1654. https://www.science.org/doi/10.1126/science.1176731.

Rusman, N., and Dahari, M. 2016. A Review of the Current Progress of Metal Hydrides Material for Solid-State Hydrogen Storage Applications. *International Journal of Hydrogen Energy*, 41(28): 12108–12126.

Ryckebosch, E., Drouillon, M., and Vervaeren, H. 2011. Techniques for Transformation of Biogas to Biomethane. *Biomass Bioenergy*, 35(5): 1633–1645. https://www.sciencedirect.com/science/article/abs/pii/S0961953411001085.

Sakai, T., Matsumoto, H., Kudo, T., Yamamoto, R., Niwa, E., Okada, S., Hashimoto, S., Sasaki, K., and Ishihara, T. 2008. High Performance of Electroless-Plated Platinum Electrode for Electrochemical Hydrogen Pumps Using Strontium-Zirconate-Based Proton Conductors. *Electrochimica Acta*, 53(28): 8172–8177.

Sankir, M., and Sankir, N.D. (Editors) 2018. *Hydrogen Storage Technologies*. Scrivener Publishing, Beverly, MA.

Sasikumar, B., Arthanareeswaran, G., and Ismail, A. 2018. Recent Progress in Ionic Liquid Membranes for Gas Separation. *Journal of Molecular Liquids*, 266: 330–341. https://www.sciencedirct.com/scien. ce/article/abs/pii/S0167732218314107.

Sazali, N., Ghazali, M.F., Siregar, J.P., Gunawan, T., Salleh, W.N.W., and Yusof, N. 2020. A Review of Carbon Membranes for Hydrogen Purification. Proceedings. *IOP Conference Series: Material Science and Engineering*, 736(2): 022004. https://iopscience.iop.org/article/10.1088/1757-899X/736/2/022004/pdf.

Schorer, L., Schmitz, S., and Weber, A. 2019. Membrane Based Purification of Hydrogen System (MEMPHYS). *International Journal of Hydrogen Energy*, 44(25): 12708–12714.

Shamsudin, I., Abdullah, A., Idris, I., Gobi, S., and Othman, M. 2019. Hydrogen Purification from Binary Syngas by PSA with Pressure Equalization Using Microporous Palm Kernel Shell Activated Carbon. *Fuel*, 253: 722–730.

Sircar, S., and Golden, T.C. 2000. Purification of Hydrogen by Pressure Swing Adsorption. *Separation Science and Technology*, 35(5): 667–687.

Sircar, S., and Golden, T.C. 2010. Pressure Swing Adsorption Technology for Hydrogen Production. In *Hydrogen and Syngas Production and Purification Technologies*. K. Liu, C. Song, and V. Subramani (Editors.). John Wiley & Sons, Inc., Hoboken, NJ. Page 414–450.

Song, C., Liu, Q., Deng, S., Li, H., and Kitamura, Y. 2019. Cryogenic-Based CO2 Capture Technologies: State-of-the-Art Developments and Current Challenges. *Renewable and Sustainable Energy Reviews*, 101: 265–278.

Speight, J.G. 2013a. *The Chemistry and Technology of Coal*, 3rd Edition. CRC Press, Taylor & Francis Group, Boca Raton, FL.

Speight, J.G. 2013b. *Coal-Fired Power Generation Handbook*. Scrivener Publishing, Salem, MA.

Speight, J.G. 2014. *The Chemistry and Technology of Petroleum*, 5th Edition. CRC Press, Taylor & Francis Group, Boca Raton, FL.

Speight, J.G. 2019. *Natural Gas: A Basic Handbook*, 2nd Edition. Gulf Publishing Company, Elsevier, Cambridge, MA.

Speight, J.G. 2020. *Synthetic Fuels Handbook: Properties, Process, and Performance*, 2nd Edition. McGraw-Hill, New York.

Strugova, D., Zadorozhnyy, M.Y., Berdonosova, E., Yablokova, M.Y., Konik, P., Zheleznyi, M., Semenov, D., Milovzorov, G., Padaki, M., and Kaloshkin, S. 2018. Novel Process for Preparation of Metal-Polymer Composite Membranes for Hydrogen Separation. *International Journal of Hydrogen Energy*, 43(27): 12146–12152.

Thomassen, M., Sheridan, E., and Kvello, J. 2010. Electrochemical Hydrogen Separation and Compression Using Polybenzimidazole (PBI) Fuel Cell Technology. *Journal of Natural Gas Science and Engineering*, 2(5): 229–234.

Vermaak, L., Neomagus, H.W.J.P., and Bessarabov, D.G. 2021. Hydrogen Separation and Purification from Various Gas Mixtures by Means of Electrochemical Membrane Technology in the Temperature Range 100–160oC. *Membranes*, 11(4): 282. https://www.mdpi.com/2077-0375/11/4/282.

Wang, Q., Xue, M., Lin, B.-L., Lei, Z., and Zhang, Z. 2020. Well-to-Wheel Analysis of Energy Consumption, Greenhouse Gas and Air Pollutants Emissions of Hydrogen Fuel Cell Vehicle in China. *Journal of Cleaner Production*, 275: 123061. https://www.sciencedirect.com/science/article/abs/pii/S0959652620331061.

Wu, T., Prasetya, N., and Li, K. 2020. Recent Advances in Aluminum-Based Metal-Organic Frameworks (MOF) and Its Membrane Applications. *Journal of Membrane Science*, 615: 118493.

Xu, R., He, L., Li, L., Hou, M., Wang, Y., Zhang, B., Liang, C., and Wang, T. 2020. Ultraselective Carbon Molecular Sieve Membrane for Hydrogen Purification. *Journal of Energy Chemistry*, 50: 16–24.

Zhao, C., Sun, B., Jiang, J., and Xu, W. 2020. Hydrogen Purification Process with Double Layer Bcc-PdCu Alloy Membrane at Ambient Temperature. *International Journal of Hydrogen Energy*, 45(35): 17540–17547.

7 Hydrogen Management

7.1 INTRODUCTION

In the refinery, hydrogen usage has increased from being a by-product to the role of a major commodity that is necessary for the production of saleable products. In fact, hydrogen is a necessity for refining operations and is used to (1) eliminate sulfur, nitrogen, and other impurities from crude oil products such as naphtha, kerosene, middle distillates, gas oil, and other feedstocks in hydrotreating units, (2) upgrade product values in processes such as the production of ultra-low sulfur diesel fuel, (3) crack higher molecular weight, higher boiling hydrocarbon products, such as gas oils, into lower molecular weight, lower boiling hydrocarbon products such as naphtha and other distillates, and (4) provide additional hydrogen along with the C1 to C4 hydrocarbon derivatives that can be used in refinery fuel gas systems.

Briefly, and by way of recall (Speight, 2014, 2017, 2023), hydrotreating is a process that removes constituents (such as nitrogen, sulfur, oxygen, and metals) that are considered contaminants in the refinery feedstock. The hydrotreating process is also used in advance of catalytic cracking to improve product yields and to upgrade the quality of the product. Put simply, the process works by mixing the feedstock with hydrogen, heating the mixture, and then passing the heated feedstock-hydrogen mixture through a catalytic reactor. The reactor converts the contaminants to other forms that can be separated; the result is a gas stream that, after treatment, can be used to fire the furnaces at the refinery, and a liquid stream that can be blended or used as feedstock. As an example, practically all the naphtha that is fed to catalytic reforming units is hydrotreated to remove the aforementioned contaminants that would damage the catalyst. The resulting product (typically referred to as 'reformate') is used as a blend stock to produce gasoline. By-products of this process include hydrogen, which is recycled within the refinery and used in hydrotreating or hydrocracking (Parkash, 2003; Gary et al., 2007; Speight, 2014; Hsu and Robinson, 2017; Speight, 2017).

On the other hand, hydrocracking is a two-stage process that combines catalytic cracking and hydrogenation in which higher boiling feedstocks are cracked in the presence of hydrogen. A hydrocracking unit typically involves a reactor section, gas separator, scrubber for sulfur removal, and a product fractionator. The hydrocracking process was developed in the 1960s to increase production of naphtha (and, hence, gasoline) and forms the basis for the modern petrochemical industry. The process is used for feedstocks that are difficult to process by either catalytic cracking or reforming because they contain substances that are considered 'poisons' for the catalyst.

Some refineries depend solely on catalytic reforming units as the source of hydrogen for hydrotreating. Often, they are semi-regenerative reformers where off-gas hydrogen quantity, purity, and availability change with feed naphtha quality, as octane requirements change seasonally, and when the reformer catalyst progresses from start-of-run to end-of-run conditions and then goes offline for regeneration. Typically, during some portions of the year, refinery margins are reduced as a result of hydrogen shortages. In fact, improved hydrogen availability through alternate sources and hydrogen purity optimization is now required to meet the needs of the future transportation product market. Many refineries developing hydrogen management programs that fit into the two general categories of either a catalytic reformer-supplied network or an on-purpose hydrogen supply.

The typical hydrogen process consists of desulfurization, steam reforming, high-temperature 'carbon monoxide shift' (CO shift), and hydrogen purification:

$$\text{Natural gas (methane)} \rightarrow \text{steam reforming} \rightarrow \text{CO shift} \rightarrow \text{hydrogen purification}$$

DOI: 10.1201/9781003186694-7

While the steam reforming process remained in principle the same, except that the size of the steam reformers increased significantly, the purification process changed from a carbon dioxide removal unit, followed by a methanation unit, to the pressure swing adsorption (PSA) unit, which can achieve a much purer hydrogen product and a more efficient hydrogen plant.

Thus, to provide the hydrogen for the hydrotreating and hydrocracking processes, refineries have seen the necessity of constructing an on-site hydrogen plant. Also, as a result of the increased use of hydrotreating processes and hydrocracking processes, hydrogen management has become a priority in current refinery operations and when planning to produce lower sulfur gasoline and low sulfur diesel fuel (Zagoria et al., 2003; Davis and Patel, 2004; Luckwal and Mandal, 2009). Along with increased hydrogen consumption for deeper hydrotreating, additional hydrogen is needed for processing heavier and high-sulfur feedstocks.

Multiple hydrotreating units compete for hydrogen – either by selectively reducing throughput, managing intermediate tankage logistics, or running the catalytic reformer sub-optimally just to satisfy downstream hydrogen requirements. Part of the operating year still runs in hydrogen surplus, and the network may be operated with relatively low hydrogen utilization (consumption/production) at 70 to 80%. Catalytic reformer off gas hydrogen supply may swing from 75 to 85% hydrogen purity. Hydrogen purity upgrade can be achieved through some hydrotreaters by absorbing heavy hydrocarbons. But without supplemental hydrogen purification, critical control of hydrogen partial pressure in hydroprocessing reactors is difficult, which can affect catalyst life, charge rates, and/or yields of naphtha.

More complex refineries, especially those with hydrocracking units, also have on-site hydrogen production facilities, typically with a steam-methane reforming unit (or units) that utilizes natural gas with supplemental amounts of refinery off-gas as the feedstock. The steam methane reformer plant provides the swing hydrogen requirements at higher purities (92 to more than 99% v/v hydrogen) and serves a hydrogen network configured with several purity and pressure levels. Multiple purities and existing purification units allow for a more optimized hydroprocessing operation by controlling hydrogen partial pressure for maximum benefit. Typical hydrogen utilization is on the order of 85% to 95%.

Prior to the 1970s, refineries operated under quite different conditions than the more modern (post-1980) refineries. The main feedstock was a light and sweet crude, and this was available in sufficient quantity and easy to handle. As crude oil prices were low most of the time (prior to the so-called 'oil shocks' of the 1970s), there was a balance between investment costs and the overall production costs. Also, as the decade of the 1980s evolved, the environmental regulations in various countries became more stringent with limitations on sulfur, nitrogen oxides (NOx), volatile organic compounds (VOCs), or carbon dioxide occurring.

Furthermore, refineries are currently facing strong competition, which compels them to lower operating costs wherever possible and to utilize all process purge gas and off-gas streams. More stringent environmental regulations and the need to process heavier crude significantly increase hydrogen demand. The profitability of a refinery depends on the method by which hydrogen is produced and how hydrogen-containing gases are recovered.

7.2 HYDROGEN PRODUCTION AND UTILIZATION

The use of hydrogen in refineries is widespread and was made necessary by the increase in hydrotreating and hydrocracking, including the treatment of progressively higher viscosity feedstocks. The continued increase in hydrogen demand over the last several decades is a result of the conversion of refinery feedstocks to match changes in product slate and the supply of heavy, high-sulfur oil and to make lower-boiling, cleaner, and more salable products. There are also many reasons other than product quality for using hydrogen in processes, adding to the need to add hydrogen at relevant stages of the refining process and, most importantly, according to availability.

There are two forms of hydrogen production in refineries and they are (1): the on-site hydrogen production using steam methane reformers, typically known as SMR, which has a high efficiency and is on the order of 90% in producing hydrogen, and (2) hydrogen production as a by-product of other chemical processes. Natural gas is used almost exclusively as feedstock for on-purpose hydrogen production in the steam methane reforming process, which is on the order of 90% efficient in producing hydrogen.

Thus, in addition to the conventional hydrogen production methods of steam methane reforming (SMR) and the use of by-product hydrogen from catalytic reforming units, new production options are emerging (Chapter 4, Chapter 5). These include using (1) renewable power directly for electrolysis, (2) various options that produce biogas that employ gasification or pyrolysis processes or biomass fermentation with microorganisms, and (3) newly developed photo-electrochemical and thermo-chemical processes, including using microbial electrolysis cells, as well as tailored molecules that can facilitate the splitting of water molecules into hydrogen and oxygen with lower energy requirements than conventional electrolysis.

Furthermore, the modern refinery operates in an environment in which the demand for refinery products is growing continuously, while the availability of light and sweet crude is decreasing. Several possibilities exist for the recovery of hydrogen from off-gases and the processing of different sources of refinery gases through the hydrogen plant, and as a result, refineries are forced to process more viscous feedstocks. Additionally, stronger environmental regulations and higher requirements on the products are already resulting in a change to the operating conditions.

The demand for hydrogen has increased significantly, and it is increasingly recognized that hydrogen is a valuable gas that needs special consideration. Utilizing all hydrogen-containing streams is therefore essential for economical refinery operations. Increased hydrogen demand.

7.2.1 Hydrogen Production

Hydrogen can be produced from many different on-site or off-site sources in different ways for use in the refinery (Chapter 3, Chapter 4, and Chapter 5). The two most common methods for producing hydrogen are (1) the steam-methane reforming process and (2) an electrolysis process in which electricity is used to dissociate water into hydrogen and oxygen. The first step in meeting the growing demand for hydrogen was the optimization of plant processes which was followed by enhancing the efficiency of existing hydrogen plants and increasing the capacity of these plants. The latest developments in processing the viscous feedstocks and producing ultra-low-sulfur products (such as gasoline and diesel fuel) have created a considerably higher demand for hydrogen which cannot be covered solely by any of the former methods. Also, many existing refinery hydrogen plants produce a medium purity hydrogen product stream (on the order of 94 to 97% v/v) by removing the carbon dioxide in a solvent absorption system accompanied by a methanation step, while the more modern plants use the pressure swing adsorption technology to recover and purify the hydrogen to a purity level in excess of 99.9% v/v.

Moreover, a major point related to hydrogen production and use is that hydrogen is a substance that does not occur naturally, similar to natural gas or crude oil, but must be manufactured from one of many different sources (Chapter 3, Chapter 4, and Chapter 5). In addition, the environmental impacts of hydrogen production and use vary significantly with the production method and the accompanying value of hydrogen during use in the refinery.

Thus, current refineries have no other choice but to look for new hydrogen sources, and the most obvious way is to build a new hydrogen plant. The actual new demand drives the capacity consideration of new hydrogen plants and can be reached by a single-train unit with a modern top-fired steam reformer with a single-train pressure swing adsorption unit.

When evaluating the process for hydrogen storage (Chapter 6), the major process steps can be defined as the separation of hydrocarbon derivatives into hydrogen and carbon. For example, steam reforming units are viewed from the perspective that they still play a major role as the major hydrogen production unit. For this step, a steam reformer or partial oxidation unit is used, depending on the feedstocks.

In fact, the partial oxidation process is the ideal process for obtaining hydrogen from viscous feedstocks, and the most common (currently used) method is the steam reforming process (Chapter 3) in which a low-boiling hydrocarbon stream (such as natural gas) is used as the feedstock. In the steam reformer, steam and natural gas are catalytically cracked at temperatures on the order of 850°C (1,560°F) to produce synthesis gas, which is a mixture of hydrogen, carbon monoxide (CO), and carbon dioxide (CO_2). The carbon monoxide is converted into carbon dioxide with steam by a downstream carbon monoxide shift reaction, during which more hydrogen is produced. After this, the hydrogen-rich gas stream is purified to the desired quality level by means of (1) a pressure swing adsorption process or (2) a membrane separation process, or (3) a cryogenic separation process (Chapter 6).

An advantage of the partial oxidation process is the fact that refineries can put the hydrogen they recover from viscous feedstocks to desulfurize any sulfur-containing products. With the trend toward stricter environmental regulations, demand for hydrogen will continue to increase.

The conversion of carbon monoxide with steam into hydrogen and carbon dioxide occurs in a shift conversion unit. Depending on the required hydrogen purity, carbon dioxide scrubbing followed by the use of a methanation process or a pressure swing adsorption unit. The refinery gases sent to a hydrogen plant for hydrogen recovery or production can be fed into the plant upstream of each of these process steps. The major portion of refinery gases will be fed into the plant upstream of the reformer. These gases typically contain a higher amount of hydrocarbons; unfortunately, not only lighter paraffins, which can be mixed with the 'main' feed gas, but also olefins and aromatics. Impurities such as halogen compounds, heavy metals, and sulfur do not make their processing easier. For special applications, the refinery gas may be fed into the carbon monoxide shift. Such a gas should contain only a negligible amount of hydrocarbons but a high amount of CO, which is to be converted to carbon dioxide and hydrogen.

Gases already containing a high proportion of hydrogen may be fed directly into the pressure swing adsorption unit. The hydrogen content of the gas stream should not be less than 60% v/v, and special care needs to be taken if the gas contains impurities or unsaturated hydrocarbons. They may require pretreatment or be fed into the steam reformer, even with a high hydrogen content. Another issue when feeding gas directly into a pressure swing adsorption unit is the increased pressure swing adsorption off-gas.

A standard hydrogen plant can use all the pressure swing adsorption of gas as a fuel for the steam reforming process. However, due to the additional gas being fed directly to the pressure swing adsorption unit, the amount of pressure swing adsorption off-gas increases without generating a higher heat demand in the steam reformer. Depending on the amount of additional gas and the composition of this gas, the pressure swing adsorption of gas flow is too high to be fed completely into the steam reformer fuel gas system. In this case, an additional consumer needs to be found.

The gas streams produced during the operation of the various refinery units typically do not match the requirements to be fed into the hydrogen plant without pretreatment. The gas streams need to be compressed first in case the supply pressure is below the plant pressure. In addition, unsaturated hydrocarbons must be hydrogenated and impurities will be removed either by suitable adsorbents like zinc oxide or promoted zinc oxide catalysts. In case the sulfur content is higher than desirable, a two-stage sulfur-removal process may be necessary. For example, the bulk of the sulfur will be removed in a stripper and the final purity will be reached over a zinc oxide bed. If the feed gas to the reformer exceeds a certain content of higher boiling hydrocarbon derivatives, a pre-reformer needs to be connected in series to the steam reformer. Design objectives A modern hydrogen plant has to process one or several refinery gases in addition to its major feedstock, natural gas, liquefied petroleum gas, or naphtha. On occasion, it may even be necessary to consider the use of refinery gases as the sole feedstock for the hydrogen production unit and the former typical feedstocks as backup only.

The hydrogen yield should be maximized insofar as all of the hydrogen atoms in the gases fed to a hydrogen plant are recovered as far as technically possible. As the hydrogen plant becomes the major hydrogen supplier for the refinery and the refinery operations depend on this hydrogen, any disturbance of the hydrogen production due to unplanned fluctuations, or even a sudden loss of a refinery gas, must be avoided. The main goal of the hydrogen plant is to achieve the highest level of reliability and availability, in addition to providing a maximum level of flexibility regarding plant load variations and changes in feed gas flows and composition. These requirements can be met with an advanced design and integrated team approach. The design needs to take into consideration the aspects of various disciplines.

The process design remains the basic element. However, due to the necessity of the refinery to operate economically, the plant controls and possible integration of the hydrogen plant into an overall advanced control system are of growing importance. Whereas the process designer has to develop several operating scenarios (sometimes more than 20 such scenarios), there is only 1 piece of equipment that must handle all operating cases.

Nearly all process steps are catalytic reactions, calling for the early involvement of the catalyst supplier in the planning team. A careful analysis of the refinery hydrogen network is the basis for the development of the hydrogen plant that best meets the requirements. Cooperation between the refinery planner and hydrogen plant designer in the early design phase ensures that the full potential of the hydrogen plant can be utilized without asking for the impossible. For the hydrogen plant designer, it is essential to have a full understanding of all aspects and the know-how of all disciplines available, which are required to design and build a hydrogen plant and integrate it into a complex refinery network.

The hydrogen production unit in a refinery not only provides on-purpose hydrogen production but is also the recycling unit for many off-gases. These gases normally contain impurities, which can poison catalysts used in the steam reformer and shift section. Removal of all impurities (for example, desulfurization) to the lowest possible level is therefore the first priority as feedstock is introduced through the series of catalysts in the hydrogen plant. Classical desulfurization systems typically consist of one reactor loaded with a hydrotreating catalyst (typically a cobalt-molybdenum catalyst, CoMo), followed by a two-reactor system loaded with a zinc oxide (ZnO) catalyst to absorb the hydrogen sulfide. Both zinc oxide reactors are in series and are interchangeable to achieve optimum purification and catalyst lifetime. In addition, this high-surface-area ZnO catalyst absorbs all of the hydrogen sulfide and, therefore, all of the reactions occur on this one catalyst in the same reactor.

The increasing demand for hydrogen and rising raw material prices make hydrogen a valuable product and also an important cost factor. The modern hydrogen plant becomes an integral part of the refinery network and can no longer be seen as a simple utility unit at the refinery's periphery. As the main supplier of hydrogen to refinery operations, the reliability of the hydrogen plant is essential to overall hydrogen management and economical refinery operations. The integration of the hydrogen plant into the refinery network in the early design phase is essential, requiring the refinery planner, hydrogen plant designer, and catalyst supplier to work as a team. Only this team approach can ensure the development of the most cost-effective solution with respect to overall hydrogen network investment and operating costs.

7.2.2 Hydrogen Utilization

Once produced and purified (Chapter 6), the hydrogen is distributed throughout the refinery complex via a network of pipelines or dedicated distribution systems to the process units where the gas is needed.

In fact, hydrogen is an indispensable commodity in the refining industry for refining operations and is used, for example, to remove sulfur from many products. In this process (commonly referred to as hydrodesulfurization), the hydrogen bonds with the sulfur to form hydrogen sulfide (H_2S), which is captured and further processed in another treatment step (Parkash, 2003; Gary et al., 2007; Speight, 2014; Hsu and Robinson, 2017; Speight, 2017). Moreover, the sulfur content of the refinery feedstock is an indicator of the quality of the feedstock along with other properties (such as, for example, the API gravity, the viscosity, and the boiling range) (Speight, 2014, 2017, 2021). This determines the process steps that are required in a refinery to produce the marketable products.

Thus, hydrogen is used in several refining processes, primarily for the purpose of removing impurities and upgrading the quality of crude oil products. Some of the key applications of hydrogen are processes such as naphtha or kerosene hydrotreating units, gas oil or residuum hydrocracking, naphtha isomerization, and hydrodealkylation. For hydrotreating of lighter crude oil streams, if high-purity hydrogen is used on the order of >90%, then gases normally are used for maintaining the pressure of the system.

Recall that for use in the modern refinery, hydrogen can be produced from diverse, domestic resources, including natural gas reforming, gasification, electrolysis, biomass-derived liquid reforming, microbial biomass conversion, thermochemical water splitting, photobiological water splitting, and photo-electrochemical water splitting (Chapter 3, Chapter 4, and Chapter 5).

In the natural gas reforming process or the gasification process, the initial product is synthesis gas, which is a mixture of hydrogen, carbon monoxide, and a small amount of carbon dioxide that is produced by reacting natural gas with high-temperature steam. Synthesis gas can also be created by reacting a carbonaceous feedstock (such as coal, a viscous feedstock such as tar sand bitumen, or biomass) with high-temperature steam and oxygen in a pressurized gasifier. This converts the feedstock into gaseous components and the resulting synthesis gas contains hydrogen and carbon monoxide, which is reacted with steam to separate the hydrogen.

In the electrolysis process, an electric current splits water into hydrogen and oxygen. If the electricity is produced by renewable sources, such as solar or wind, the resulting hydrogen will be considered renewable and has numerous emission benefits.

In the biomass-derived liquid reforming process, the liquid feedstock (usually termed 'bio-oil') is reacted with high-temperature steam to produce hydrogen near the point of end use. On the other hand, the conversion of biomass using microbes produces sugar-rich feedstocks that can be fermented to produce hydrogen. The thermochemical water splitting process uses a high temperature that is generated by solar concentrators or nuclear reactors to drive chemical reactions that split water to produce hydrogen. The photobiological process uses microbial species, such as green algae, to consume water in the presence of sunlight and produce hydrogen as a by-product.

However, the most common method currently used to produce hydrogen is the steam-methane reforming process in which a low-boiling hydrocarbon derivative (such as methane, i.e., natural gas) is used as the feedstock. In the steam reformer, the feedstock (in the presence of steam) is catalytically cracked at a temperature on the order of 850°C (1,560°F) to yield a product known as 'synthesis gas' which is a mixture of hydrogen, carbon monoxide (CO), and carbon dioxide (CO_2). The carbon monoxide is converted into carbon dioxide with steam in a downstream carbon monoxide shift reaction, during which more hydrogen is produced (Chapter 3). Thus:

Steam-methane reforming reaction:

$$CH_4 + H_2O \text{ (+ heat)} \rightarrow CO + 3H_2$$

Water-gas shift reaction

$$CO + H_2O \rightarrow CO_2 + H_2 \text{ (+ a small amount of heat)}$$

The hydrogen-rich gaseous product is finally purified to the desired quality level by means of pressure swing adsorption (PSA), membranes, or cryogenic processes (Chapter 6).

Thus, the distribution network delivers hydrogen to various hydroprocessing units (i.e., hydrotreating and hydrocracking units) that require it according to the pressure and purity requirements. These units typically require higher purity hydrogen, after which the purge gases (from the units) are sent to units that do not require such high purity hydrogen and, ultimately, the purge gas streams (containing any residual hydrogen) are used as fuel gas.

7.3 HYDROGEN MANAGEMENT

Refineries use large quantities of hydrogen in various process operations to produce products that are environmentally acceptable. In fact, the production of hydrogen is essential for the successful operation of a refinery, and recently demand has been increasing. In fact, refineries are under increasing pressure to meet environmental legislation and reduce the emissions of their processes, so it is desirable to produce and use hydrogen in higher quantities than ever before.

Because of this, the refining industry has been the subject of several major forces that affect most industries and which have hastened the development of new refining processes of which the most prominent are: (1) the increased demand for products such as gasoline, diesel, jet fuel, and fuel oil, (2) feedstock supply, specifically the changing quality of crude oil and geopolitics between different countries and the emergence of alternate feed supplies such as heavy crude oil, extra heavy crude oil, bitumen from tar sand, (3) environmental regulations that include more stringent regulations in relation to sulfur in gasoline and diesel fuel, and (4) technology development such as new catalysts and processes.

Also, increased market competition and tighter environmental legislation lead to the increasing demand for hydrogen in the refinery, which require effective hydrogen management strategies within the refinery in order to satisfy the increasing hydrogen requirements.

In addition, the hydrogen requirement for product improvement, which is the hydrotreatment of crude oil products, to ensure that they meet utility and performance specifications, is also increasing. Product improvement can not only involve hydrotreatment but also changes in molecular shape (reforming and isomerization) or changes in molecular size (alkylation and polymerization), and hydrotreating can play a major role in product improvement.

The use of hydrogen in thermal processes as an aid to the conversion of the feedstock is perhaps the single most significant advance in refining technology during the twentieth century (Dolbear, 1998). In fact, hydrogenation processes are the principal processes used in the manufacture of naphtha and gasoline (Parkash, 2003; Gary et al., 2007; Speight, 2014; Hsu and Robinson, 2017; Speight, 2017). Indeed, with the influx of the more viscous feedstocks into refineries, hydroprocessing will assume a greater role in the refinery of the future.

While the lighter, sweet crude oils (i.e., low density, high API gravity, low sulfur crude oils) require less processing, the viscous refinery feedstocks contain higher levels of sulfur as well as other contaminants. Processing the viscous feedstocks may also begin with the same distillation process as used for the conventional crude oils to produce intermediate products, but additional steps are necessary. One such process is the hydrotreating process that was originally (and still is) introduced to remove sulfur (a downstream pollutant) and other undesirable compounds, such as unsaturated hydrocarbon derivatives and any nitrogen-containing constituents from the process stream.

In fact, to emphasize the above paragraphs, hydrogen is one of the most important elements in crude oil refineries and is used in various refining processes to upgrade the quality of crude oil products and meet sales specifications. The demand for hydrogen in crude oil refineries has increased significantly due to stricter environmental regulations, the processing of heavier crude oils, and the conversion of residues into lighter, more valuable products. However, the increasing demand for hydrogen requires advanced hydrogen management and on-purpose production of more hydrogen through various processes.

Hydrogen management in refineries has become a priority in current refinery operations and when planning to produce lower sulfur gasoline and low sulfur diesel fuel. Along with the increased consumption of hydrogen for hydrotreating the various feedstocks, additional hydrogen is needed for processing heavier and higher sulfur crude slates.

In many refineries, the hydroprocessing capacity and the associated hydrogen network can cause a major limit to refinery throughput. Thus, many refineries have developed or are developing hydrogen management programs that fit into the two general categories of either (1) a catalytic reformer-supplied network or (2) an on-purpose hydrogen supply. Some refineries depend solely on catalytic reformer(s) as their source of hydrogen for hydrotreating. Often, they are semi-regenerative reformers where off-gas hydrogen quantity, purity, and availability change with feed naphtha quality, as octane requirements change seasonally, and when the reformer catalyst progresses from start-of-run (SOR) to end-of-run (EOR) conditions and then goes offline for regeneration. Typically, during some portions of the year, refinery margins are reduced as a result of hydrogen shortages.

Multiple hydrotreating units compete for hydrogen – either by selectively reducing throughput, managing intermediate tankage logistics, or running the catalytic reformer sub-optimally just to satisfy downstream hydrogen requirements. Part of the operating year still runs in hydrogen surplus, and the network may be operated with relatively low hydrogen utilization (consumption/production) at 70 to 80%. Catalytic reformer off-gas hydrogen supply may swing from 75 to 85% hydrogen purity. A hydrogen purity upgrade can be achieved through some hydrotreaters by absorbing heavy hydrocarbons. But without supplemental hydrogen purification, critical control of hydrogen partial pressure in hydroprocessing reactors is difficult, which can affect catalyst life, charge rates, and/or the yield of naphtha.

More complex refineries, especially those with hydrocracking units, also have on-purpose hydrogen production, typically with a steam methane reformer (SMR) that utilizes refinery off-gas and supplemental natural gas as feedstock. The SMR plant provides the swing hydrogen requirements at higher purities (92–99+% hydrogen) and serves a hydrogen network configured with several purity and pressure levels. Multiple purities and existing purification units allow for a more optimized hydroprocessing operation by controlling hydrogen partial pressure for maximum benefit. Typical hydrogen utilization is on the order of 85% to 95%.

Many refineries developing hydrogen management programs that fit into the two general categories of either a catalytic reformer-supplied network or an on-purpose hydrogen supply. Some refineries depend solely on catalytic reformer(s) as their source of hydrogen for hydrotreating. Often, they are semi-regenerative reformers where off-gas hydrogen quantity, purity, and availability change with feed naphtha quality, as octane requirements change seasonally, and when the reformer catalyst progresses from start-of-run to end-of-run conditions and then goes offline for regeneration.

Multiple hydrotreating units compete for hydrogen – either by (1) selectively reducing throughput or (2) running the catalytic reformer sub-optimally just to satisfy downstream hydrogen requirements. Part of the operating year still runs in hydrogen surplus, and the network may be operated with relatively low hydrogen utilization (i.e., consumption/production). Catalytic reformer off-gas hydrogen supply may swing from 75 to 85% hydrogen purity. A hydrogen purity upgrade can be achieved through some hydrotreaters by absorbing heavy hydrocarbons. But without supplemental hydrogen purification. Critical control of hydrogen partial pressure in hydroprocessing reactors is difficult, which can affect catalyst life, charge rates, and/or gasoline yields.

When the refinery steam-methane reforming plant begins reaching its nameplate hydrogen capacity during hydrogen production swings from catalytic reforming units. Hydroprocessing bottlenecks again reduce refinery throughput and operating margins. Furthermore, the energy bills associated with the efficiency of the steam-methane reforming plant should be benchmarked to find cost savings that go right to the refinery's bottom line. The on-stream reliability of the steam-methane reforming plant also takes on a higher priority to maintain high margins.

In summary, the hydrogen plant in the modern refinery is an integral part of the refinery network and, as the main supplier of hydrogen for the refinery, the reliability of the plant is essential to overall hydrogen management and refinery operations. The integration of the hydrogen plant into the refinery network in the early design phase is a necessary part of the design. This approach (and acceptance of the major role of hydrogen in the refinery) can ensure the development of the overall hydrogen network investment.

The final aspect of hydrogen management is hydrogen storage, which has been presented in an earlier chapter (Chapter 6) but is worthy of a brief mention here because of the relationship to the hydrogen network.

Briefly, hydrogen can be stored using several different methods which involve (1) high-pressure gas cylinders at pressures up to 12,000 psi, (2) low-temperature storage as liquid hydrogen in cryogenic tanks, (3) low-temperature adsorption on materials with a large specific surface area, (4) absorption on interstitial sites in a host metal at ambient pressure and temperature, (5) at ambient pressure as chemically bonded hydrogen in covalent and ionic compounds, or (6) by oxidation of reactive metals, such as lithium, Li, sodium, Na, magnesium, Mg, aluminum, Al, and zinc, Zn, with water. The common storage systems involve the use of high-pressure gas cylinders with a maximum pressure on the order of 3,000 psi. Typically, hydrogen that has been converted to the liquid state is stored in cryogenic tanks at low temperature and ambient pressure.

Many metals and alloys are capable of reversibly absorbing hydrogen using molecular hydrogen gas or hydrogen atoms from an electrolyte. The group 1, 2, and 3 light metals (e.g., lithium, Li, magnesium, Mg, Boron, B, and aluminum, Al) can combine with hydrogen to form a large variety of metal-hydrogen complexes (Chapter 6). These bonded species are especially interesting because of the light weight and because of the number of hydrogen atoms (typically 2 hydrogen atoms) per metal atom. Hydrogen can also be stored indirectly in reactive metals (such as lithium, Li, sodium, Na, aluminum, Al, or zinc, Zn) which react with water to form the corresponding hydroxide and liberate the hydrogen from the water. Because water is the product of the combustion of hydrogen with either oxygen or with air, it can be recycled in a closed loop and reacts with the metal. Finally, the metal hydroxides can be thermally reduced to the respective metals.

7.4 THE HYDROGEN NETWORK

The term 'hydrogen network' as it exists in a crude oil refinery refers to a system that produces, distributes, and utilizes hydrogen within the refinery complex. High-pressure pure hydrogen is sent to one of the consumers, and the purge gas of that consumer is treated and diverted to the downstream unit using the same pressure. The network covers from generation to consumption and then finally purging to the refinery fuel gas system. Here's an overview of the hydrogen network in a crude oil refinery.

Refinery hydrogen network typically interconnect many producers, consumers, and purification units with different pressures, purity, and operating objectives. The network grows with each subsequent refinery project, modified in ways that minimize complexity and the interruption of existing units rather than for refinery-wide optimization. Hydrogen production costs and constraints on availability are typically much greater than when the network was first envisioned.

Hydrogen is an increasingly important component of refining, particularly in view of the increased demand for clean fuels. There are opportunities to optimize the use of hydrogen in a refinery to maximize profits, which include (1) hydrogen network studies and flow scheme optimization, (2) pressure swing adsorption and membrane technologies to recover and purify hydrogen, (3) reforming process and catalyst technologies to produce more hydrogen, and (4) hydroprocessing and catalyst technologies to consume less hydrogen.

7.4.1 Optimizing the Hydrogen Network

The first step in improving the hydrogen network is to clarify the objective. Operating costs, capital costs, and refinery margins are all part of the picture. At a high level, the process of hydrogen network optimization is: (1) identify the constraint that is most limiting profitability, (2) identify ways to alleviate the constraint and select the most cost-effective approach, (3) consider all the options listed above. Many improvements can be implemented quickly without capital projects, (4) repeat these steps until a constraint is reached that cannot be cost-effectively relaxed or alleviated, (5) utilize a broad range of tools to make comprehensive optimization possible, and (6) in many hydrogen optimization schemes, it often occurs that the greater the number of degrees of freedom, the larger the improvement that is possible.

The most successful programs for improving the hydrogen network draw the largest possible envelope and take advantage of all the 'knobs' that are available to turn, including network connectivity, increased hydrogen production capacity, target hydrogen partial pressures, process changes in producers and consumers, catalysts, operating procedures, revamped and new purification capacity, pressure swing absorption unit feed to product bypass, feed to the hydrogen plant, compressor modifications, ability of LP models to accurately represent hydrogen availability constraints, and header pressure control system improvements.

Every refinery is different, and from time to time, the active constraints in a refinery can change with different crudes or operating objectives. A refinery network may be constrained by the total moles of hydrogen available, hydrogen purity, hydraulics, purifier capacity, compression, hydrogen scrubbing, fuel system constraints, or other refinery-specific issues.

What is ultimately constrained is the profitability of the refinery. When there is insufficient quantity or purity of hydrogen, charge rates, the processing of more difficult feeds, or product quality is limited and refinery margins are reduced. When sufficient hydrogen is available, the effect of inefficiencies is higher operating (hydrogen production) costs.

7.4.2 Hydrogen and Process Performance

Hydrogen has a significant effect on process performance and profitability. Hydrogen partial pressure, a variable completely under the operator's control, can be utilized to increase catalyst life in hydroprocessing units, increase throughput, increase conversion, improve product quality, or process more profitable feeds. The potential benefits of affecting refinery profitability through hydrogen management are much greater than those from simply reducing hydrogen production or purchase costs. Of course, the benefits must be considered in concert with the capital and operating costs of increasing the hydrogen partial pressure in order to determine the most profitable targets.

For refiners who want to maximize the effectiveness of their hydrogen network, further optimization is possible when targets for recycle hydrogen purity are modified through major operating changes, such as variation in charge rate, feed properties, and severity. The measured variable that represents hydrogen partial pressure is recycle purity. Make-up purity and purge rate can both affect the hydrogen partial pressure, but they do not determine it.

7.4.3 Network Analysis

Hydrogen network pinch analysis is a valuable analytical method to identify the theoretical minimum hydrogen requirements for a given refinery. For example, in the design work for a major new refinery for Petrobras, we had full freedom to route and recover streams, since design pressures and makeup purities were not yet fixed. We set the separator pressure of one hydrotreater such that the flash gas could be sent to the suction of the makeup compressor of another hydrotreater, and the revised makeup hydrogen purity could be taken into account in the design of the consuming unit. The flash drum in another unit was set at a pressure such that its flash gas could easily be purified and recovered in an existing pressure swing adsorption unit.

If it is not possible to purchase an external fuel for the refinery, the fuel balance is not only critical from an economic standpoint but it determines the feed selection for the hydrogen plant. In our early estimates, we expected that an internally generated LPG stream would have to be burned to meet the fuel balance. After energy optimization of the preliminary design, the fuel gas balance shifted to positive, and we were able to utilize lower-value fuel gas as hydrogen plant feed. As an additional benefit, we could isolate hydrogen-rich fuel gas streams and utilize them as hydrogen plant feed. Not only is fuel gas less valuable than LPG, but the hydrogen content enables a significant reduction in firing and energy consumption in the hydrogen plant furnace compared to other hydrocarbon feeds.

7.4.4 Optimizing Beyond the Hydrogen Network

Broadening the optimization provides a target and a pinch purity that can guide the analyst in proposing changes to the existing network. However, pinch analysis is limited in a number of ways. Its key limitations are that it is not concerned with the pressures of sources and sinks, it treats all light hydrocarbon components the same (in reality, methane has a much greater tendency to build up in the recycle loop than propane), it does not consider the significant effects on compressors and hydraulics of changing a hydrogen recycle hydroprocessing unit into a once-through unit, nor does it consider sulfur levels.

To evaluate potential network improvements, a detailed refinery-wide model of the hydrogen network should be employed. One of the critical features of this model (as opposed to a spreadsheet model) is that it simulates the non-linear relationships between charge rate, make-up rate, make-up purity, and recycle purity (hydrogen partial pressure) and purge rate. With this model, potential modifications can be tested, such as changing the make-up source, trading off purity against recovery in an existing purifier, adding new purification, or changing the target hydrogen partial pressure in a hydrogen-consuming process unit. The same model can be used by the refiner for operations and project planning.

7.4.5 Process and Purification Models

Process models of catalytic reformers enable the analyst to understand the effects of operating, mechanical, or catalyst changes on hydrogen yield and purity.

A hydroprocessing model can be assembled to predict the effect of changing the hydrogen partial pressure on process performance (catalyst life, product quality, conversion) and hydrogen consumption. Purifier (which typically may include a pressure swing adsorption unit and membrane unit) models are used to predict the impact of changing a purifier's feed composition and operating conditions on hydrogen product purity and recovery. These models are also used to identify the potential for debottlenecking existing purifiers.

7.4.6 Hydrogen Network Improvements and Optimizing the Network

The potential for improving hydrogen efficiency is estimated by evaluating the hydrogen-containing streams currently going to fuel, flare, and hydrogen plant feed. The potential financial benefits will also be a function of the value of hydrogen in a refinery. The potential benefit of improvements in process performance through hydrogen optimization can be estimated roughly by evaluating current constraints to process performance, refinery drivers, and refinery economics.

In many hydrogen optimization schemes, it often occurs that the greater the number of degrees of freedom, the larger the improvement that is possible. The most successful programs for improving the hydrogen network draw the largest possible envelope and take advantage of all the 'knobs' that are available to turn, including network connectivity, increased hydrogen production capacity, target hydrogen partial pressures, process changes in producers and consumers, catalysts, operating procedures, revamped and new purification capacity, pressure swing absorption (PSA) unit feed to product bypass, feed to the hydrogen plant, compressor modifications, ability of LP models to accurately represent hydrogen availability constraints, and header pressure control system improvements.

Every refinery is different, and from time to time, the active constraints in a refinery can change with different crudes or operating objectives. When there is insufficient quantity or purity of hydrogen, charge rates, the processing of more difficult feeds, or product quality is limited and refinery margins are reduced. When sufficient hydrogen is available, the effect of inefficiencies is higher operating (hydrogen production) costs.

7.4.7 Hydrogen and Process Performance

Hydrogen has a significant effect on process performance and profitability because it can be employed to increase catalyst life in hydroprocessing units, increase throughput, increase conversion, improve product quality, or process more profitable feeds. The potential benefits of affecting refinery profitability through hydrogen management are much greater than those from simply reducing hydrogen production or purchase costs. Of course, the benefits must be considered in concert with the capital and operating costs of increasing the hydrogen partial pressure in order to determine the most profitable targets.

For refiners who want to maximize the effectiveness of their hydrogen network, further optimization is possible when targets for recycle hydrogen purity are modified through major operating changes, such as variation in charge rate, feed properties, and severity. The measured variable that represents hydrogen partial pressure is recycle purity. Make-up purity and purge rate can both affect the hydrogen partial pressure, but they do not determine it.

7.5 NETWORK OPTIMIZATION

Hydrogen network optimization is at the forefront of options to address clean fuel trends, meet growing transportation fuel demands, and help maintain profitability from refining crude oils.

The hydrogen distribution system in the modern refinery cascades through multiple hydroprocessing units, where higher hydrogen purity and higher-pressure consumers send their purge gases to lower hydrogen purity and lower pressure consumers. Ultimately, purge gases containing residual hydrogen are employed as fuel for the refinery.

Many makeup and recycle compressors drive the hydrogen through this complex network of hydrotreater and hydrocracker units. The various hydroprocessing units consume hydrogen to (1) remove sulfur and nitrogen, (2) reduce the boiling point of the products, and/or (3) saturate aromatic derivatives that occur in the liquid products.

The characteristics of the refinery feedstock, the configuration of the refinery, and new production requirements of each refinery are different, so various hydrogen network optimization considerations must be considered. These considerations include (1) recovering hydrogen from refinery gas streams, (2) expanding or adding a new on-purpose hydrogen plant, or (3) outsourcing to an industrial gas company.

7.5.1 Recovery Technologies

Hydrogen from off-gas sources, typically ranging from 50 to 90%, with a few of them as low as 10%, is an increasingly important alternative to refiners (Davis and Patel, 2004; Düker and Basse, 2006). The wide range of off-gas hydrogen content makes choosing a recovery technology critical to obtain an economically attractive hydrogen product. In addition, a recovery solution should consider effective utilization of existing compression and minimizing new compression requirements.

Each of the hydrogen recovery technologies has its unique capabilities and constraints. Table 3 summarizes important considerations when determining a technology fit to provide economically attractive hydrogen for a specific refinery application.

Adsorption (using the pressure swing adsorption system) is the best choice when ultra-high purity, 99%+, hydrogen product is required. Any impurities and unrecovered hydrogen are delivered at low pressure and are available as a refinery fuel. Hydrogen recoveries for a pressure swing adsorption system are typically lower than those available from other technologies.

Pressure swing adsorption system designs for feeds must account for variability in feed gas conditions. For example, catalytic reformer off-gas hydrogen can change by 10 to 15% v/v from start to end of run catalyst conditions. Hydrotreater off-gas composition changes as a function of makeup hydrogen purity and specific purge rate set points. When multiple refinery off-gas streams are mixed, different operating scenarios can significantly change pressure swing adsorption feed composition. Refinery off-gas feed contaminants can include a long 'tail' of heavy C5+ hydrocarbons, including aromatics. Some heavy hydrocarbons and aromatics can create problems if they are irreversibly adsorbed. Therefore, it is advantageous to design refinery off-gas hydrogen pressure swing adsorption systems using various available adsorbents and tailoring the adsorbent quantities and mix to each application to provide feed flexibility while striving for the best achievable recoveries and high operating bed utilization. Hydrogen recoveries (typically greater than ~80%) are dependent on feed conditions, hydrogen product specifications, and system configuration.

Prism® membranes can provide the lowest cost solution for high-pressure refinery off-gas feeds, especially when integration with existing hydrogen compression is possible. Membrane systems are bulk separators that produce refinery-quality hydrogen purities of 90 to 98% at high recoveries (85% or better). Since pressure drives the separation, they are best suited for high-pressure feeds (>300 psi) such as hydroprocessing purge gases. A common disadvantage of membranes is that the hydrogen product stream is produced at pressures much lower than the feed pressure and may need recompression. The hydrogen product may be sent to existing make-up or recycle compression in hydroprocessing loops. An important consideration for refinery off-gas applications is that hydrogen is also a 'fast gas' through the membrane polymer and will end up in the hydrogen product. Prism® membrane systems have been utilized in almost 50 refinery installations with feed capacities of up to 39 thousand Nm3/hr (or 35 MM scfd) and design pressures of up to 2,500 psi. In refinery operation, they are flexible to changing refinery off-gas feed conditions and provide a robust separation system when upstream units are designed to handle various feed contaminants such as water, particulates, amines, and hydrocarbon liquids. A detailed review of potential impurities in the feed stream, particularly during upsets, must be completed prior to selecting the technology.

Cryogenic Systems provides an excellent vehicle to purify or recover refinery-quality hydrogen when by-product liquid hydrocarbons, especially ethylene and propylene, can subsidize the higher capital expenditure. Even low pressure (>75 psig) and dilute hydrogen refinery off-gas feeds, such as FCC off-gas, are good candidates for a cryogenic separation system.

7.5.2 Expansion of an Existing Hydrogen Plant

Generally, complex refineries source 30~60% of their total hydrogen requirements from on-purpose hydrogen capacity.

The endothermic steam reforming reaction is accomplished by flowing feed gas through catalyst-filled tubes that are located in a furnace. The resulting hydrogen and carbon oxides are further processed in shift reactors to convert carbon monoxide to additional hydrogen. Many existing refinery hydrogen plants produce a medium purity hydrogen product stream (on the order of 94 to 97% v/v) by removing the carbon dioxide in a solvent absorption system accompanied by a methanation step. Newer hydrogen production plants use pressure swing adsorption technology to recover and purify the hydrogen to purities in excess of 99.9% v/v. These pressure swing adsorption-based hydrogen plants have a higher efficiency than the conventional low purity plants due to additional export steam credits.

Existing hydrogen plants can be debottlenecked and expanded through various methods. Current limitations can include: (1) reformer furnace firing capacity – reformer tubes/convection section coils overheating, fuel system control limits, or burner firing limitations, (2) system pressure drop limitations – maximum pressure through the catalyst bed or full use of available feed pressure from the supplier or feed compressor, (3) induced-draft fan capacity limiting available draft for the reformer furnace firing, and (4) purification equipment limitations – flooding in the CO_2 absorber, CO_2 stripper capacity, overheating in the methanator, bed fluidization in pressure swing adsorption, or insufficient adsorption capacity in pressure swing adsorption.

The refiner must be aware of not only the major bottleneck but also other areas of the plant that are near their respective limits. If the first bottleneck is resolved, another is often right behind it.

Secondary limitations are usually difficult to identify. In many plants, multiple bottlenecks may need to be addressed to accomplish a meaningful expansion in capacity. This could be very costly and may require significant downtime.

Increasing hydrogen production from 5% to 30% may be possible depending on the age of the plant, the extent of previous debottlenecking work, utility availability, constraints of emission permits, and flexibility of the refinery with respect to product hydrogen pressure, steam production, and design margins.

The optimum solution for an overall refinery hydrogen network can involve one or more of several potential options. These options can involve: (1) new or improved hydrogen recovery/purification equipment properly located within the refinery flowsheet; (2) refinery off-gas and/or liquefied petroleum gas integration with on-site hydrogen production; (3) integration with petrochemical operations; (4) improving hydrogen plant efficiency; (5) expanding or replacing the existing hydrogen plant capacity; and (6) providing new hydrogen via pipeline supply.

Thus, to ensure the long-term production of hydrogen, it is crucial to conduct a critical assessment of various production routes and their environmental impacts, as well as storage options and utilization options. Hydrogen is produced from either fossil-based or renewable feedstocks (Chapter 3, Chapter 4, and Chapter 5).

7.6 HYDROGEN DISTRIBUTION

The use of hydrogen requires an infrastructure that enables the hydrogen to be delivered from the point of production to the point of end use, such as a refinery or any industrial facility. The delivery technology for hydrogen (from the production point to the point of use) is currently available commercially and the growth in the demand for hydrogen will require regional expansion of this infrastructure and development of new technologies, such as chemical carriers to transport hydrogen at high density and high-throughput technologies for transportation.

In many countries, hydrogen is produced at, or at least as close to the plant where it is used as possible. On the other hand, if the hydrogen has to be transported to the point of use, there are three methods of transportation: (1) pipelines, (2) high-pressure tube trailers, and (3) liquefied hydrogen tankers.

7.6.1 Pipeline Transportation

Transportation by pipeline (in a similar manner to the transportation of natural gas) is a convenient method for the transportation of hydrogen, the least-expensive way to deliver large volumes of hydrogen, but the capacity may be limited because of the limited mileage of the available pipeline systems. Typically, these pipelines are located near large crude oil refineries and chemical (or petrochemical) plants.

If not, the site where the hydrogen is produced can have a major impact on the method of delivery. For example, if the hydrogen is produced at a large, centrally located production facility, it is possible that the hydrogen can be produced at a lower cost because the facility is producing more, but there may be a disadvantage insofar as there could be higher costs to deliver the hydrogen because of the distance to the point of delivery. In comparison, distributed production facilities produce hydrogen at locations such as a refinery or a petrochemical plant (Chapter 3), so delivery costs are relatively low, but the cost to produce the hydrogen is likely to be higher because production volumes are less.

Transporting gaseous hydrogen via existing pipelines (such as natural gas pipelines) is a low-cost option for delivering large volumes of hydrogen. The high initial capital costs of new pipeline construction constitute a major barrier to expanding hydrogen pipeline delivery infrastructure. Technical issues that are related to pipeline transmission include: (1) the potential for hydrogen to embrittle the steel and welds used to fabricate the pipelines; (2) the need to control hydrogen permeation and leaks; (3) the need for reliable and more durable hydrogen compression technology.

Potential solutions to these issues include using fiber reinforced polymer (FRP) pipelines for hydrogen distribution. The installation costs for fiber reinforced polymer pipelines are about 20% less than those of steel pipelines because the fiber reinforced polymer can be obtained in sections that are much longer than steel, thereby minimizing welding requirements.

One possibility for rapidly expanding the hydrogen delivery infrastructure is to adapt part of the natural gas delivery infrastructure to accommodate hydrogen. Converting natural gas pipelines to carry a blend of natural gas and hydrogen (up to about 15% v/v hydrogen) may require only modest modifications to the pipeline. However, converting existing natural gas pipelines to be suitable for the delivery of pure hydrogen may require more substantial modifications.

7.6.2 Transportation by Tube Trailers

Transporting compressed hydrogen gas in high-pressure tube trailers by truck, railcar, ship, or barge is used primarily for distances of 200 miles or less. Gaseous hydrogen is most commonly delivered either by trucks or through pipelines. On the other hand, hydrogen in the gaseous phase is typically produced at relatively low pressures (300 to 540 psi) and prior to truck transportation, the hydrogen must be compressed to pressures on the order of 2,600 psi or higher into long cylinders which are stacked on the trailer that the truck hauls, which gives the appearance of long tubes to the onlooker, hence the name tube trailer.

Tube trailers are currently limited to pressures of 3,750 psi by the regulations of the United States Department of Transportation (US DOT) but exemptions have been granted to enable operation at higher pressures (on the order of 7,500 psi or higher). Steel tube trailers can transport on the order of 840 pounds of hydrogen but the carrying capacity of these types of trailers is limited by the weight of the steel tubes. As a result, storage vessels made of composite materials have been developed that have capacities of 1,235 to 1,985 pounds of hydrogen per trailer.

Briefly and by way of definition, a composite material is a combination of two materials with different physical and chemical properties which, when combined, create a material that is specialized to perform a particular task such as, for instance, to become stronger and lighter with an improvement in.

7.6.3 Transportation by Liquefied Hydrogen Tankers

Cryogenic liquefaction is a process in which the hydrogen is cooled to a temperature where liquefaction of the hydrogen occurs, allowing hydrogen to be transported more efficiently (compared with high-pressure tube trailers) over longer distances by truck, railcar, ship, or barge. If the liquefied hydrogen is not used at a sufficiently high rate at the point of consumption, the hydrogen can boil off (or evaporate) from the containment vessels. As a result, hydrogen delivery and consumption rates must be carefully matched and the balance maintained.

Gaseous hydrogen is liquefied by cooling it to below −253°C (−423°F) and, once liquefied, the hydrogen can be stored in the liquid state at the liquefaction plant in insulated tanks. Caution is advised because some amount of stored hydrogen will be lost through evaporation, or 'boil off' of liquefied hydrogen, especially when using small tanks with large surface-to-volume ratios.

Currently, hydrogen is transported over longer distances as a liquid in super-insulated, cryogenic tanker trucks. After the liquefaction process, the liquid hydrogen is dispensed to delivery trucks and transported to distribution center sites where it is vaporized to a high-pressure gaseous product for the distribution process. Over long distances, transporting the liquid hydrogen by truck may be more economical than trucking gaseous hydrogen. Put simply, a liquid tanker truck can hold a much larger amount (i.e., mass) of hydrogen than a gaseous tube trailer.

7.6.4 The Transportation Infrastructure

Creating an infrastructure for distribution (and delivery) does present challenges, not the least of which is the cost of constructing a new hydrogen pipeline network involves high initial capital costs, and the properties of the hydrogen (Chapter 6) present challenges to pipeline materials and compressor design. The possible solution is that because hydrogen can be produced from a wide variety of resources, the production of hydrogen at regional production centers or even at local hydrogen production can maximize the use of local resources, which is an aid to resolving and minimizing distribution challenges.

The infrastructure at the production site includes (in addition to the hydrogen production equipment), the transportation infrastructure. At the point of hydrogen use, additional infrastructure components that are necessary deployed include (1) compression, (2) storage facilities, (3) dispensers, (4) contaminant detection monitors, and (5) purification technologies.

Most compressors employed for the compression of gaseous hydrogen are either positive displacement compressors (which can be reciprocating or rotary) or centrifugal compressors.

By way of explanation, reciprocating compressors (sometimes referred to as 'recips') are the most commonly used compressors for applications that require a very high compression ratio and use a motor with a linear drive to move a piston or a diaphragm back and forth to compress the hydrogen by reducing the volume it occupies. On the other hand, rotary compressors compress through the rotation of gears, lobes, screws, vanes, or rollers. On the other hand, ionic compressors are similar to reciprocating compressors but use ionic liquids in place of the piston. Also, rotary compressors do not require bearings and seals, two of the common sources of failure in reciprocating compressors. An ionic compressor is a compression system in which compression is achieved by a column of ionic fluid (a fluid composed solely of ions and has a melting point lower than 100°C/212°F) is in direct contact with the gaseous hydrogen, allowing for the continuous transportation of the gas.

Centrifugal compressors are often the compressor of choice for pipeline applications due to their high throughput and moderate compression ratio. They operate by rotation of a turbine at very high speeds to compress the gas. Hydrogen centrifugal compressors must operate at tip speeds three times faster than that of natural gas compressors to achieve the same compression ratio because of the low molecular weight of hydrogen.

In summary, hydrogen can be transported over the distance from the point of production to the point of use. Producing hydrogen in a production center can decrease production costs but may increase distribution costs. In contrast, producing hydrogen at the point of end-use stations will decrease distribution costs but increase production costs.

In the future, other hydrogen carriers may be sought and used. For example, hydrogen carriers store hydrogen in some other chemical state rather than as free hydrogen molecules (Chapter 6). Chemicals that can act as hydrogen carriers include metal hydrides, carbon or other nanostructures, and reversible reactors. Using such types of carriers would be a significant departure from the way hydrogen is currently delivered.

REFERENCES

Davis, R.A., and Patel, N.M. 2004. Refinery Hydrogen Management. *Petroleum Technology Quarterly*, Spring: 28–35.

Düker, A., and Basse, R. 2006. Hydrogen Management. *Petroleum Technology Quarterly*, Q1: 1–6. https://www.digitalrefining.com/article/1000631; https://www.researchgate.net/publication/293516074_Hydrogen_management.

Gary, J.H., Handwerk, G.E., and Kaiser, M.J. 2007. *Petroleum Refining: Technology and Economics*, 5th Edition. CRC Press, Taylor & Francis Group, Boca Raton, FL.

Hsu, C.S., and Robinson, P.R. (Editors) 2017. *Handbook of Petroleum Technology*. Springer AG, Cham.

Luckwal, K., and Mandal, K.K. 2009. Improve Hydrogen Management of Your Refinery. *Hydrocarbon Processing*, 88(2): 55–61.

Parkash, S. 2003. *Refining Processes Handbook*. Gulf Professional Publishing, Elsevier, Amsterdam.

Speight, J.G. 2014. *The Chemistry and Technology of Petroleum*, 5th Edition. CRC Press, Taylor & Francis Group, Boca Raton, FL.

Speight, J.G. 2017. *Handbook of Petroleum Refining*. CRC Press, Taylor & Francis Group, Boca Raton, FL.

Speight, J.G. 2019. *Natural Gas: A Basic Handbook*, 2nd Edition. Gulf Publishing Company, Elsevier, Cambridge, MA.

Speight, J.G. 2021. *Petroleum Refining Technology Series Volume 1: Refinery Feedstocks*. CRC Press, Taylor & Francis Group, Boca Raton, FL.

Speight, J.G. 2023. *Petroleum Refining Technology Series Volume 4: Hydrotreating and Hydrocracking in Petroleum Refining*. CRC Press, Taylor & Francis Group, Boca Raton, FL.

Zagoria, A., Huychke, R., and Boulter, P.H. 2003. Refinery Hydrogen Management – The Big Picture. Proceedings. DGMK Conference on Innovation in the Manufacture and Use of Hydrogen. Dresden. October 15–17. Page 95.

8 Hydrogen in the Refinery of the Future

8.1 INTRODUCTION

The projections for the continued use of fossil fuels indicate that there will be at least another five decades of fossil fuel use (especially in the cases of natural gas and crude oil) before biomass and other forms of alternate energy take hold and are able to provide sufficient energy not only to subsidize the production of energy from fossil fuels but to replace the fossil fuels (Speight, 2011a, 2011b, 2013a, 2013b). Furthermore, estimations that the era of fossil fuels (natural gas, crude oil, coal, and biomass, and carbonaceous wastes) will be almost over when the cumulative production of the fossil resources reaches 85% of their initial total reserves may or may not have some merit. In fact, the relative scarcity (compared to a few decades ago) of crude oil was real, but it seems that the remaining reserves make it likely that there will be an adequate supply of energy for several decades (Martin, 1985; MacDonald, 1990; Banks 1992; Krey et al., 2009; Speight, 2011c, 2013a, 2013b, 2014a). The environmental issues are very real and require serious and continuous attention.

Technologies which ameliorate the effects of fossil fuel combustion on acid rain deposition, urban air pollution, and global warming must be pursued vigorously (Mohnen, 1988; Vallero, 2008). There is a challenge that must not be ignored, and the effects of acid rain on soil and water leave no doubt about the need to control its causes (Mohnen, 1988). Indeed, recognition of the need to address these issues is the driving force behind recent energy strategies, as well as a variety of research and development programs (Stigliani and Shaw, 1990; United States Department of Energy, 1990; United States General Accounting Office, 1990).

While regulations on the greenhouse gas (GHG) carbon dioxide (CO_2) would be an immediate hurdle to deployment of coal plants, gasification plants are in the best position compared to other coal-based alternatives to capture carbon dioxide. However, with the continued uncertainty of carbon dioxide regulation, there is industry reluctance to make large investments in projects with high emissions of carbon dioxide since a cost-effective solution for reducing such emissions is not yet available. Nevertheless, the reduction in greenhouse gas emissions can be an enhancing factor for gasification in the long run because the carbon dioxide from a gasification plant is more amenable to capture.

As new technology is developed, emissions may be reduced by repowering in which aging equipment is replaced by more advanced and efficient substitutes. Such repowering might, for example, involve an exchange in which an aging unit is exchanged for a newer combustion chamber, such as the atmospheric fluidized-bed combustor (AFBC) or the pressurized fluidized-bed combustor (PFBC).

Indeed, recognition of the production of these atmospheric pollutants in considerable quantities every year has led to the institution of national emission standards for many pollutants. Using sulfur dioxide as an example, the various standards are not only very specific but will become more stringent with the passage of time. Atmospheric pollution is being taken very seriously and there is also the threat, or promise, of heavy fines and/or jail terms for any pollution-minded miscreants who seek to flaunt the laws (Vallero, 2008). Nevertheless, a trend toward the increased use of fossil fuels will require more stringent approaches to environmental protection issues than we have ever known at any time in the past. The need to protect the environment is strong.

DOI: 10.1201/9781003186694-8

Only a few decades ago, viscous feedstocks being utilized in the modern refinery today would not have even been a consideration to produce mainstream products and were used mainly as bunker fuels. Hydrogen is therefore critical to convert this poor-quality crude oil into modern-day products, and to comply with strict environmental mandates.

Although these heavy crudes are cheaper, refineries are faced with the additional expense of upgrading to sophisticated processes to refine them to the required standards and product slate meeting demand. The alternative is to pay a premium for the lighter crudes.

The challenge is made more complex by the fact that no two refineries are alike and that the naturally occurring hydrocarbon distribution in crude does not always match customer demand. Various additional processing steps are required to re-adjust the molecules, reshape them, and remove contaminants to ensure the refinery products meet the requirements for end-use and the product demand profile, as well as environmental performance.

Hydrogen is key to refinery operations, allowing refineries to comply with the latest product specifications and environmental requirements for fuel production mandated by the market and governments and helping to reduce the carbon footprint of their plants.

Margins are tight in the highly competitive refinery business, a situation exacerbated by the costs of creating low sulfur fuels from heavier, more sour crude, as the world's crude oil resources dwindle. Product sulfur levels are lower than ever before – for instance, the tolerance level is 30 parts per million (ppm) in gasoline and 15 ppm in diesel fuels, above which these fuels (by regulation in the United States and many other countries) must not exceed this level of sulfur.

8.2 GROWTH IN DEMAND

The growth in demand for hydrogen will continue to increase, and a new source of hydrogen must be developed. Such development would well (must) involve the use of the always available biomass resources, which may also offer a ready source of specialty chemicals. Thus, in addition to using traditional feedstocks such as coal and crude oil coke, gasifiers can be designed to utilize biomass, such as yard and crop waste, bio-solids, energy crops, such as switch grass, and waste and residual pulp/paper plant materials as feedstock.

Like many solid feedstocks, biomass can come in a range of sizes. In many biomass gasification systems, the biomass must be processed to a uniform size or shape to feed into the gasifier at a consistent rate and to ensure that as much of the biomass is gasified as possible. However, beyond the issue of biomass availability, including the seasonal factors associated with many of the biomass feedstocks, another major concern is that more energy is expended in the collection and preparation stages than is generated through processing the biomass, and although technical hurdles to biomass use remain, in general, there is the general perception in many countries that increased use of biomass feedstocks is driven more by environmental and regulatory factors, if anything, than by free-market forces. Without tax credits or similar incentives, biomass is unlikely to be used as a base-load feedstock, and market entry is likely to occur as a blend with other feedstocks (Clayton et al., 2002).

In general, biomass gasification plants are much smaller than the typical coal or crude oil coke gasification plants used in the power, chemical, fertilizer, and refining industries. As such, they are less expensive to build and have a smaller *facility footprint*. While a large industrial gasification plant may take up 150 acres of land and process 2,500 to 15,000 tons per day of feedstock (such as coal or crude oil coke), the smaller biomass plants typically process 25 to 200 tons of feedstock per day and take up less than 10 acres.

Vast amounts of corn (and land, water, and fertilizer) are needed to produce the ethanol. As more corn is being used, there is an increasing concern about less corn being available for food. Gasifying biomass, such as corn stalks, husks, and cobs, and other agricultural waste products to produce ethanol and synthetic fuels such as diesel and jet fuel can help break this energy-food competition.

Biomass, such as wood pellets, yard and crop wastes, and energy crops such as switch grass and waste from pulp and paper mills, can be used to produce ethanol and synthetic diesel. The biomass is first gasified to produce synthesis gas, and then converted via catalytic processes to these downstream products.

Each year, municipalities spend millions of dollars collecting and disposing of waste, such as yard waste (grass clippings and leaves) and construction and demolition debris. While some municipalities compost yard waste, this takes a separate collection by a city – an expense many cities just can't afford. Thus, instead of paying to dispose of and manage waste for years in a landfill, using it as a feedstock reduces disposal costs and landfill space, and converts that waste to power and fuels.

Thus, the benefits of biomass gasification include (1) the conversion of a waste product into high-value products, (2) the reduced need for landfill space for disposal of solid wastes, (3) the decrease in the emission of methane from landfills, (4) the reduced risk of contamination of the groundwater by runoff from landfills, and (5) the production of ethanol from non-food sources.

While still relatively new, biomass gasification shows a great deal of promise. A key advantage of biomass gasification is that the process can convert non-food biomass materials, such as corn stalks and wood wastes, to useful products such as alcohol derivatives that can be used as sources of useful chemicals (Speight, 2014, 2017, 2019). Furthermore, biomass gasification does not diminish the availability of removing food-based biomass, such as corn, from the economy, unlike typical fermentation processes for making alcohol derivatives.

In fact, a critical issue facing the refineries at present and in the future is the changing slate of crude oil feedstocks and the conversion of these feedstocks into refined transportation fuels under an environment of increasingly more stringent clean fuel regulations, decreasing heavy fuel oil demand, and increasing supply of heavy, high-sulfur crude oils. Innovative hydrogen network optimization will be a critical factor influencing future refinery operating flexibility and profitability in a shifting world of crude feedstock supplies and the need for low-sulfur naphtha and low-sulfur kerosene from which ultra-low-sulfur (ULS) gasoline and diesel fuel are produced.

Upgrading feedstocks such as heavy oils and residua evolved after the introduction of hydrodesulfurization processes (Rana et al., 2007; Ancheyta and Speight, 2007). In the early days, the goal was desulfurization but, in later years, the processes were adapted to a 10 to 30% partial conversion operation, intended to achieve desulfurization, and obtain low-boiling fractions simultaneously, by increasing severity in operating conditions.

Hydrogen, one of the two major components of synthesis gas, is used to produce high-quality fuels (gasoline, diesel fuel, and jet fuel), thereby meeting the requirements for clean fuels according to state and federal clean air regulations (in the United States). Hydrogen is also used to upgrade heavy crude oil and tar sand bitumen. Refineries can gasify low-value residuals, such as crude oil coke, asphalts, tars, and some oily wastes from the refining process, to generate both the required hydrogen and the power and steam needed to run the refinery.

Thus, the gasification of crude oil residua, crude oil coke, and other feedstocks such as biomass (Speight, 2011a, 2011b, 2014a, 2020) to produce hydrogen and/or power may become an attractive option for refiners. The premise that the gasification section of a refinery will be the *garbage can* for deasphalter residues, high-sulfur coke, as well as other refinery wastes is worthy of consideration. Other processes such as ammonia dissociation, steam-methanol interaction, or electrolysis are also available for hydrogen production, but economic factors and feedstock availability assist in the choice between processing alternatives.

Biomass gasification can be considered as a form of pyrolysis, which takes place at higher temperatures and produces a mixture of gases with a hydrogen content on the order of 6 to 6.5% v/v. The synthetic gas produced by the gasification of biomass is made up of hydrogen, carbon monoxide, methane, nitrogen, carbon dioxide, oxygen, and volatile tar. The product distribution and gas composition depend on many factors, including the gasification temperature and the reactor type.

From a global perspective, hydrogen is demonstrating significant growth. Large heavy crude oil reserves, still under development, may increase the demand for hydrogen even further. Two examples are the extra heavy crude oil in the Orinoco Belt in southern Venezuela and the deposits of bitumen in Alberta, Canada. Hydrogen is utilized in several refining processes, all aiming to obtain better product qualities. The main processes include the hydrotreating of various refinery streams and hydrocracking of heavy products.

Processing crude oil typically begins with the same distillation process as for the sweet crudes to produce intermediate products, but additional steps are necessary.

Hydrotreating is one such process, introduced to remove sulfur, a downstream pollutant, and other undesirable compounds, such as unsaturated hydrocarbons and nitrogen from the process stream. Hydrogen is added to the hydrocarbon stream over a bed of catalyst that contains molybdenum with nickel or cobalt at intermediate temperature, pressure, and other operating conditions. This process causes sulfur compounds to react with hydrogen to form hydrogen sulfide, while nitrogen compounds form ammonia. Aromatics and olefins are saturated by hydrogen and lighter products are created. The final product of the hydrotreating process is typically the original feedstock free of sulfur and other contaminants. Single or multiple product streams (fractionated) are possible, depending on the process configuration.

High boiling gas oils, viscous residues, or similar boiling-range heavy distillates react with hydrogen in the presence of a catalyst at high temperature and pressure. The heavy feedstocks are converted (cracked) into light distillates – for example, naphtha, kerosene, and diesel – or base stocks for lubricants. In fact, the hydrocracking unit is often the major hydrogen consumer in the refinery insofar as hydrogen is the key enabler of the hydrocracking to reduce the product boiling range appreciably by converting most of the feed to lower-boiling products, especially when applied to the viscous feedstocks and products from these feedstocks. Other refinery processes including isomerization, alkylation, and tail gas treatment also consume small amounts of hydrogen.

In addition to its mainstream uses, hydrogen is also utilized in refineries for smaller, but no less critical, applications. However, whenever hydrogen is employed in processes that also require a stage of the hydrogen, there are certain cautions that need to be in place (Chapter 6).

Fuel gases, such as hydrogen, are needed if flames are part of the system, as they are in gas chromatography (GC) and in gas chromatography-flame ionization detection systems (GC-FID systems) which require a carrier gas that is often hydrogen. All of the gases used in these systems must be of an exceptionally high and controlled quality in order to obtain the very accurate analytical results required in modern refineries.

Hydrogen is one of the smallest and lightest molecules but is also one of the most flammable and explosive gases processed at a refinery. Any leakage poses a potential risk of fire or explosion, potentially resulting in damage to capital equipment, production downtime and, at the extreme end, potential loss of life. Since hydrogen is used on such a refinery-wide scale, it is critical to detect any leaks and, as such, there is a vital need for sophisticated detection equipment.

Gas detectors are used to monitor and warn of the existence of specific toxic or combustible gases and oxygen deficiency hazards. Typical measurement levels will be in the low parts per million for toxic gas, while combustible gases are measured in terms of the lower explosive limit (LEL).

A fixed system refers to a monitoring system permanently installed in the workplace. The detecting sensor may be hardwired or use wireless signals to a central reporting station. Most will come with an audible alarm.

All refinery gas detection equipment also needs to be tested and calibrated regularly and accurately, and this requirement must also be addressed by, for example, the use of environmentally friendly portable and small refillable gas cylinders that have a completely integrated valve, pressure regulator, and flow meter for direct use.

8.3 HYDROGEN PRODUCTION AND THE ENVIRONMENT

In the modern technologically advanced refining industry, it is becoming increasingly common to find synergies between hydrogen and other processes, such as the recovery or sequestration of carbon dioxide, wastewater treatment, and enhanced oil recovery (EOR), among other uses.

The hydrogen production process can be harnessed to sequester carbon dioxide, which is a by-product of this process. Instead of venting carbon dioxide into the atmosphere, the plant design can add additional steps to the hydrogen process to remove carbon dioxide from the gaseous stream and recover it for other industrial uses. An interesting and pertinent application for this carbon dioxide is EOR, where carbon dioxide is pumped at high pressure into deep oil wells to improve the recovery of the crude oil trapped at these low levels.

Hydrogen is also used to enable the production of biofuels, produced from biological matter instead of fossil fuels. This matter can range from corn or soybeans to crop oils and animal fat, depending on the type of fuel being made and the production method.

The combustion of fossil fuels can account for most of the sulfur oxides and nitrogen oxides released into the atmosphere. Whichever technologies succeed in reducing the amounts of these gases in the atmosphere should also succeed in reducing the amounts of urban smog, those notorious brown and gray clouds that are easily recognizable at some considerable distances from urban areas, not only by their appearance but also by their odor.

$$SO_2 + H_2O \rightarrow H_2SO_4 \text{ (sulfurous acid)}$$

$$2SO_2 + O_2 \rightarrow 2SO_3$$
$$SO_3 + H_2O \rightarrow H_2SO_4 \text{ (sulfuric acid)}$$

$$2NO + H_2O \rightarrow HNO_2 + HNO_3 \text{ (nitrous acid + nitric acid)}$$

$$2NO + O_2 \rightarrow 2NO_2$$
$$NO_2 + H_2O \rightarrow HNO_3 \text{ (nitric acid)}$$

Current awareness of these issues by various levels of government has resulted, in the United States, in the institution of the Clean Fossil Fuels Program to facilitate the development of pollution abatement technologies. And it has led to successful partnerships between government and industry (United States Department of Energy, 1993). In addition, there is the potential that new laws, such as the passage in 1990 of the Clean Air Act Amendments in the United States (United States Congress, 1990; Stensvaag, 1991) will be a positive factor and supportive of the controlled clean use of fossil fuels. However, there will be a cost, but industry is supportive of the measure and confident that the goals can be met.

Gasification plants can readily capture carbon dioxide, the leading greenhouse gas, much more easily and efficiently than coal-fired power plants.

Carbon dioxide captured during the gasification process can be used to help recover oil from otherwise depleted oil fields. The Dakota Gasification plant in Beulah, North Dakota, captures its carbon dioxide while making substitute natural gas and sells it for enhanced oil recovery. Since 2000, this plant has captured and sent the carbon dioxide via pipeline to EnCana's Weyburn oil fields in Saskatchewan, Canada, where it is used for enhanced oil recovery. More than five million tons of carbon dioxide have been sequestered.

8.3.1 Carbon Dioxide

Converting the carbon monoxide to carbon dioxide and capturing it prior to combustion is more economical than removing carbon dioxide after combustion, effectively 'de-carbonizing' or, at least, reducing the carbon in the synthesis gas.

Gasification plants manufacturing ammonia, hydrogen, fuels, or chemical products routinely capture carbon dioxide as part of the manufacturing process. According to the Environmental Protection Agency, the higher thermodynamic efficiency of the IGCC cycle minimizes carbon dioxide emissions relative to other technologies. IGCC plants offer a least-cost alternative for capturing carbon dioxide from a coal-based power plant. In addition, IGCC will experience a lower energy penalty than other technologies if carbon dioxide capture is required. While carbon dioxide capture and sequestration will increase the cost of all forms of power generation, an IGCC plant can capture and compress carbon dioxide at one-half the cost of a traditional pulverized coal plant.

8.3.2 Air Emissions

Coal-based IGCC offers the lowest emissions of sulfur dioxide, nitrogen oxides, and particulate matter (PM) of any coal-based power production technology. In fact, a coal IGCC plant is able to achieve low air-emissions rates that approach those of a natural gas combined cycle (NGCC) power plant. In addition, mercury emissions can be removed from an IGCC plant at one-tenth the cost of removal from a coal combustion plant. Technology exists today to remove more than 90% w/w of the volatile mercury from the synthesis gas in a coal-based gasification-based plant.

8.3.3 Solids Generation

The mineral material in the feedstock separates from the gaseous products, and the ash and other inert materials melt and fall to the bottom of the gasifier as a non-leachable, glass-like solid, or other marketable material. This material can be used for many construction and building applications. In addition, more than 99% w/w of the sulfur can be removed using commercially proven technologies and converted into marketable elemental sulfur or sulfuric acid.

8.3.4 Water Use

Gasification uses approximately 14 to 24% v/v less water to produce electric power from coal compared to other coal-based technologies, and water losses during operation are about 32 to 36% v/v less than other coal-based technologies. This is a major issue in many countries – including the United States – where water supplies have already reached critical levels in certain regions.

Finally, for this section, one issue that cannot (and must not) be ignored relates to plant safety insofar as the gasification process produces a highly flammable gaseous mixture, including hydrogen, and carbon monoxide, which is extremely toxic. In plant sections where pressure build-up exists, there is a risk of gas escaping to atmosphere and precautions are necessary to prevent such escape of toxic or environmentally destructive gas. The areas outside the equipment must be adequately ventilated to prevent the buildup of an explosive atmosphere, but also to ensure that there is no buildup of a toxic atmosphere that could cause carbon monoxide poisoning. Carbon monoxide detection equipment should be provided to detect possible leaks. Applying the relevant techniques to the gasification process is necessary to ensure safety and that the plant meets environmental standards.

8.4 THE HYDROGEN ECONOMY

Hydrogen will play an important role in a future energy economy mainly as a storage and transportation medium for renewable sources, which are nuclear energy, solar energy including photovoltaic electrolyzer, photo-electrocatalysis in water splitting, photovoltaic-support fuel cell, wind energy, hydro power, geothermal energy, and biomass energy.

The goal of a hydrogen economy is to become able to compete with fossil fuels in the near future. A variety of methods for obtaining, storing, and delivering hydrogen have been developed since no single method has yet proven superior.

Alternative thermochemical and biological processes can be practically applied to produce hydrogen. Bio-hydrogen production from food industry wastewaters and agricultural wastes is a two-stage, anaerobic dark and photo-fermentation process.

The future prospects for the hydrogen economy or economic hydrogen production are the basic point of many articles. Biomass, in particular, could be a low-cost option for some countries. An assessment of hydrogen production technologies concludes that biomass gasification is the most economical process for renewable hydrogen production. Other analyses have shown that biomass gasification/shift conversion is economically unfavorable compared to natural gas steam reforming, except for very low-cost biomass and potential environmental incentives (Demirbaş, 2009). Pyrolysis with a valuable co-product approach yields hydrogen in the price range of US$6–8/GJ, which is promising for near-term application.

Raw material resources of fossil fuels are not geographically distributed homogeneously around the world. However, the most important source of hydrogen is water, which is almost independent of geography. This advantage in the development of the hydrogen economy is its greatest driving force. However, the hydrogen economy would require a huge investment in new infrastructure to produce, store, and deliver hydrogen to end users, as well as in transportation and establishing of hydrogen stationary systems. Improving the hydrogen-powered system could take several decades due to the slow turnover of the existing stock of capital. The transition to hydrogen is likely to begin later in most developing economies compared to industrialized countries. The earlier industrialized countries/economies begin the transition to hydrogen, the quicker they could achieve their energy sustainability.

The pyrolysis-based technology, in particular, because of the co-products opportunity, has the most favorable economics. An added advantage of biomass as a renewable feedstock is that it is not intermittent but can be used to produce hydrogen as and when required. With scientific and engineering advancements, biomass can be viewed as a key and economically viable component to a renewable-based hydrogen economy.

The ever-increasing global energy demand and the fast-depleting fossil fuels have shifted focus to sustainable energies such as biomass and waste in the recent past. The importance of the gasification of such alternate feedstocks cannot be ignored as potential sources of sustainable energy to meet the energy demand of future generations. The various technologies that are in practice at the commercial and pilot scale, with respect to bubbling, circulating fluidized beds, and dual fluidized beds, are being developed for feedstocks other than coal (Stanislaus et al., 2000; Bridgwater, 2002; Speight, 2011a, 2011b, 2020).

In fact, gasification could now be proposed as a viable alternative solution for biomass waste treatment with energy recovery. On the other hand, the gasification process still faces some technical and economic problems, mainly related to the highly heterogeneous nature of feeds, such as municipal solid waste, which has continuous operating experience under commercial conditions. However, the issues associated with biomass are not insurmountable. The high moisture content and the high mineral matter content can cause ignition and combustion problems, but high-moisture content biomass is more appropriate for a wet conversion process, such as fermentation, i.e., processes involving biochemically mediated reactions, whereas low moisture content biomass is more economically suited to a conversion process, such as gasification (Oliveira and Franca, 2009).

In the aggressive working environment of municipal solid waste management, with its uncompromising demand for reasonable cost, high reliability, and operational flexibility, it could be premature to indicate gasification as the thermal processing strategy of the future or even as a strong competitor for combustion systems, at least for any size of waste-to-energy plants.

The success of any advanced thermal technology is determined by its technical reliability, environmental sustainability, and economic convenience. The first, and then mainly the on-line availability, is supported by years of successful continuous operations of about 100 gasification-based WtE plants, mainly in Japan but now also in Korea and Europe.

Economic aspects are probably the crucial factor for relevant market penetration, since gasification-based waste-to-energy (often referred to as WtE) tends to have operating and capital costs higher than those of conventional combustion-based waste-to-energy (in the order of about 10%), mainly as a consequence of the ash melting system or, in general, the added complexity of the technology.

The greatest technical challenge to overcome for a wider market penetration of commercial advanced gasification technologies appears still to be that of improved and cheaper synthesis gas cleaning, able to conveniently meet defined specifications and to obtain higher electric energy conversion efficiencies. It is essential that the performance and experience from several commercial waste gasifiers in operation will point to the gasification process as a strong competitor of conventional moving grate or fluidized-bed combustion systems.

Environmental concerns and newly enacted rules and regulations mean the feedstock that crude oil products are expected to meet lower and lower limits on contaminants, such as sulfur and nitrogen. New regulations require the removal of sulfur compositions from liquid hydrocarbon derivatives, such as those used in gasoline, diesel fuel, and other transportation fuels.

In the hydrodesulfurization process, high temperatures and pressures may be required to obtain the desired low levels of sulfur. High-temperature processing of olefinic naphtha, however, may result in a lower grade fuel due to saturation of olefins, leading to an octane loss. Low octane gasoline (or low octane naphtha, a blend stock for low octane gasoline) may require additional refining, isomerization, blending, and the like to produce higher quality fuels suitable for use in saleable products. Such extra processing adds additional cost, expense, and complexity to the process and may result in other undesirable changes in the products. As a result, the future will see processes that are more chemically precise in hydrotreating and will offer higher efficiency and conversion to selected products on the base that is not always practiced currently.

Five of the most common approaches to upgrading hydrotreaters for clean-fuels production (in order of increasing capital cost) are currently and will continue (at least for the next two decades) to be: (1) developing higher-activity and more resilient catalysts, (2) replacing reactor internals for increased efficiency, (3) adding reactor capacity to accommodate high-boiling feedstocks and increase the production of gasoline and diesel fuel, (4) increasing hydrogen partial pressure, (5) designing specialized process design and hardware, and (6) focusing on process schemes that effectively reduce hydrogen consumption.

Finally, severe hydrotreating requires high-purity (99% v/v) hydrogen, while less severe hydrotreating can employ lower purity (90% v/v) hydrogen. Refiners will continue to optimize hydrogen use by cascading hydrogen through the refinery. High purity hydrogen will continue to be used only where required, and the lower purity hydrogen purged from these applications is used for services that do not require high purity hydrogen.

8.5 THE FUTURE

The worldwide refining industry has been impacted by several trends that have increased hydrogen demand significantly. First, in aggregate, crude oil has been getting heavier and contains more sulfur and nitrogen; second, decreasing heavy fuel oil demand requires more 'bottoms' upgrading; and third, increasingly stringent environmental regulations require cleaner transportation fuels production. These factors have led to higher hydrogen consumption for upgrading crude oil into light transportation fuels and removing sulfur and nitrogen compounds. This long-term trend is expected to continue through the next several decades as the viscous feedstocks are used by refineries and the clean fuels specifications are progressively tightened.

Hydrogen is required in large volumes – typically 10–200,000 Nm3/hr in a refinery but is needed for a variety of applications in several different scales of supply. Due to hydrogen representing a significant percentage of a refinery's total investment, a pivotal decision confronting operators is the supply method. There are essentially three options for large-scale hydrogen supply.

Firstly, the refiner can build an on-site hydrogen production plant, which is owned by the refinery and operated by on-site refinery personnel, thereby allowing hydrogen production operation(s) to become fully integrated with the other refinery processes. While this enables the refinery to keep control of its own hydrogen supply, this option requires more capital and demands skilled attention from the refinery labor force for efficient operation, maintenance, and repair.

The second option is to outsource hydrogen supply to a world-class, highly experienced industrial gas service provider. Compared to on-site hydrogen production, the advantages here are that no capital investment is required and the risks in terms of operation, maintenance, capital overrun, raw material overconsumption, schedule, and labor issues are all transferred to the service provider in exchange for a monthly service charge. The potential disadvantage, from the refiner's perspective, is the reliance on a third party to supply a critical reactant for their processes. However, this 'over the fence' concept has been widely embraced by refiners in developed countries and is given serious consideration in many other parts of the world.

Lower purity hydrogen can be recovered and purified from catalytic reforming, purge stream, or other refinery sources such as refinery off gas via methods such as Pressure Swing Adsorption and cryogenic systems. This option is limited by the availability and quality of hydrogen-containing streams in the refinery. However, it may supplement the first two options.

Hydrogen can also be used on a smaller scale in the refinery – that is, less than 1000 Nm3/hr. In these cases, hydrogen can be brought in and stored on site in a liquefied form and further vapourized to be used as gaseous hydrogen. Other forms of supply include a hydrogen tube trailer delivered by a third-party supplier, hooked to the back of a truck, or 'bundle' delivery, which involves a 'bundle' of 12 to 15 cylinders being delivered to the refinery site. Typically, bundle supply is used to support hydrogen needs during start-up and possibly an on-site pilot plant where R&D work is being conducted, such as experimental hydrogenation studies. Individual cylinders also have their place in refineries to keep on-site laboratories supplied with hydrogen and some of these laboratories also make use of bench-top generators that produce hydrogen by electrolysis of distilled water.

The configuration of each refinery, the crude slate, and any new production requirements is different, so various hydrogen network optimization considerations must be considered. These considerations should include (1) recovering hydrogen from fuel streams, (2) expanding or adding a new on-purpose hydrogen plant, or (3) outsourcing to an industrial gas company.

The recovery technologies used for the recovery of refinery off-gases. Hydrogen is recoverable from these sources, typically ranging from 50 to 90% v/v, with some process gases as low as 10% v/v hydrogen. It is an increasingly important alternative to refiners for hydrogen recovery. The wide range of the hydrogen content in the off-gases makes choosing a recovery technology critical to obtaining an economically attractive and relatively pure hydrogen product. In addition, a recovery solution should consider effective utilization of existing compression and minimizing new compression requirements.

Each of these technologies has its unique capabilities and constraints, and there are important considerations when determining a technology fit to provide economically attractive hydrogen for a specific refinery application. Adsorption on (PSA system) is the best choice when ultra-high purity, 99%+, hydrogen product is required. Impurities and unrecovered H_2 are delivered at low pressure and available for fuel. Hydrogen recoveries for an Adsorption PSA system are typically lower than those available from the other technologies.

The pressure swing adsorption system designs that can be applied to refinery off-gas streams must account for variability in feed gas conditions. For example, catalytic reformer off-gas hydrogen can change by 10–15% v/v from the start of a run to the end of run catalyst conditions. Hydrotreater off-gas composition changes as a function of makeup hydrogen purity and specific purge rate set points. When multiple refinery off-gas streams are mixed, different operating scenarios can significantly change the composition of the gas stream to the pressure swing adsorption unit. In fact, the contaminants that can occur in refinery off-gas streams can include a long 'tail' of higher boiling (heavy C_5+) hydrocarbon derivatives, including aromatic derivatives. Moreover, some higher boiling hydrocarbon derivatives and aromatic derivatives can create problems if they are irreversibly adsorbed.

Therefore, it is advantageous to design pressure swing adsorption systems for refinery off-gases by selecting the various available adsorbents that are tailored for the application. The quantities of the adsorbent in the adsorbent mix allow feedstock flexibility while striving for the best achievable recoveries and high utilization of the adsorbent bed. Hydrogen recoveries (typically greater than 80% v/v) depend on feed conditions, hydrogen product specifications, and system configuration. Prism® membranes can provide the lowest cost solution for high-pressure refinery off-gases, especially when integration with existing hydrogen compression is possible. Membrane systems are bulk separators that produce refinery-quality hydrogen purities of 90 to 98% v/v at high recoveries (85% or better). Since pressure drives the separation, they are best suited for high-pressure feeds (in excess of 300 psi), such as hydroprocessing purge gases.

A common disadvantage of membranes is that the hydrogen product stream is produced at pressures much lower than the feed pressure and may need recompression. The hydrogen product may be sent to existing make-up or recycle compression in hydroprocessing loops. An important consideration for refinery off-gas applications is that hydrogen sulfide is also a 'fast gas' that can pass through the membrane polymer and end up in the hydrogen product. However, membrane systems have been utilized in many refineries with feed capacities up to 35 million scfd and design pressures up to 2,500 psi.

In refinery operations, it is flexible to changing refinery off-gas feed conditions and provides a robust separation system when upstream units are designed to handle various feed contaminants such as water, particulates, amines and hydrocarbon liquids. A detailed review of potential impurities in the feed stream, particularly during upsets, must be completed prior to selecting the technology. Cryogenic Systems provide an excellent vehicle to purify or recover refinery-quality hydrogen when by-product liquid hydrocarbons, especially ethylene and propylene, can subsidize the higher capital expenditure. Even low pressure (>75 psi) and dilute hydrogen refinery off-gas feeds, such as the off-gas from fluid catalytic cracking units, are good candidates for a cryogenic separation system.

Without liquid recovery to offset the higher capital costs, cryogenic separation systems cannot generally be justified over other separation technologies. However, the refiner, working with suppliers of these separation technologies, must determine which technology to use and which off-gas streams can provide the most economical hydrogen product. Many off-gas streams can be eliminated from consideration because the pressure or hydrogen concentration is too low, rendering the recovered hydrogen uneconomical relative to other sourcing options.

In addition, these off-gas recovery options may not provide sufficient quantities of hydrogen to meet the refiner's clean-fuel project requirements. Therefore, the refiner must evaluate other methods of obtaining the necessary hydrogen. Generally, complex refineries source 30 to 60% of their total hydrogen requirements from on-site hydrogen production systems.

The endothermic steam reforming reaction is accomplished by flowing feed gas through catalyst-filled tubes that are located in a furnace. The resulting hydrogen and carbon oxides are further processed in shift reactors to convert carbon monoxide to additional hydrogen. Many existing refinery hydrogen plants produce a medium purity (94~97%) hydrogen product stream by removing the carbon dioxide in a solvent absorption system accompanied by a methanation step. Newer plants use PSA technology to recover and purify the hydrogen to purities in excess of 99.9%.

These hydrogen production plants that are based on the pressure swing adsorption system have higher efficiency than the conventional low purity plants due to additional export steam credits. Existing hydrogen plants can be debottlenecked and expanded through various methods. Current limitations can include: (1) reformer furnace firing capacity because of overheating in the reformer tubes/convection section coils, (2) fuel system control limits, (3) burner firing limitations, (4) system pressure drop limitations in which there is maximum pressure through the catalyst bed or full use of available feed pressure from supplier or feed compressor, (5) induced-draft fan capacity limiting available draft for the reformer furnace firing, (6) purification equipment limitations that can be affected by flooding in the carbon dioxide adsorber through exceeding the carbon dioxide stripper capacity, (7) overheating in the methanator, (8) bed fluidization in the pressure swing adsorption, or (8) insufficient adsorption capacity in the pressure swing adsorption system.

The refiner must be aware not only of the potential of a major bottleneck in the system due to these events but also of other areas of the plant that are near their respective limits. If the first bottleneck is resolved, another is often right behind. Secondary limitations are usually difficult to identify. In many plants, multiple bottlenecks may need to be addressed to accomplish a meaningful expansion in capacity. This could be very costly and may require significant downtime. Revamping and debottlenecking an existing hydrogen plant can be a cost-effective way of fulfilling incremental hydrogen needs.

Increasing hydrogen production from 5% to 30% may be possible depending on (1) the age of the plant, (2) the extent of any previous debottlenecking work, (3) utility availability, (4) the constraints of emission permits, and (5) the flexibility of the refinery with respect to product hydrogen pressure, steam production, and design effects. However, debottlenecking options are available for the steam methane reforming unit.

Increasing plant throughput by increasing reformer firing is usually the starting point for most users. Increasing flow through the reformer beyond design conditions will require an increase in reformer firing rate. As flow rate through the tubes and the firing increase, the metal temperature of the reformer tube also increases. Since the tube life is a function of tube metal temperature at a given wall thickness, tube life will be reduced. This problem is solved by reducing the reformer outlet temperature (and conversion) or by accepting a lower tube life. Other equipment limits need to be examined such as the induced draft (ID) fan and valve sizes.

Lowering the steam to carbon ratio can be incorporated as a standalone option or in conjunction with other debottlenecking options. Non-naphtha feed plants should have steam to carbon ratios between 2.5 and 3.5. Plants have the potential to easily increasing hydrogen production if they are operating at steam to carbon ratios higher than 3.5. Lowering the steam to carbon ratio in a conventional (non-pressure swing adsorption based) plant will lower the hydrogen purity and reduce the energy available for the CO_2 unit.

In a pre-reformer, adiabatic steam-hydrocarbon reforming is performed upstream of the fired reformer. The heat required for the endothermic reaction is provided by hot flue gas from the reformer convection section. Since the feed to the fired reformer is now partially reformed, the reformer is unloaded and can then operate at an increased feed rate. An additional advantage of the pre-reformer is that it allows flexibility in feedstock composition (refinery off-gas streams, liquefied petroleum gas, butane, or low-boiling naphtha), while maintaining relatively constant operating conditions within the fired reformer.

An oxygen secondary reformer is installed immediately downstream from the primary reformer, allowing the shifting of a portion of the reforming load from the primary to the secondary. It provides low methane slip and the ability to lower the outlet temperature of the primary, providing a means of processing more feed gas in the primary reformer without increasing the reformer-firing rate. The secondary reformer vessel is a refractory-lined carbon steel vessel housing an oxygen burner in its top neck and a fixed catalyst bed.

Installation of a secondary reformer usually requires significant changes to the carbon dioxide removal system leading to an increase in hydrogen purity on the order of 98% v/v. A number of options are available to increase hydrogen production as a result of modifications to or around the pressure swing adsorption system which include intermediate hydrogen off-gas feed, adsorbent replacement, cycle time adjustment, operating at atmospheric purge gas pressure, and purge gas recycle.

Adding a carbon dioxide (CO_2) wash system to a pressure swing adsorption hydrogen plant can improve the hydrogen recovery of the pressure swing adsorption by reducing the total volume and carbon dioxide content of the feed gas. Improvements in hydrogen recovery from the pressure swing adsorption of 3 to 5% are possible. The incremental hydrogen benefit is not sufficient to make this option economically attractive unless the recovered carbon dioxide has value and there must be a market for liquid or gaseous carbon dioxide. When considering a new on-purpose hydrogen plant, the refiner must identify which hydrocarbon feedstocks are available to produce hydrogen so that the right generation technology can be determined.

The lighter fractions, such as methane (natural gas) through naphtha and various refinery off-gases, can be converted into hydrogen by either steam reforming technology or partial oxidation technology (POX technology; also known as gasification technology). The higher boiling fractions, such as vacuum residue, deasphalted bottoms, refinery sludge, and crude oil coke can only be processed into hydrogen using partial oxidation technology. The refiner may have preferences for using specific hydrocarbons as feeds depending on the overall refinery configuration and operation.

Refinery off-gases that contain unrecoverable hydrogen can be used as feedstocks for the conversion of the hydrocarbon derivatives to hydrogen that can be recovered along with any hydrogen contained in the off-gas stream.

The 'bottom of the barrel' materials (i.e., the non-volatile residue) normally sold as high sulfur fuel oil, and/or crude oil coke, may present a growing disposal problem. The final selection of the appropriate feedstock and technology requires a detailed analysis to determine which combination provides the lowest cost hydrogen. The primary factors to be considered include: the costs of feedstock and utilities; by-product credits (e.g., for steam, carbon monoxide, syngas, carbon dioxide, sulfur, and power as appropriate); the cost of capital; and fixed operating expenses. The required investment and operating costs are typically obtained from the various suppliers of the technology. It is critical, however, to ensure that the scope basis and hydrogen supply reliability are consistent when evaluating vendor proposals. The integration of refinery off-gas streams with hydrogen supply streams that are hydrocarbon-rich can be used as part of the reformer feedstock network after being pretreated for unacceptable levels of sulfur-containing derivatives, olefin derivatives, and/or chloride-containing derivatives.

Refinery off-gases such as coker gas, isomerization vent gas, Flexicoker gas, butane, propane, and hydroprocessing purge gas are some of the feeds that have recently been used to supplement natural gas. The range of feedstock stream compositions in some hydrogen plants includes 5 to 60% v/v hydrogen, up to 10% olefin derivatives, and higher boiling hydrocarbon derivatives that can be 5% v/v C5+ derivatives.

For the low-boiling refinery hydrocarbon derivatives, steam reforming technology is expected to continue to be the technology of choice for hydrogen production, much as it has been during the last 40 years. The steam reforming process is a simple, efficient, well-proven technology with significant flexibility in operating with different light hydrocarbon feedstocks. These plants can also be configured to produce steam, power, and/or carbon dioxide as by-products. Valuing by-products at the appropriate pricing for the particular locale can reduce the hydrogen cost by as much as 10%. The partial oxidation (gasification) of viscous feedstocks and crude oil coke can produce hydrogen at a lower cost than steam reforming under certain conditions.

The steam reforming plant, which is principally designed for the production of hydrogen, can also produce steam and electric power. One of the key benefits is the ability to use the hot exhaust from a gas turbine in the radiant section of the steam reformer. This hot exhaust gas, at a temperature on the order of 540°C (approximately 1,000°F) still contains oxygen (on the order of 13% v/v) and serves as combustion air to the reformer. Since this stream is hot, fuel consumption in the reformer furnace is reduced. The reformer convection section is utilized as a heat recovery steam generator (HRSG) in a cogeneration design. Steam raised in the convection section can be put through either a topping or condensing turbine for additional power generation.

All the high-pressure steam generated is sent to the topping turbine for power recovery before process use for the reforming reaction and export to the refiner. Appropriate bypasses can be incorporated into the plant design to decouple hydrogen and steam production from power production. The system can also be designed such that production of hydrogen and steam can continue in the event of loss of the power generation units through the incorporation of fresh air intakes into the reformer furnace. This permits outages of the gas turbine for maintenance and repair. Separately, a gas turbine exhaust bypass system allows the gas turbine to be de-coupled from the reformer. In the event of a reformer trip, the gas turbine can operate and continue to produce power and some export steam. Depending on the natural gas cost, economics for this integration scheme can result in a very favorable power cost.

The optimization of the hydrogen network is increasing in importance in order to maximize higher value transportation fuels production, minimize clean fuels investments, and overall refinery profitability. In fact, the value of the hydrogen that occurs in process off-gases can improve hydro-processing network purities and consider new hydrogen production options. The optimum solution for an overall refinery hydrogen network can be complex and involve the screening of numerous options.

These options can involve: (1) new or improved hydrogen recovery/purification equipment properly located within the refinery flowsheet; (2) refinery off-gas and/or integration of liquefied petroleum gas with on-purpose hydrogen production; (3) integration with petrochemical operations; (4) improving hydrogen plant efficiency; (5) expanding or replacing the existing hydrogen plant capacity; and (6) providing new hydrogen via pipeline supply.

8.5.1 Catalyst Technology

Catalyst synthesis will attempt to harness the intrinsic activity of various metals and remedy the deficiencies that currently plague low metal loading and thermal instability by using mixed oxides.

Use of ultra-deep desulfurization of liquid hydrocarbon fuels such as gasoline, diesel, and jet fuel (or the blend stocks that are the precursors to these fuels) to satisfy new environmental regulations and fuel cell applications is receiving increased attention and will continue to receive attention. These ultra-deep desulfurization technologies include (1) biodesulfurization and (2) adsorptive desulfurization. Both technologies will continue to receive attention and have a high potential for being incorporated into future refineries. In fact, the biodesulfurization technology is likely to see application during microbial enhanced oil recovery processes.

In respect to the growing demand for ultra-low-sulfur diesel (ULSD), low-boiling cycle oil hydrotreating is receiving much attention. Feeds such as these are typically high in heavy metals, which will require additional unit modifications and/or the installation of guard beds/reactors. In fact, many fluid catalytic cracking units incorporate pretreaters (in the form of hydrotreating the feedstocks or guard beds/reactors) to meet the naphtha-gasoline sulfur requirements.

Currently, the process of choice is the hydrotreater, in which hydrogen is added to the fuel to remove the sulfur from the fuel. Some hydrogen may be lost to reduce the octane number of the fuel, which is undesirable.

Selection of catalyst types is one of the important decisions and depends on (1) feedstock, (2) operating conditions, (3) hydrogen availability, and product properties play a role (Parkash, 2003; Gary et al., 2007; Speight, 2014a; Hsu and Robinson, 2017; Speight, 2017). In this respect, the continued innovation could well involve (in part) the use of specific guard beds or specific reactors or the use of scavengers (such as calcium oxide, alone or supported on the catalyst) that will remove sulfur and coke formers during the hydrotreating process.

The catalysts will have multiple functions, such as (1) optimized hydrodesulfurization and (2) minimized hydrogen consumption.

Areas such as (1) catalyst supports, (2) catalyst morphology, and (3) reaction pathways will continue to provide new opportunities for the development of improved commercial hydrotreating catalysts.

Breakthrough technology is essential if catalysts are to be developed that are able to exhibit high activity to produce the desired products. In addition, hydrotreating, as applied to the production of products from non-fossil fuel feedstocks (Chapter 5) into renewable diesel, utilizes specially designed catalysts compatible with the type of catalysts used for hydrotreating fossil diesel streams to meet environmental specifications. The hydrotreating options may also take place in a dedicated stand-alone unit that produces 100% renewable drop-in diesel.

8.5.2 The Gasification Refinery

The concept of a refinery in which one or more gasification units have been installed (each unit could be devoted to a specific type of feedstock) is not a new concept but now deserves more serious consideration than at any time in the history of crude oil refinery operations because of the large number of different feedstocks that can be processed (Table 8.1).

Gasification differs from more traditional energy-generating schemes in that it is not a combustion process, but rather, it is a conversion process. Instead of the carbonaceous feedstock being wholly burned in air to create heat to raise steam, which is used to drive turbines, the feedstock to be gasified is combined with steam and limited oxygen in a heated, pressurized vessel. The atmosphere inside the vessel is deficient in oxygen, leading to a complex series of reactions of the feedstocks to produce synthesis gas. Moreover, using current technologies, the synthesis gas can be cleaned beyond environmental regulatory requirements (current and proposed), as demonstrated by currently commercial chemical production plants that require ultra-clean synthesis gas to protect the integrity of expensive catalysts. The clean synthesis gas can be combusted in turbines or engines using higher temperatures (more efficient) cycles than the conventional steam cycles associated with burning carbonaceous fuels, allowing possible efficiency improvements. Synthesis gas can also be used in fuel cells and fuel cell-based cycles with even higher efficiencies and exceptionally low emissions of pollutants.

Furthermore, one of the major challenges of the 21st century is finding a way to meet national and global energy needs while minimizing the impact on the environment. There is extensive debate

TABLE 8.1
Examples of Feedstocks for, and Products from, the Gasification Process

Feedstocks*	Products*
Biomass	Acetic acid
Black liquor	Acetic anhydride
Coal	Ammonia
Crude oil coke	Ammonium nitrate
Municipal solid waste	Carbon Dioxide
Natural gas	Carbon Monoxide
Refinery gas	Dimethyl ether (DME)
Resids	Electric Power
Sewage sludge	Ethanol
Waste oil	Hot Water
	Hydrogen
	Industrial chemicals
	Methanol
	Steam
	Sulfur
	Sulfuric Acid
	Urea

*Listed alphabetically rather than by preference

surrounding this issue, but there are certain areas of focus: (1) production of cleaner energy, both from conventional fuel sources and alternative technologies, (2) use of energy sources that are not only environmentally sound, but also economically viable, (3) investment in a variety of technologies, and (4) resources to produce clean energy to meet energy needs. Gasification technologies will help to answer these challenges.

8.5.2.1 The Gasification Process

Gasification, a time-tested, reliable, and flexible technology, will be an increasingly important component of this new energy equation, even to the point of the evolution of the crude oil refinery as more gasification units are added to current refineries (Speight, 2011a). Any investment in gasification will yield valuable future returns in clean, abundant, and affordable energy from a variety of sources (Speight, 2011b, 2020).

Gasification is an environmentally sound way to transform any carbon-based material, such as coal, refinery by-products, biomass, or even waste, into energy by producing synthesis gas that can be converted into electricity and valuable products, such as transportation fuels, fertilizers, substitute natural gas, or chemicals (Clayton et al., 2002; Chadeesingh, 2011; Speight, 2013a).

Gasification is currently playing an important role in meeting energy needs in many countries and will continue to play an increasingly important role as one of the economically attractive manufacturing technologies that will allow the production of clean, abundant energy. And, while gasification has typically been used in industrial applications, the technology is increasingly being adopted in smaller-scale applications to convert biomass and waste to energy and products.

Gasification can convert low-value materials into high-value products, such as chemicals and fertilizers, substitute natural gas, transportation fuels, electric power, steam, and hydrogen. The process can be used to convert biomass, municipal solid waste, and other materials (that are normally burned as fuel) into a clean gas. Most importantly, for many countries dependent upon high-cost imported crude oil and natural gas from politically unstable regions of the world, gasification allows the use of domestic resources to generate energy.

In fact, gasifiers can be designed to use a single material or a blend of feedstocks: (1) solids, such as coal, crude oil coke, biomass, wood waste, agricultural waste, household waste, and hazardous waste, (2) liquids, such as crude oil resids, including used/recovered road asphalts, tar sand bitumen, and liquid wastes from chemical plants and other processing plants, and (3) natural gas, or (4) refinery off-gas (also known as 'process off-gas').

The major sought-after products of gasification – synthesis gas and hydrogen – are dependent upon the specific gasification technology; smaller quantities of methane, carbon dioxide, hydrogen sulfide, and water vapor are also produced. Typically, 70 to 85% of the carbon in the feedstock is converted into the synthesis gas. The ratio of carbon monoxide to hydrogen depends in part upon the hydrogen and carbon content of the feedstock and the type of gasifier used but can also be adjusted downstream of the gasifier through the use of catalysts.

In fact, carbon dioxide is routinely removed in gasification-based ammonia, hydrogen, and chemical manufacturing plants. Gasification-based ammonia plants already capture/separate approximately 90% v/v of the carbon dioxide, and gasification-based methanol plants capture approximately 70% v/v of the carbon dioxide.

Other by-products include slag – a glass-like product – composed primarily of sand, rock, and minerals originally contained in the gasifier feedstock. This slag is non-hazardous and can be used in roadbed construction, cement manufacturing, or in roofing materials. Also, in most gasification plants, more than 99% w/w of the feedstock sulfur is removed and recovered either as elemental sulfur or sulfuric acid.

Plasma is an ionized gas that is formed when an electrical charge passes through a gas. Plasma torches generate extremely high temperatures which, when used in a gasification plant, initiate and intensify the gasification reaction, increasing the rate of those reactions and making gasification more efficient.

8.5.2.2 Gasification in the Refinery of the Future

With the entry into the 21st century, crude oil refining technology is experiencing its great innovation driven by the increasing supply of heavy oils with decreasing quality and the fast increases in the demand for clean and ultra-clean vehicle fuels and petrochemical raw materials.

To meet the challenges of restructuring from simple crude trends in product slate and the distillation operations into increasingly stringent specifications imposed by environmental chemical operations involving legislation, the refining industry in the near future will become increasingly flexible and innovative with new processing schemes, transforming crude oil into a variety of refined products with specifications that meet user requirements.

During the next 20 to 30 years, the future evolution of crude oil refining and the current refinery configuration (Speight, 2014a) will be primarily on process modification with some new innovations coming on-stream. The industry will move predictably on to (1) deep conversion of heavy feedstocks, (2) higher hydrocracking and hydrotreating capacity, and (3) more efficient processes currently under development.

High conversion refineries will move to gasification of feedstocks for the development of alternative fuels and to enhance equipment usage. A major trend in the refining industry market demand for refined products will be in synthesizing fuels from simple basic reactants (e.g., synthesis gas) when it becomes uneconomical to produce super clean transportation fuels through conventional refining processes. Fischer-Tropsch plants, together with IGCC systems, will be integrated with or even into refineries, which will offer the advantage of high-quality products (Chadeesingh, 2011; Speight, 2014b).

A gasification refinery would have, as the centerpiece, gasification technology as is the case of the Sasol refinery in South Africa (Couvaras, 1997). The refinery would produce synthesis gas (from the carbonaceous feedstock) from which liquid fuels would be manufactured using the Fischer-Tropsch synthesis technology.

Inorganic components of the feedstock, such as metals and minerals, are trapped in an inert and environmentally safe form as char, which may have use as a fertilizer. Biomass gasification is therefore one of the most technically and economically convincing energy possibilities for a potentially carbon-neutral economy.

Partial oxidation processes using oxygen instead of steam also found wide application for synthesis gas manufacture, with the special feature that they could utilize low-value feedstocks such as heavy crude oil residue.

As crude oil supplies decrease, the desirability of producing gas from other carbonaceous feedstocks will increase, especially in areas where natural gas is in short supply.

The conversion of the gaseous products of gasification processes to synthesis gas, a mixture of hydrogen (H_2) and carbon monoxide (CO), in a ratio appropriate to the application, needs additional steps after purification.

8.5.2.3 Economic Aspects

While a gasification plant is capital intensive (like any manufacturing unit), its operating costs can be lower than those of many other manufacturing processes or coal combustion plants. Because a gasification plant can use low-cost feedstocks, such as crude oil coke or high-sulfur coal, converting them into high-value products, it increases the use of available energy in the feedstocks while reducing disposal costs.

In addition, gasification has a number of other significant economic benefits such as (1) the principal gasification by-products – sulfur, sulfuric acid, and slag – are marketable, (2) the gasification process can produce a number of high-value products at the same time thereby helping a facility to offset the capital cost and the operating costs by diversifying the risks, (3) the gasification process offers wide feedstock flexibility – a gasification plant can be designed to vary the mix of the solid feedstocks or operate on natural gas or liquid feedstocks when desirable, and (4) the gasifiers require less emission control equipment because they generate fewer emissions, thereby further reducing the operating costs.

Investment in gasification injects capital into the economy (by building large-scale plants using domestic labor and suppliers) and creates domestic jobs (construction to build, well-paid jobs to run) that cannot be outsourced to overseas workers. While a gasification plant is capital intensive (like any manufacturing unit), its operating costs can be lower than those of many other manufacturing processes or coal combustion plants. Because a gasification plant can use low-cost feedstocks, such as crude oil coke or high-sulfur coal, converting them into high-value products, it increases the use of available energy in the feedstocks while reducing disposal costs.

8.5.2.4 Market Outlook

The forecast for growth of gasification capacity focuses on two areas: large-scale industrial and power generation plants and the smaller-scale biomass and waste-to-energy area.

The industrial and power gasification industry in the United States faces a number of challenges, including rising construction costs and uncertainty about policy incentives and regulations. Despite these challenges, the industrial and power gasification capacity in the United States is expected to grow.

A number of factors will contribute to this expansion and include (1) the price of natural gas and conventional crude oil, which will make low-cost and abundant domestic resources with stable prices increasingly attractive as feedstocks and (2) gasification processes will be able to comply with more stringent environmental regulations by having emission profiles that are substantially lower than the emission profiles of the more conventional technologies.

In fact, there is a growing consensus that carbon dioxide management will be required in power generation and energy production. Since the gasification process allows carbon dioxide to be captured in a cost-effective and efficient manner, it will be an increasingly attractive choice for the continued use of fossil fuels. Because they are smaller in scale, these plants are easier to finance, easier to permit, and take less time to construct. In addition, municipal and state restrictions on landfills and incineration and a growing recognition that these materials contain valuable sources of energy are driving the demand for these plants.

Furthermore, a number of factors will contribute to the growing interest in waste and biomass gasification and include (1) the restrictions on landfill space, (2) the efforts to reduce costs associated with waste management, (3) the growing recognition that waste and biomass contain unused energy that can be captured and converted into energy and valuable products, and (4) the ability to use non-food biomass materials and convert them into valuable energy products.

In summary, the future of the crude oil refinery is dependent not only (1) upon the availability of the feedstocks but also (and predominantly) (2) on the availability of hydrogen to produce the desired products from these feedstocks (Table 8.1).

8.6 CONCLUSIONS

The most obvious issue with fossil fuel use relates to the effects on the environment. As technology evolves, the means to reduce the damage done by fossil fuel use also evolves, and the world is on the doorstep of adapting to alternative energy sources. In the meantime, gasification offers alternatives to meet the demand for fuels of the future and to reduce potentially harmful emissions.

Recent policies to tackle climate change and resource conservation, such as the Kyoto Protocol, the deliberations at Copenhagen in 2009, and the Landfill Directive of the European Union, have stimulated the development of renewable energy and landfill diversion technology, thus providing gasification technology development with renewed impetus. However, even though renewable sources of energy are the fastest-growing source of energy, they will still represent 15% of the world's energy requirements in 2035 (up from the current estimation of 10%). Divesting from fossil fuels does not mean an end to environmental emissions. Crude oil, tar sand bitumen, coal, natural gas, and perhaps oil shale will still be dominant energy sources – and will grow at a relatively robust rate over, at least, the next two decades. These estimates serve as a reality check on the challenge ahead for clean technologies if they are to make an impact in reducing greenhouse gas emissions and satisfying future energy demands (EIA, 2013).

Gasification could now be proposed as a viable alternative solution for waste treatment with energy recovery. In the aggressive working environment of municipal solid waste management, with its uncompromising demand for reasonable cost, high reliability, and operational flexibility, it could be premature to indicate gasification as the thermal processing strategy of the future or even as a strong competitor for combustion systems, at least for any size of waste-to-energy plants.

The success of any advanced thermal technology is determined by its technical reliability, environmental sustainability, and economic convenience. The first, and then mainly the on-line availability, is supported by years of successful continuous operations of about one hundred gasification-based WtE plants, mainly in Japan but now also in Korea and Europe.

Economic aspects are probably the crucial factor for relevant market penetration, since gasification-based WtE tends to have operating and capital costs higher than those of conventional combustion-based waste-to-energy (in the order of about 10%), mainly as a consequence of the ash melting system or, in general, the added complexity of the technology.

The greatest technical challenge to overcome for a wider market penetration of commercial advanced gasification technologies appears still to be that of improved and cheaper synthesis gas cleaning, able to conveniently meet defined specifications and to obtain higher electric energy conversion efficiencies. It is essential that the performance and experience from several commercial waste gasifiers in operation will point to the gasification process as a strong competitor of conventional moving grate or fluidized bed combustion systems.

The refineries have been implementing advances and technological innovations such as those we have described to satisfy their markets and the specifications applicable to their products. The choice of a refining technology is based on the specific circumstances of its operation and depends on the type of crude oil, the demand and quality of the products, and economic factors such as the cost of crude oil and products, availability of services such as steam, water, electricity, etc., type and cost of catalysts, and the cost of the necessary equipment for its operation.

Considering our current knowledge on the past and present of oil refining technology, it is worth asking whether the refinery will remain as a supplier of essential products for society or there are risks of the industry being overcome by new technological advances or even disappearing.

The crude oil refining industry began in 1859, but it took more than a century for coal to cease to be the main source of energy. Crude oil, in turn, represented 5% of energy consumption in 1840, and its growth was slow until 1900 when it represented 50% of the energy supply, sharing the leadership with fuels as exotic as whale oil. In general, history indicates that energy transitions energy are slow phenomena and also that usually the preceding energy sources do not disappear and remain in niches of consumption in considerable quantities; for example, around 1960, oil was the main source of energy surpassing coal, but since then, coal consumption has tripled despite being a secondary source of energy.

Historically, the transitions from one energy source to another have been due to the emergence of new technologies, better prices and lower energy costs, and often to policies looking for more secure supplies. Currently, it is climate change that is causing the transition from low-cost energy sources to energies whose costs are currently higher but seek to reduce the emission of greenhouse gases (GHG). The difference in costs is covered by various types of subsidies, incentives, and regulations, until investments in research and development make those energies competitive. However, it should not be forgotten that the growth in wealth growth in the world has always been associated with an efficient and cheap source of energy.

Crude oil refining technologies, as we have seen, have been responding to the environmental challenges that have been presented, but the last one we have mentioned, decarbonization, affects the very nature of its products that are made up of carbon atoms. Next, we will make a brief analysis of where it is more feasible to replace crude oil products and which of them will be maintained over time.

Vehicles with an internal combustion engine can be replaced by electric vehicles, hydrogen, hybrids, natural gas, etc. Car manufacturers have understood this trend and, regardless of the legislation, they are making large investments in developing electric vehicles. As in the past, the automobile industry can dictate the demand for fuels and contribute to a significant decrease in consumption of gasoline and diesel fuel.

Current advances that allow the commercialization of electric heavy transport vehicles or the replacement of aviation fuel seems less viable. Therefore, the demand for diesel and especially aviation kerosene should be maintained for the next decade. Marine fuels, as we have already described, have the alternative of natural gas that would have a significant bunker market share in the coming years. Hydrogen also appears as a long-term option for decarbonized bunker fuels. The future of transport fuels is therefore a challenge, perhaps the most complicated one that the refining industry has to face.

Road paving is another market where the use of asphalt mixtures seems to be the most economically and technically viable alternative, and therefore, the demand for asphalt can be maintained over time. The future of lubricants, already threatened by synthetic lubricants, is associated with that of the internal combustion engine. If this disappears, a large part of the manufacture of lubricants will also disappear, although some demand will still exist for other lubrication applications such as oils for machines, greases, waxes, etc.

An area where oil seems irreplaceable is petrochemicals. This market demands 12% of the crude oil production and maintains sustained annual growth, especially in areas such as Asia where the standard of living is growing rapidly. The refineries have increased their integration with petrochemical plants starting with 5–10% of their products being petrochemicals to more than 20–30%.

In recent years, chemical products based on renewable raw materials have been developed. Although technically possible, a disadvantage is the low production yield that makes large quantities of raw material necessary, and often leads to irrational demands on land availability and logistics. The recycling of plastics is also a booming trend.

The high demand for petrochemical products is giving rise to a new refinery configuration known as 'Crude to Chemicals', which is a good example of what the future could be. In these refineries, existing technologies and new developments are combined to directly convert crude into petrochemical products with conversions that exceed 50%. We have examples of these refineries in Saudi Arabia and China where the direct conversion of crude oil is the technological challenge; in that vein, Exxon at its Jurong refinery, launched in 2014, a process of direct crude cracking. The number of refineries that might follow this production strategy is a concern due to the small size of the petrochemical market in comparison to the fuel market.

REFERENCES

Ancheyta, J., and Speight, J.G. 2007. *Hydroprocessing of Heavy Oils and Residua*. CRC Press, Taylor & Francis Group, Boca Raton, FL.

Banks, F.E. 1992. Some Aspects of Natural Gas and Economic Development – A Short Note. *OPEC Bulletin*, 16(2): 235–240.

Bridgwater, A.V. 2002. The Future for Biomass Pyrolysis and Gasification: Status, Opportunities and Policies for Europe. Contract No: 4.1030/S/01-009/2001. Energy Framework Program, European Union, Directorate-General for Transport and Energy, Brussels, Belgium.

Clayton, S.J., Stiegel, G.J., and Wimer, J.G. 2002. Gasification Technologies: Gasification Markets and Technologies – Present and Future, an Industry Perspective. Report No. DOE/FE-0447. Office of Fossil Energy, United States Department of Energy, Washington, DC.

Couvaras, G. 1997. Sasol's Slurry Phase Distillate Process and Future Applications. Proceedings. Monetizing Stranded Gas Reserves Conference. Houston, December.

Davis, R.A., and Patel, N.M. 2004. Refinery Hydrogen Management. *Petroleum Technology Quarterly*, Spring: 29–35.

Gary, J.H., Handwerk, G.E., and Kaiser, M.J. 2007. *Petroleum Refining: Technology and Economics*, 5th Edition. CRC Press, Taylor & Francis Group, Boca Raton, FL.

Krey, V., Canadell, J.G., Nakicenovic, N., Abe, Y., Andruleit, H., Archer, D., Grubler, A., Hamilton, N.T.M., Johnson, A., Kostov, V., Lamarque, J.-F., Langhorne, N., Nisbet, E.G., O'Neill, B., Riahi, K., Riedel, M., Wang, W., and Yakushev, V. 2009. Gas Hydrates: Entrance to a Methane Age or Climate Threat? *Environmental Research Letters*, 4(3): 034007.

Mohnen, V.A. 1988. The Challenge of Acid Rain. *Scientific American*, 259(2): 30–38.

Oliveira, L.S., and Franca, A.S. 2009. From Solid Bio-waste to Liquid Biofuels. In *Agricultural Wastes*. G.S. Ashworth and P. Azevedo (Editors). Nova Science Publishers Inc., Hauppauge, New York.

Hsu, C.S., and Robinson, P.R. (Editors) 2017. *Handbook of Petroleum Technology*. Springer AG, Cham.

MacDonald, G.J. 1990. The Future of Methane as an Energy Resource. *Annual Review of Energy*, 15(1): 53–83.

Martin, A.J. 1985. The Prediction of Strategic Reserves. In *Prospects for the World Oil Industry*. T. Niblock and R. Lawless (Editors). Croom Helm Publishers, Beckenham, Kent. Chapter 1.

Rana, M.S., Sámano, V., Ancheyta, J., and Diaz, J.A.I. 2007. A Review of Recent Advances on Process Technologies for Upgrading of Heavy Oils and Residua. *Fuel*, 86(9): 1216–1231.

Speight, J.G. 2011a. *The Refinery of the Future*. Gulf Professional Publishing, Elsevier, Oxford.

Speight, J.G. 2011b. *The Biofuels Handbook*. Royal Society of Chemistry, London.

Speight, J.G. 2011c. *An Introduction to Petroleum Technology, Economics, and Politics*. Scrivener Publishing, Salem, MA.

Speight, J.G. 2013a. *The Chemistry and Technology of Coal*, 3rd Edition. CRC Press, Taylor & Francis Group, Boca Raton, FL.

Speight, J.G. 2013b. *Coal-Fired Power Generation Handbook*. Scrivener Publishing, Salem, MA.

Speight, J.G. 2014a. *The Chemistry and Technology of Petroleum*, 5th Edition. CRC Press, Taylor & Francis Group, Boca Raton, FL.

Speight, J.G. 2014b. *Gasification of Unconventional Feedstocks*. Gulf Professional Publishing Company, Elsevier, Oxford.

Speight, J.G. 2017. *Handbook of Petroleum Refining*. CRC Press, Taylor & Francis Group, Boca Raton, FL.

Speight, J.G. 2019. *Handbook of Petrochemical Processes*. CRC Press, Taylor & Francis Group, Boca Raton, FL.

Speight, J.G. 2020. *Synthetic Fuels Handbook: Properties, Process, and Performance*, 2nd Edition. McGraw-Hill, New York.

Stanislaus, A., Qabazard, H., and Absi-Halabi, M. 2000. Refinery of the Future. Proceedings. 16th World Petroleum Congress, Calgary, Alberta, Canada. June 11–15.

Vallero, D. 2008. *Fundamentals of Air Pollution*, 4th Edition. Elsevier, London.

Glossary

ABN separation: a method of fractionation by which crude oil is separated into acidic, basic, and neutral constituents.

Absorber: see absorption tower.

Absorption gasoline: gasoline extracted from natural gas or refinery gas by contacting the absorbed gas with an oil and subsequently distilling the gasoline from the higher-boiling components.

Absorption oil: oil used to separate the heavier components from a vapor mixture by absorption of the heavier components during intimate contacting of the oil and vapor; used to recover natural gasoline from wet gas.

Absorption plant: a plant for recovering the condensable portion of natural or refinery gas, by absorbing the higher boiling hydrocarbons in an absorption oil, followed by separation and fractionation of the absorbed material.

Absorption tower: a tower or column which promotes contact between a rising gas and a falling liquid so that part of the gas may be dissolved in the liquid.

Acetone-benzol process: a dewaxing process in which acetone and benzol (benzene or aromatic naphtha) are used as solvents.

Acid catalyst: a catalyst having acidic character; the alumina minerals are examples of such catalysts.

Acid deposition: acid rain; a form of pollution depletion in which pollutants, such as nitrogen oxides and sulfur oxides, are transferred from the atmosphere to soil or water; often referred to as atmospheric self-cleaning. The pollutants usually arise from the use of fossil fuels.

Acid gas removal: a process for the removal of hydrogen sulfide, other sulfur species, and some carbon dioxide from syngas by absorption in a solvent with subsequent solvent regeneration and production of a hydrogen sulfide-rich (H_2S-rich) rich stream for sulfur recovery.

Acidity: the capacity of an acid to neutralize a base such as a hydroxyl ion (OH^-).

Acidizing: a technique for improving the permeability of a reservoir by injecting acid.

Acid number: a measure of the reactivity of crude oil with a caustic solution and given in terms of milligrams of potassium hydroxide that are neutralized by one gram of crude oil.

Acid rain: the precipitation phenomenon that incorporates anthropogenic acids and other acidic chemicals from the atmosphere to the land and water (see Acid deposition).

Acid sludge: the residue left after treating crude oil of a crude oil product with sulfuric acid for the removal of impurities; a black, viscous substance containing the spent acid and impurities.

Acid treating: a process in which unfinished crude oil products, such as gasoline, kerosene, and lubricating-oil stocks, are contacted with sulfuric acid to improve their color, odor, and other properties.

Acoustic log: see sonic log.

Acre-foot: a measure of bulk rock volume where the area is one acre and the thickness is one foot.

Additive: a material added to a reaction mixture (usually in small amounts) in order to (1) enhance the reaction, (2) enhance the desirable properties of the products, or (3) to suppress undesirable properties of the products. Different additives, even when added for identical purposes, may be incompatible with each other, for example react and form new compounds. See also incompatibility, instability.

Add-on control methods: the use of devices that remove refinery process emissions after they are generated but before they are discharged to the atmosphere.

Adsorption: a surface process that leads to the transfer of a molecule from a fluid bulk to a solid surface and adsorption occurs when a particle adheres to the surface of a solid substance without entering it and is held there by inter-molecular forces; the transfer of a substance from a solution to the surface of a solid resulting in relatively high concentration of the substance at the place of contact. See also Chromatographic adsorption.

Adsorption gasoline: natural gasoline obtained by the adsorption process from wet gas.

Afterburn (after-burn): the combustion of carbon monoxide (CO) to carbon dioxide (CO_2); usually in the cyclones of a catalyst regenerator.

After flow: flow from the reservoir into the wellbore that continues for a period after the well has been shut in; after-flow can complicate the analysis of a pressure transient test.

Agricultural waste: unwanted or unsalable materials produced wholly from agricultural operations directly related to the growing of crops or raising of animals for the primary purpose of making a profit or for a livelihood; plant residues from agriculture which originate from arable land and horticulture; agricultural waste are all parts of crops that are not used for human or animal food; crop residues consist mainly of stems, branches, and leaves. See also waste.

Air-blown asphalt: asphalt produced by blowing air through residua at elevated temperatures.

Air injection: an oil recovery technique using air to force oil from the reservoir into the wellbore.

Airlift Thermofor catalytic cracking: a moving-bed continuous catalytic process for conversion of heavy gas oils into lighter products; the catalyst is moved by a stream of air.

Air pollution: the discharge of toxic gases and particulate matter introduced into the atmosphere, principally as a result of human activity.

Air separation unit (ASU): a plant that separates oxygen and nitrogen from air, usually by cryogenic distillation.

Air sweetening: a process in which air or oxygen is used to oxidize lead mercaptan derivatives (RSH) to disulfide derivatives (RSSR) instead of using elemental sulfur.

Air toxics: hazardous air pollutants.

Albertite: a black, brittle, natural hydrocarbon possessing a conchoidal fracture and a specific gravity of approximately 1.

Alcohol: the family name of a group of organic chemical compounds composed of carbon, hydrogen, and oxygen; the molecules in the series vary in chain length and are composed of a hydrocarbon plus a hydroxyl group; alcohol includes methanol and ethanol.

Alicyclic hydrocarbon: a compound containing carbon and hydrogen only which has a cyclic structure (e. g., cyclohexane); also collectively called naphthenes.Aliphatic hydrocarbon: a compound containing carbon and hydrogen only which has an open-chain structure (e. g., as ethane, butane, octane, butene) or a cyclic structure (e.g., cyclohexane).Aliquot: that quantity of material of proper size for measurement of the property of interest; test portions may be taken from the gross sample directly, but often preliminary operations such as mixing or further reduction in particle size are necessary.

Alkaline: a high pH usually of an aqueous solution; aqueous solutions of sodium hydroxide, sodium orthosilicate, and sodium carbonate are typical alkaline materials used in enhanced oil recovery.

Alkaline flooding: see EOR process.

Alkalinity: the capacity of a base to neutralize the hydrogen ion (H^+).

Alkali treatment: see caustic wash.

Alkali wash: see caustic wash.

Alkanes: hydrocarbons that contain only single carbon-hydrogen bonds. The chemical name indicates the number of carbon atoms and ends with the suffix 'ane'.

Alkenes: hydrocarbons that contain carbon-carbon double bonds. The chemical name indicates the number of carbon atoms and ends with the suffix 'ene'.

Alkylate: the product of an alkylation process.

Alkylate bottoms: residua from fractionation of alkylate; the alkylate product which boils higher than the aviation gasoline range; sometimes called heavy alkylate or alkylate polymer.

Alkylation: a refinery process by which an olefin (e.g., ethylene) is combined with a branched-chain hydrocarbon (e.g., *iso*-butane); alkylation may be accomplished as a thermal or as a catalytic reaction.Alkyl groups: a group of carbon and hydrogen atoms that branch from the main carbon chain or ring in a hydrocarbon molecule. The simplest alkyl group, a methyl group, is a carbon atom attached to three hydrogen atoms.

Alpha-scission: the rupture of the aromatic carbon-aliphatic carbon bond that joins an alkyl group to an aromatic ring.

Alumina (Al_2O_3): a chemical used in separation methods as an adsorbent and in refining as a catalyst.

American Society for Testing and Materials (ASTM): See ASTM International.

Amine washing: a method of gas cleaning whereby acidic impurities such as hydrogen sulfide and carbon dioxide are removed from the gas stream by washing with an amine (usually an alkanolamine).

Anaerobic digestion: decomposition of biological wastes by micro-organisms, usually under wet conditions, in the absence of air (oxygen), to produce a gas comprising mostly methane and carbon dioxide.

Analytical equivalence: the acceptability of the results obtained from the different laboratories; a range of acceptable results.

Analyte: the chemical for which a sample is tested, or analyzed. *Antibody* a molecule having chemically reactive sites specific for certain other molecules.

Antibody: a molecule having chemically reactive sites specific for certain other molecules.

Aniline point: the temperature, usually expressed in °F, above which equal volumes of a crude oil product are completely miscible; a qualitative indication of the relative proportions of paraffin derivatives in a crude oil product which are miscible with aniline only at higher temperatures; a high aniline point indicates low aromatics.

Annual removals: the net volume of growing stock trees removed from the inventory during a specified year by harvesting, cultural operations such as timber stand improvement, or land clearing.

Antiknock: resistance to detonation or pinging in spark-ignition engines.

Antiknock agent: a chemical compound such as tetraethyl lead which, when added in small amount to the fuel charge of an internal-combustion engine, tends to lessen knocking.

Antistripping agent: an additive used in an asphaltic binder to overcome the natural affinity of an aggregate for water instead of asphalt.

API gravity: a measure of the lightness (low density) or heaviness (high density) of crude oil which is related to density and specific gravity.

$$\text{API} = (141.5/\text{sp gr @ }60°\text{F}) - 131.5$$

Apparent bulk density: the density of a catalyst as measured; usually loosely compacted in a container.

Apparent viscosity: the viscosity of a fluid, or several fluids flowing simultaneously, measured in a porous medium (rock), and subject to both viscosity and permeability effects; also called effective viscosity.

Aquifer: a subsurface rock interval that will produce water; often the underlay of a crude oil reservoir.

Areal sweep efficiency: the fraction of the flood pattern area that is effectively swept by the injected fluids.

Aromatic hydrocarbon: a hydrocarbon characterized by the presence of an aromatic ring or condensed aromatic rings; benzene and substituted benzene, naphthalene and substituted naphthalene, phenanthrene and substituted phenanthrene, as well as the higher condensed ring systems; compounds that are distinct from those of aliphatic compounds or alicyclic compounds.

Aromatics: see aaromatic hydrocarbon.

Aromatization: the conversion of non-aromatic hydrocarbons to aromatic hydrocarbons by: (1) rearrangement of aliphatic (noncyclic) hydrocarbons into aromatic ring structures; and (2) dehydrogenation of alicyclic hydrocarbons (naphthenes).

Arosorb process: a process for the separation of aromatic derivatives from non-aromatic derivatives by adsorption on a gel from which they are recovered by desorption.

Asphalt: the nonvolatile product obtained by distillation and treatment of an asphaltic crude oil; a manufactured product.

Asphalt cement: asphalt especially prepared as to quality and consistency for direct use in the manufacture of bituminous pavements.

Asphalt emulsion: an emulsion of asphalt cement in water containing a small amount of emulsifying agent.

Asphaltene (asphaltenes, asphaltene fraction): The brown to black powdery material produced by treatment of crude oil, crude oil residua, or bituminous materials with a low-boiling liquid hydrocarbon, e. g., pentane or heptane; soluble in benzene (and other aromatic solvents), carbon disulfide, and chloroform (or other chlorinated hydrocarbon solvents). Asphalt flux: an oil used to reduce the consistency or viscosity of hard asphalt to the point required for use.

Asphalt primer: a liquid asphaltic material of low viscosity which, upon application to a non-bituminous surface to waterproof the surface and prepare it for further construction.

Asphaltene association factor: the number of individual asphaltene species which associate in non-polar solvents as measured by molecular weight methods; the molecular weight of asphaltenes in toluene divided by the molecular weight in a polar non-associating solvent, such as dichlorobenzene, pyridine, or nitrobenzene.

Asphaltic pyrobitumen: see Asphaltoid.

Asphaltic road oil: a thick, fluid solution of asphalt; usually a residual oil; see also Nonasphaltic road oil (Non-asphaltic road oil).

Asphaltite: a variety of naturally occurring, dark brown to black, solid, nonvolatile bituminous material that is differentiated from bitumen primarily by a high content of material insoluble in n-pentane (asphaltene) or other liquid hydrocarbons.

Asphaltoid: a group of brown to black, solid bituminous materials of which the members are differentiated from asphaltites by their infusibility and low solubility in carbon disulfide.

Asphaltum: see asphalt.

Associated molecular weight: the molecular weight of asphaltenes in an associating (non-polar) solvent, such as toluene.

ASTM International: the official organization in the United States for designing standard tests for crude oil and other industrial products; formerly the American Society for Testing and Materials.

Atmospheric residuum: a residuum obtained by distillation of a crude oil under atmospheric pressure and which boils above 350°C (660°F).

Atmospheric equivalent boiling point (AEBP): a mathematical method of estimating the boiling point at atmospheric pressure of non-volatile fractions of crude oil.

Attainment area: a geographical area that meets NAAQS for criteria air pollutants (non-attainment area).

Attapulgus clay: see Fuller's earth.

Autofining process: a catalytic process for desulfurizing distillates; the process differs from other hydrorefining processes in that an external source of hydrogen is not required and sufficient hydrogen to convert sulfur-containing constituents to hydrocarbon derivatives and hydrogen sulfide is obtained by dehydrogenation of naphthene derivatives that are present in the feedstock.

Average particle size: the weighted average particle diameter of a catalyst.

Aviation gasoline: any of the special grades of gasoline suitable for use in certain airplane engines.

Aviation turbine fuel: see Jjet fuel.

Back mixing: the phenomenon observed when a catalyst travels at a slower rate in the riser pipe than the vapors.

BACT: best available control technology.

Baghouse: a filter system for the removal of particulate matter from gas streams; so-called because of the similarity of the filters to coal bags.

Bank: the concentration of oil (oil bank) in a reservoir that moves cohesively through the reservoir.

Bari-Sol process: a dewaxing process which employs a mixture of ethylene dichloride and benzol as the solvent.

Barrel (bbl): the unit of measure used by the crude oil industry; equivalent to approximately forty-two US gallons or approximately thirty four (33. 6) Imperial gallons or 159 liters; 7.2 barrels are equivalent to one tonne of oil (metric).

Barrel of oil equivalent (boe): the amount of energy contained in a barrel of crude oil, i. e. approximately 6.1 GJ (5.8 million Btu), equivalent to 1,700 kWh.

Base number: the quantity of acid, expressed in milligrams of potassium hydroxide per gram of sample that is required to titrate a sample to a specified end-point.

Base stock: a primary refined crude oil fraction into which other oils and additives are added (blended) to produce the finished product.

Basic nitrogen: nitrogen (in crude oil) which occurs in pyridine form

Basic sediment and water (BS&W, BSW): the material which collects in the bottom of storage tanks, usually composed of oil, water, and foreign matter; also called bottoms, bottom settlings.

Batch blending: a blending process (also called the in-tank process) in which specific (calculated and measured) volumes of the different crude oils that have been stored in separate tanks are loaded into a blending tank where they are mixed (by a mechanical stirring operation).

Battery: a series of stills or other refinery equipment operated as a unit.

Baumé gravity: the specific gravity of liquids expressed as degrees on the Baumé (°Bé) scale; for liquids lighter than water:Sp gr 60°F = 140/(130 + °Bé)For liquids heavier than water:Sp gr 60°F = 145/(145 – °Bé)

Bauxite: a mineral that is used as a treating agent; hydrated aluminum oxide formed by the chemical weathering of igneous rock.

Bbl: see barrel.

Bell cap: a hemispherical or triangular cover placed over the riser in a (distillation) tower to direct the vapors through the liquid layer on the tray; see bubble cap.

Bender process: a chemical treating process using lead sulfide catalyst for sweetening light distillates by which mercaptan derivatives (RSH) are converted to disulfide derivatives (RSSR) by oxidation.

Bentonite: montmorillonite (a magnesium-aluminum silicate); used as a treating agent.

Benzene: a colorless aromatic liquid hydrocarbon (C_6H_6).

Benzin: a refined light naphtha used for extraction purposes.

Benzine: an obsolete term for low boiling crude oil distillates covering the gasoline and naphtha range; see ligroine.

Benzol: the general term which refers to commercial or technical (not necessarily pure) benzene; also the term used for aromatic naphtha.

Beta-scission: the rupture of a carbon-carbon bond two bonds removed from an aromatic ring.

Billion: 1×10^9

Biocide: a chemical capable of killing bacteria and biorganisms.

Biochemical conversion: the use of fermentation or anaerobic digestion to produce fuels and chemicals from organic sources.

Biodiesel: a fuel derived from biological sources that can be used in diesel engines instead of crude oil-derived diesel; through the process of transesterification, the triglycerides in the biologically derived oils are separated from the glycerin, creating a clean-burning, renewable fuel.

Bioenergy: useful, renewable energy produced from organic matter - the conversion of the complex carbohydrates in organic matter to energy; organic matter may either be used directly as a fuel, processed into liquids and gasses, or be a residual of processing and conversion.

Bioethanol: ethanol produced from biomass feedstocks; includes ethanol produced from the fermentation of crops, such as corn, as well as cellulosic ethanol produced from woody plants or grasses.

Biofuels: a generic name for liquid or gaseous fuels that are not derived from crude oil based fossils fuels or contain a proportion of non-fossil fuel; fuels produced from plants, crops such as sugar beet, rape seed oil or re-processed vegetable oils or fuels made from gasified biomass; fuels made from renewable biological sources and include ethanol, methanol, and biodiesel; sources include, but are not limited to: corn, soybeans, flaxseed, rapeseed, sugarcane, palm oil, raw sewage, food scraps, animal parts, and rice.

Biogas: a combustible gas mixture that is derived from decomposing biological waste under anaerobic conditions. Biogas normally consists of 50 to 60%methane. See also landfill gas.

Biogenic: material derived from bacterial or vegetation sources.

Biological lipid: any biological fluid that is miscible with a nonpolar solvent. These materials include waxes, essential oils, chlorophyll, etc.

Biological oxidation: the oxidative consumption of organic matter by bacteria by which the organic matter is converted into gases.

Biomass: any organic matter that is available on a renewable or recurring basis, including agricultural crops and trees, wood and wood residues, plants (including aquatic plants), grasses, animal manure, municipal residues, and other residue materials. Biomass is generally produced in a sustainable manner from water and carbon dioxide by photosynthesis. There are three main categories of biomass - primary, secondary, and tertiary.

Biopower: the use of biomass feedstock to produce electric power or heat through direct combustion of the feedstock, through gasification and then combustion of the resultant gas, or through other thermal conversion processes; power is generated with engines, turbines, fuel cells, or other equipment.

Biorefinery: a facility that processes and converts biomass into value-added products. These products can range from biomaterials to fuels such as ethanol or important feedstocks for the production of chemicals and other materials.

Biomass to liquid (BTL): the process of converting biomass to liquid fuels. Hmm, that seems painfully obvious when you write it out.

Biopolymer: a high molecular weight carbohydrate produced by bacteria.

Bitumen: also, on occasion, referred to as native asphalt, and extra heavy crude oil; a naturally occurring material that has little or no mobility under reservoir conditions and which cannot be recovered through a well by conventional oil well production methods including currently used enhanced recovery techniques; current methods involve mining for bitumen recovery.

Bituminous: containing bitumen or constituting the source of bitumen.

Bituminous rock: see Bituminous sand.

Bituminous sand: a formation in which the bituminous material (see Bitumen) is found as a filling in veins and fissures in fractured rock or impregnating relatively shallow sand, sandstone, and limestone strata; a sandstone reservoir that is impregnated with a heavy, viscous black crude oil-like material that cannot be retrieved through a well by conventional production techniques.

Black acid(s): a mixture of the sulfonates found in acid sludge which are insoluble in naphtha, benzene, and carbon tetrachloride; very soluble in water but insoluble in 30 per cent sulfuric acid; in the dry, oil-free state, the sodium soaps are black powders.

Black liquor: a solution of lignin-residue and the pulping chemicals used (in the Kraft process) to extract lignin during the manufacture of paper. See also: Brown liquor, Red liquor, Sulfite liquor, Thick liquor.

Black oil: a term applied to any of the dark-colored oils that has little scientific or technical meaning which is often applied to heavy crude oil and extra heavy crude oil as well as distillation residua; the term does not give any indication of the composition of the oil.

Black soap: see Black acid.

Black strap: the black material (mainly lead sulfide) formed in the treatment of sour light oils with doctor solution and found at the interface between the oil and the solution.

Blended crude: a mixture of crude oils, blended in the pipeline to create a crude with specific physical properties. For example, heavy crude oil, extra heavy crude oil, and tar sand bitumen cannot flow from the field to the refinery in their original state and at normal surface temperatures, they are blended with lighter crude oils primarily to reduce the viscosity, thereby enabling transportation to a refinery. In addition, the blend can provide a blended refinery feedstock crude oil that has significantly higher value than the raw heavy feedstocks in which case the blend is usually constructed so that the value of the overall blended volume is greater than the summed value of the initial volumes of the individual heavy and light components of the blend.

Blending: the process of mixing two or more crude oils without producing a chemical reaction or a physical reaction (i. e. phase separation of asphaltene constituents) to create a composite feedstock that is still suitable for refining.

Blown asphalt: the asphalt prepared by air blowing a residuum or an asphalt.

Blue hydrogen: hydrogen that is produced by reforming natural gas through the steam methane reforming process (the SMR process), and then capturing the carbon dioxide emissions created by this process; the captured carbon can then be utilized or stored to form a part of a circular carbon economy in which carbon is considered a resource that can be captured, reused, and recycled; if the carbon dioxide emissions are then captured and stored underground, the process is considered carbon-neutral, and the resulting hydrogen is known as 'blue hydrogen'. See also: green hydrogen, gray hydrogen.

Bogging: a condition that occurs in a coking reactor when the conversion to coke and light ends is too slow causing the coke particles to agglomerate.

Boiling point: a characteristic physical property of a liquid at which the vapor pressure is equal to that of the atmosphere and the liquid is converted to a gas.

Boiling range: the range of temperature, usually determined at atmospheric pressure in standard laboratory apparatus, over which the distillation of an oil commences, proceeds, and finishes.

Bone dry: having 0% moisture content; wood heated in an oven at a constant temperature of 100°C (212°F) or above until its weight stabilizes is considered bone dry or oven dry.

Bottled gas: usually butane or propane, or butane-propane mixtures, liquefied and stored under pressure for domestic use; see also Liquefied petroleum gas.

Bottoming cycle: A cogeneration system in which steam is used first for process heat and then for electric power production.

Bottoms: the liquid which collects in the bottom of a vessel (tower bottoms, tank bottoms) either during distillation; also the deposit or sediment formed during storage of crude oil or a crude oil product; see also Residuum; Basic sediment and water.

Bright stock: refined, high-viscosity lubricating oils usually made from residual stocks by processes such as a combination of acid treatment or solvent extraction with dewaxing or clay finishing.

British thermal unit: see Btu.

Bromine number: the number of grams of bromine absorbed by 100 g of oil which indicates the percentage of double bonds in the material.

Brown acid: oil-soluble crude oil sulfonates found in acid sludge which can be recovered by extraction with naphtha solvent. Brown-acid sulfonates are somewhat similar to mahogany sulfonates but are more water-soluble. In the dry, oil-free state, the sodium soaps are light-colored powders.

Brown liquor: the spent cooking liquor from sulfite pulping process; the terms red liquor, thick liquor and sulfite liquor are also used (compared to black liquor in the kraft process). Pulp washers, using countercurrent flow, remove the spent cooking chemicals and degraded lignin and hemicellulose. See: Black liquor.

Brown soap: see Brown acid.

Brønsted acid: a chemical species which can act as a source of protons.

Brønsted base: a chemical species which can accept protons.

BS&W: see Basic sediment and water.

BTEX: benzene, toluene, ethylbenzene, and the xylene isomers.

Btu (British thermal unit): the energy required to raise the temperature of one pound of water one degree Fahrenheit.

Bubble cap: an inverted cup with a notched or slotted periphery to disperse the vapor in small bubbles beneath the surface of the liquid on the bubble plate in a distillation tower.

Bubble plate: a tray in a distillation tower.

Bubble point: the temperature at which incipient vaporization of a liquid in a liquid mixture occurs, corresponding with the equilibrium point of 0 per cent vaporization or 100 per cent condensation.

Bubble tower: a fractionating tower so constructed that the vapors rising pass up through layers of condensate on a series of plates or trays (see Bubble plate); the vapor passes from one plate to the next above by bubbling under one or more caps (see Bubble cap) and out through the liquid on the plate where the less volatile portions of vapor condense in bubbling through the liquid on the plate, overflow to the next lower plate, and ultimately back into the reboiler thereby effecting fractionation.

Bubble tray: a circular, perforated plates having the internal diameter of a bubble tower, set at specified distances in a tower to collect the various fractions produced during distillation.

Buckley-Leverett method: a theoretical method of determining frontal advance rates and saturations from a fractional flow curve.

Bumping: the knocking against the walls of a still occurring during distillation of crude oil or a crude oil product which usually contains water.

Bunker: a storage tank.

Bunker C oil: see No. 6 fuel oil.Burner fuel oil: any crude oil liquid suitable for combustion.

Burning oil: an illuminating oil, such as kerosene (also written as kerosine) suitable for burning in a wick lamp.

Burning point: see Fire point.

Burning-quality index: an empirical numerical indication of the likely burning performance of a furnace or heater oil; derived from the distillation profile and the API gravity, and generally recognizing the factors of paraffin character and volatility.

Burton process: an older thermal cracking process in which oil was cracked in a pressure still and any condensation of the products of cracking also took place under pressure.

Butane dehydrogenation: a process for removing hydrogen from butane to produce butenes and, on occasion, butadiene.

Butane vapor-phase isomerization: a process for isomerizing n-butane to *iso*-butane using aluminum chloride catalyst on a granular alumina support and with hydrogen chloride as a promoter.

Butanol: a c4 alcohol that is generally produced from fossil fuels; also produced through bacterial fermentation of alcohol.

C_1, C_2, C_3, C_4, C_5 fractions: a common way of representing fractions containing a preponderance of hydrocarbons having 1, 2, 3, 4, or 5 carbon atoms, respectively, and without reference to hydrocarbon type.

CAA: Clean Air Act; this act is the foundation of air regulations in the United States

Calcining: heating a metal oxide or an ore to decompose carbonates, hydrates or other compounds often in a controlled atmosphere.

Capillary forces: interfacial forces between immiscible fluid phases, resulting in pressure differences between the two phases.

Capillary number: N_c, the ratio of viscous forces to capillary forces, and equal to viscosity times velocity divided by interfacial tension.

Carbene: the pentane- insoluble material or the heptane-insoluble material in crude oil or processed crude oil that is insoluble in benzene or toluene but which is soluble in carbon disulfide (or pyridine); a type of rifle used for hunting bison and for use by the United States cavalry.

Carboid: the pentane-insoluble material or the heptane-insoluble material in crude oil or processed crude oil that is insoluble in benzene or toluene and which is also insoluble in carbon disulfide (or pyridine).

Carbonaceous material: See Hydrocarbonaceous material.

Carbonate washing: processing using a mild alkali (e.g. potassium carbonate) process for emission control by the removal of acid gases from gas streams.Carbon dioxide augmented waterflooding: injection of carbonated water, or water and carbon dioxide, to increase water flood efficiency; see immiscible carbon dioxide displacement.

Carbon dioxide miscible flooding: see EOR process.

Carbon-forming propensity: see Carbon residue.

Carbonization: the conversion of an organic compound into char or coke by heat in the substantial absence of air; often used in reference to the destructive distillation (with simultaneous removal of distillate) of coal.

Carbon monoxide (CO): a lethal gas produced by incomplete combustion of carbon-containing fuels in internal combustion engines and which is colorless, odorless, and tasteless.

Carbon-oxygen log: information about the relative abundance of elements such as carbon, oxygen, silicon, and calcium in a formation; usually derived from pulsed neutron equipment.

Carbon rejection: upgrading processes in which coke is produced, e.g. coking.Carbon residue: the amount of carbonaceous residue remaining after thermal decomposition of crude oil, a crude oil fraction, or a crude oil product in a limited amount of air; also called the *coke-* or *carbon-forming propensity*; often prefixed by the terms Conradson or Ramsbottom in reference to the inventor of the respective tests.

Carbon sink: a geographical area whose vegetation and/or soil soaks up significant carbon dioxide from the atmosphere. Such areas, typically in tropical regions, are increasingly being sacrificed for energy crop production.CAS: Chemical Abstract Service.

Cascade tray: a fractionating device consisting of a series of parallel troughs arranged on stair-step fashion in which liquid frown the tray above enters the uppermost trough and liquid thrown from this trough by vapor rising from the tray below impinges against a plate and a perforated baffle and liquid passing through the baffle enters the next longer of the troughs.

Casinghead gas: natural gas which issues from the casinghead (the mouth or opening) of an oil well.

Casinghead gasoline: the liquid hydrocarbon product extracted from casinghead gas by one of three methods: compression, absorption, or refrigeration; see also Natural gasoline.

Catagenesis: the alteration of organic matter during the formation of crude oil that may involve temperatures in the range 50°C (120°F) to 200°C (390°F); see also Diagenesis and Metagenesis.

Catalyst: a chemical agent which, when added to a reaction (process) will enhance the conversion of a feedstock without being consumed in the process.

Catalyst selectivity: the relative activity of a catalyst with respect to a particular compound in a mixture, or the relative rate in competing reactions of a single reactant.

Catalyst stripping: the introduction of steam, at a point where spent catalyst leaves the reactor, in order to strip, i.e., remove, deposits retained on the catalyst.Catalytic activity: the ratio of the space velocity of the catalyst under test to the space velocity required for the standard catalyst to give the same conversion as the catalyst being tested; usually multiplied by 100 before being reported.

Catalytic cracking: the conversion of high-boiling feedstocks into lower boiling products by means of a catalyst which may be used in a fixed-bed or fluid bed.

Cat cracking: see Catalytic cracking.

Catalytic reforming: a process in which there is rearrangement of hydrocarbon molecules in a gasoline-boiling-range feedstock to produce other hydrocarbons having a higher antiknock quality; isomerization of paraffins, cyclization of paraffins to naphthenes, dehydrocyclization of paraffins to aromatics.

Catforming: a process for reforming naphtha using a platinum-silica-alumina catalyst which permits relatively high space velocities and results in the production of high-purity hydrogen.

Caustic consumption: the amount of caustic lost from reacting chemically with the minerals in the rock, the oil, and the brine.

Chemical flooding: see EOR process.

Caustic wash: the process of treating a product with a solution of caustic soda to remove minor impurities; often used in reference to the solution itself.

Ceresin: a hard, brittle wax obtained by purifying ozokerite; see Microcrystalline wax and Ozokerite).

Cetane index: an approximation of the cetane number calculated from the density and mid-boiling point temperature; see also Diesel index.

Cetane number: a number indicating the ignition quality of diesel fuel; a high cetane number represents a short ignition delay time; the ignition quality of diesel fuel can also be estimated from the following formula:

CFR: Code of Federal Regulations; Title 40 (40 CFR) contains the regulations for protection of the environment.

Characterization factor: the UOP characterization factor K, defined as the ratio of the cube root of the molal average boiling point, T_B, in degrees Rankine (°R = °F + 460), to the specific gravity at 60°F/60°F:

$$K = (T_B)^{1/3}/\text{sp gr}$$

The value ranges from 12.5 for paraffin stocks to 10.0 for the highly aromatic stocks; also called the Watson characterization factor.

Cheesebox still (cheese box still): an early type of vertical cylindrical still designed with a vapor dome.

Chelating agents: complex-forming agents having the ability to solubilize heavy metals.

Chemical flooding: see EOR process.

Chemical octane number: the octane number added to gasoline by refinery processes or by the use of octane number improvers such as tetraethyl lead.

Chemical waste: any solid, liquid, or gaseous material discharged from a process and that may pose substantial hazards to human health and environment.

Chlorex process: a process for extracting lubricating-oil stocks in which the solvent used is Chlorex (ß,ß-dichlorodiethyl ether).

Chromatographic adsorption: selective adsorption on materials such as activated carbon, alumina, or silica gel; liquid or gaseous mixtures of hydrocarbons are passed through the adsorbent in a stream of diluent, and certain components are preferentially adsorbed.

Chromatographic separation: the separation of different species of compounds according to their size and interaction with the rock as they flow through a porous medium.

Chromatography: a method of separation based on selective adsorption; see also Chromatographic adsorption.

Clarified oil: the heavy oil which has been taken from the bottom of a fractionator in a catalytic cracking process and from which residual catalyst has been removed.

Clarifier: equipment for removing the color or cloudiness of an oil or water by separating the foreign material through mechanical or chemical means; may involve centrifugal action, filtration, heating, or treatment with acid or alkali.

Clay: silicate minerals that also usually contain aluminum and have particle sizes are less than 0.002 micron; used in separation methods as an adsorbent and in refining as a catalyst.

Clay contact process: see contact filtration.

Clay refining: a treating process in which vaporized gasoline or other light crude oil product is passed through a bed of granular clay such as fuller's earth.

Clay regeneration: a process in which spent coarse-grained adsorbent clay minerals from percolation processes are cleaned for reuse by de-oiling the clay minerals with naphtha, steaming out the excess naphtha, and then roasting in a stream of air to remove carbonaceous matter.

Clay treating: see gray clay treating.

Clay wash: light oil, such as kerosene (kerosine) or naphtha, used to clean fuller's earth after it has been used in a filter.

Clastic: composed of pieces of pre-existing rock.

Cleanup: a preparatory step following extraction of a sample media designed to remove components that may interfere with subsequent analytical measurements.

Closed-loop biomass: crops grown, in a sustainable manner, for the purpose of optimizing their value for bioenergy and the production of bio-products. This includes annual crops such as maize and wheat, and perennial crops such as trees, shrubs, and grasses such as switch grass.Cloud point: the temperature at which paraffin wax or other solid substances begin to crystallize or separate from the solution, imparting a cloudy appearance to the oil when the oil is chilled under prescribed conditions.

Coal: an organic rock. Coalescence: the union of two or more droplets to form a larger droplet and, ultimately, a continuous phase.

Coal tar: the specific name for the tar produced from coal.

Coal tar pitch: the specific name for the pitch produced from coal.

Coarse materials: Wood residues suitable for chipping, such as slabs, edgings, and trimmings.

COFCAW: an enhanced oil recovery process (EOR process) that combines forward combustion and water flooding.

Cogeneration: an energy conversion method by which electrical energy is produced along with steam generated for EOR use.

Coke: a gray to black solid carbonaceous material produced from crude oil during thermal processing; characterized by having a high carbon content (95%+ by weight) and a honeycomb type of appearance and is insoluble in organic solvents.

Coke drum: a vessel in which coke is formed and which can be cut oil from the process for cleaning.

Coke number: used, particularly in Great Britain, to report the results of the Ramsbottom carbon residue test, which is also referred to as a coke test.

Coker: the processing unit in which coking takes place.

Coking: a process for the thermal conversion of crude oil in which gaseous, liquid, and solid (coke) products are formed.

Cold pressing: the process of separating wax from oil by first chilling (to help form wax crystals) and then filtering under pressure in a plate and frame press.

Cold settling: processing for the removal of wax from high-viscosity stocks, wherein a naphtha solution of the waxy oil is chilled and the wax crystallizes out of the solution.

Color stability: the resistance of a crude oil product to color change due to light, aging, etc.

Combined cycle: a combustion (gas) turbine equipped with a heat recovery steam generator that produces steam for the steam turbine; power is produced from both the gas and steam turbines – hence the term combined cycle.

Combustible liquid: a liquid with a flash point in excess of 37.8oC (100°F) but below 93.3°C (200°F).

Combustion zone: the volume of reservoir rock wherein crude oil is undergoing combustion during enhanced oil recovery.

Composition: the general chemical make-up of crude oil.

Completion interval: the portion of the reservoir formation placed in fluid communication with the well by selectively perforating the wellbore casing.

Composition map: a means of illustrating the chemical make-up of crude oil using chemical and/or physical property data.

Con carbon: see carbon residue.

Condensate: a mixture of light hydrocarbon liquids obtained by condensation of hydrocarbon vapors: predominately butane, propane, and pentane with some heavier hydrocarbons and relatively little methane or ethane; see also natural gas liquids.

Conductivity: a measure of the ease of flow through a fracture, perforation, or pipe.

Conformance: the uniformity with which a volume of the reservoir is swept by injection fluids in area and vertical directions.

Connate water: water that is trapped in the pores of a rock during formation of the rock – also described as fossil water. The chemistry of connate water can change in composition throughout the history of the rock; connate water can be dense and saline compared with seawater. On the other hand, formation water or interstitial water is water found in the pore spaces of a rock, and might not have been present when the rock was formed.

Conradson carbon residue: see carbon residue.

Contact filtration: a process in which finely divided adsorbent clay is used to remove color bodies from crude oil products.

Contaminant: a substance that causes deviation from the normal composition of an environment.

Continuous contact coking: a thermal conversion process in which crude oil-wetted coke particles move downward into the reactor in which cracking, coking, and drying take place to produce coke, gas, gasoline, and gas oil.

Continuous contact filtration: a process to finish lubricants, waxes, or special oils after acid treating, solvent extraction, or distillation.

Conventional crude oil: crude oil that is pumped from the ground and recovered using the energy inherent in the reservoir; also recoverable by application of secondary recovery techniques; the proeprties are typically on the order of API gravity: >20°; sulfur: <2.0% w/w; metals: <200 ppm, mobile in the reservoir and recovered by primary recovery methods or secondary recovery.

Conventional recovery: primary and/or secondary recovery.

Conversion: the thermal treatment of crude oil which results in the formation of new products by the alteration of the original constituents.

Conversion cost: the cost of changing a production well to an injection well, or some other change in the function of an oilfield installation.

Conversion factor: the percentage of feedstock converted to light ends, gasoline, other liquid fuels, and coke, Copper sweetening: processes involving the oxidation of mercaptan derivatives (RSH) to disulfide derivatives (RSSR) by oxygen in the presence of cupric chloride ($CuCl_2$).

Cord: a stack of wood comprising 128 cubic feet (3.62 m3); standard dimensions are 4 × 4 × 8 feet, including air space and bark. One cord contains approx. 1.2 U.S. tons (oven-dry) = 2,400 pounds = 1,089 kg.

Core floods: laboratory flow tests through samples (cores) of porous rock.

Co-surfactant: a chemical compound, typically alcohol that enhances the effectiveness of a surfactant.

Cp (centipoise): a unit of viscosity.

Craig-Geffen-Morse method: a method for predicting oil recovery by water flood.

Cracked residua: residua that have been subjected to temperatures above 350°C (660°F) during the distillation process.

Cracking: the thermal processes by which the constituents of crude oil are converted to lower molecular weight products.

Cracking activity: see Catalytic activity.

Cracking coil: equipment used for cracking viscous (and high molecular weight) crude oil products consisting of a coil of heavy pipe running through a furnace so that the oil passing through it is subject to high temperature.

Cracking still: the combined equipment-furnace, reaction chamber, fractionator for the thermal conversion of heavier feedstocks to lighter products.

Cracking temperature: the temperature (350°C; 660°F) at which the rate of thermal decomposition of crude oil constituents becomes significant.

Criteria air pollutants: air pollutants or classes of pollutants regulated by the Environmental Protection Agency; the air pollutants are (including VOCs): ozone, carbon monoxide, particulate matter, nitrogen oxides, sulfur dioxide, and lead.

Cracking: a secondary process in a refinery that uses heat and/or a catalyst to break down high molecular weight chemical components into lower molecular weight products which can be used as blending components for fuels.

Cropland: total cropland includes five components: cropland harvested, crop failure, cultivated summer fallow, cropland used only for pasture, and idle cropland.

Cropland pasture: land used for long-term crop rotation. However, some cropland pasture is marginal for crop uses and may remain in pasture indefinitely. This category also includes land that was used for pasture before crops reached maturity and some land used for pasture that could have been cropped without additional improvement.Cross-linking: combining of two or polymer molecules by use of a chemical that mutually bonds with a part of the chemical structure of the polymer molecules.

Crude assay: a procedure for determining the general distillation characteristics (e.g., distillation profile, *q.v.*) and other quality information of crude oil.Crude oil: a naturally occurring mixture of gaseous, liquid, and solid hydrocarbon compounds usually found trapped deep underground beneath impermeable cap rock and above a lower dome of sedimentary rock such as shale; most crude oil reservoirs occur in sedimentary rocks of marine, deltaic, or estuarine origin.

See also: Conventional crude oil.

Crude scale wax: the wax product from the first sweating of the slack wax.

Crude still: distillation equipment in which crude oil is separated into various products.

Cull tree: a live tree, 5.0 inches in diameter at breast height (d.b.h.) or larger that is non-merchantable for saw logs now or prospectively because of rot, roughness, or species. (See definitions for rotten and rough trees.)

Cultivated summer fallow: cropland cultivated for one or more seasons to control weeds and accumulate moisture before small grains are planted.

Cumene: a colorless liquid [$C_6H_5CH(CH_3)_2$] used as an aviation gasoline blending component and as an intermediate in the manufacture of chemicals.

Cut point: the boiling-temperature division between distillation fractions of crude oil.

Cutback: the term applied to the products from blending heavier feedstocks or products with lighter oils to bring the heavier materials to the desired specifications.

Cutback asphalt: asphalt liquefied by the addition of a volatile liquid such as naphtha or kerosene which, after application and on exposure to the atmosphere, evaporates leaving the asphalt.

Cutting oil: an oil to lubricate and cool metal-cutting tools; also called cutting fluid, cutting lubricant.

Cycle stock: the product taken from some later stage of a process and recharged (recycled) to the process at some earlier stage.

Cyclic steams injection: the alternating injection of steam and production of oil with condensed steam from the same well or wells.

Cyclization: the process by which an open-chain hydrocarbon structure is converted to a ring structure, e.g., hexane to benzene.Cyclone: a device for extracting dust from industrial waste gases. It is in the form of an inverted cone into which the contaminated gas enters tangential from the top; the gas is propelled down a helical pathway, and the dust particles are deposited by means of centrifugal force onto the wall of the scrubber.Deactivation: reduction in catalyst activity by the deposition of contaminants (e.g., coke, metals) during a process.

Dealkylation: the removal of an alkyl group from aromatic compounds.

Deasphaltened oil: the fraction of crude oil after the asphaltene constituents have been removed.

Deasphaltening: removal of a solid powdery asphaltene fraction from crude oil by the addition of the low-boiling liquid hydrocarbons such as n-pentane or n-heptane under ambient conditions.

Deasphalting: the removal of the asphaltene fraction from crude oil by the addition of a low-boiling hydrocarbon liquid such as n-pentane or n-heptane; more correctly the removal asphalt (tacky, semi-solid) from crude oil (as occurs in a refinery asphalt plant) by the addition of liquid propane or liquid butane under pressure.

Debris: see waste.

Debutanization: distillation to separate butane and lighter components from higher boiling components.

Decant oil: the highest boiling product from a catalytic cracker; also referred to as slurry oil, clarified oil, or bottoms.

Decarbonizing: a thermal conversion process designed to maximize coker gas-oil production and minimize coke and gasoline yields; operated at essentially lower temperatures and pressures than delayed coking.

Decoking: removal of crude oil coke from equipment such as coking drums; hydraulic decoking uses high-velocity water streams.

Decolorizing: removal of suspended, colloidal, and dissolved impurities from liquid crude oil products by filtering, adsorption, chemical treatment, distillation, bleaching, etc.

De-ethanization: distillation to separate ethane and lighter components from propane and higher-boiling components; also called de-ethanation.

Degradation: the loss of desirable physical properties of EOR fluids, e.g., the loss of viscosity of polymer solutions.Dehydrating agents: substances capable of removing water (drying, q.v.) or the elements of water from another substance.

Dehydrocyclization: any process by which both dehydrogenation and cyclization reactions occur.

Dehydrogenation: the removal of hydrogen from a chemical compound; for example, the removal of two hydrogen atoms from butane to make butene(s) as well as the removal of additional hydrogen to produce butadiene.

Delayed coking: a coking process in which the thermal reaction are allowed to proceed to completion to produce gaseous, liquid, and solid (coke) products.

Demethanization: the process of distillation in which methane is separated from the higher boiling components; also called demethanation.

Demulsifiers: a group of chemicals or surfactants used to separate water content in the water-in-oil and oil-in-water emulsions usually at low concentrations. See: emulsion.

Density: the mass (or weight) of a unit volume of any substance at a specified temperature; see also Specific gravity.

Deoiling (De-oiling): a reduction in the quantity of liquid oil entrained in solid wax by draining (sweating) or by a selective solvent; see MEK deoiling.

Depentanizer: a fractionating column for the removal of pentane and lighter fractions from a mixture of hydrocarbons.

Depropanization: distillation in which lighter components are separated from butanes and higher boiling material; also called depropanation.

Desalting: the removal of mineral salts (mostly chlorides) from crude oils; the first refining process applied to crude oil which removes salt, water and solid particles that would otherwise lead to operational problems during refining such as corrosion, fouling of equipment, or poisoning of catalysts.

Desorption: the reverse process of adsorption whereby adsorbed matter is removed from the adsorbent; also used as the reverse of absorption.

Desulfurization: the removal of sulfur or sulfur compounds from a feedstock.

Detergent oil: lubricating oil possessing special sludge-dispersing properties for use in internal-combustion engines.

Dewaxing: see solvent dewaxing.

Devolatilized fuel: smokeless fuel; coke that has been reheated to remove all of the volatile material.

Diagenesis: the concurrent and consecutive chemical reactions which commence the alteration of organic matter (at temperatures up to 50°C (120°F) and ultimately result in the formation of crude oil from the marine sediment; see also Catagenesis and Metagenesis.

Diagenetic rock: rock formed by conversion through pressure or chemical reaction) from a rock, e.g., sandstone is a diagenetic.Diesel cycle: a repeated succession of operations representing the idealized working behavior of the fluids in a diesel engine.

Diesel engine: named for the German engineer Rudolph Diesel, this internal-combustion, compression-ignition engine works by heating fuels and causing them to ignite; can use either crude oil or bio-derived fuel.

Diesel fuel: fuel used for internal combustion in diesel engines; usually that fraction which distills after kerosene.

Diesel index: an approximation of the cetane number of diesel fuel calculated from the density and aniline point.

$$DI = (\text{aniline point } (^{\circ}F) \times \text{API gravity})\ 100$$

Diesel knock: the result of a delayed period of ignition is long and the accumulated of diesel fuel in the engine.

Differential-strain analysis: measurement of thermal stress relaxation in a recently cut well. Digester: an airtight vessel or enclosure in which bacteria decomposes biomass in water to produce biogas.

Direct-injection engine: a diesel engine in which fuel is injected directly into the cylinder.

Dispersion: a measure of the convective mi fluids due to flow in a reservoir.Displacement efficiency: the ratio of the amount of oil moved from the zone swept by the re process to the amount of oil present in the zone prior to start of the process.

Distribution coefficient: a coefficient that describes the distribution of a chemical in reservoir fluids, usually defined as the equilibrium concentrations in the aqueous phases.

Distillate: a refinery product produced by boiling crude oil and collecting the vapors produced as a condensate in a separate vessel, for example gasoline (light distillate), gas oil (middle distillate), or fuel oil (heavy distillate).

Distillation: the primary refining process which uses high temperature to separate crude oil into vapor and fluids (bulk products) which can then be fed into a variety of processes to produce saleable products.

Distillation curve: see distillation profile.

Distillation loss: the difference, in a laboratory distillation, between the volume of liquid originally introduced into the distilling flask and the sum of the residue and the condensate recovered.

Distillation range: the difference between the temperature at the initial boiling point and at the end point, as obtained by the distillation test.

Distillation profile: the distillation characteristics of crude oil or crude oil products showing the temperature and the per cent distilled.

Doctor solution: a solution of sodium plumbite used to treat gasoline or other light crude oil distillates to remove mercaptan sulfur; see also doctor test.

Doctor sweetening: a process for sweetening gasoline, solvents, and kerosene by converting mercaptan derivatives (RSH) to disulfide derivatives (RSSR) using sodium plumbite (Na_2PbO_2) and sulfur.

Doctor test: a test used for the detection of compounds in low boiling crude oil distillates which react with sodium plumbite; see also doctor solution.

Domestic heating oil: see No. 2 fuel oil.Donor solvent process: a conversion process in which hydrogen donor solvent is used in place of or to augment hydrogen.

Downcomer: a means of conveying liquid from one tray to the next below in a bubble tray column.

Downdraft gasifier: a gasifier in which the product gases pass through a combustion zone at the bottom of the gasifier.

Downhole steam generator: a generator installed downhole in an oil well to which oxygen-rich air, fuel, and water are supplied for the purposes of generating steam for it into the reservoir. Its major advantage over a surface steam generating facility is the losses to the wellbore and surrounding formation are eliminated.

Dross: See waste.

Drying: removal of a solvent or water from a chemical substance; also referred to as the removal of solvent from a liquid or suspension.

Dropping point: the temperature at which grease passes from a semisolid to a liquid state under prescribed conditions.

Dry gas: a gas which does not contain fractions that may easily condense under normal atmospheric conditions.

Dry point: the temperature at which the last drop of crude oil fluid evaporates in a distillation test.

Dualayer distillate process: a process for removing mercaptan derivatives (RSH) and oxygenated compounds from distillate fuel oils and similar products, using a combination of treatment with concentrated caustic solution and electrical precipitation of the impurities.

Dualayer gasoline process: a process for extracting mercaptan derivatives (RSH) and other objectionable acidic compounds from crude oil distillates; see also Dualayer solution.

Dualayer solution: a solution which consists of concentrated potassium or sodium hydroxide containing a solubilizer; see also Dualayer gasoline process.

Dubbs cracking: an older continuous, liquid-phase thermal cracking process formerly used.

Dutch oven furnace: one of the earliest types of furnaces, having a large, rectangular box lined with firebrick (refractory) on the sides and top; commonly used for burning wood.

Dykstra-Parsons coefficient: an index of reservoir heterogeneity arising from permeability variation and stratification.

E85: an alcohol fuel mixture containing 85%ethanol and 15% gasoline by volume, and the current alternative fuel of choice of the U.S. government.Ebullated bed: a process in which the catalyst bed is in a suspended state in the reactor by means of a feedstock recirculation pump which pumps the feedstock upwards at sufficient speed to expand the catalyst bed at approximately 35% above the settled level.

Edeleanu process: a process for refining oils at low temperature with liquid sulfur dioxide (SO_2), or with liquid sulfur dioxide and benzene; applicable to the recovery of aromatic concentrates from naphtha and high boiling more viscous crude oil distillates.

Effective viscosity: see apparent viscosity.

Effluent: any contaminating substance, usually a liquid, which enters the environment via a domestic industrial, agricultural, or sewage plant outlet.

Electric desalting: a continuous process to remove inorganic salts and other impurities from crude oil by settling out in an electrostatic field.

Electrical precipitation: a process using an electrical field to improve the separation of hydrocarbon reagent dispersions. May be used in chemical treating processes on a wide variety of refinery stocks.

Electrofining: a process for contacting a light hydrocarbon stream with a treating agent (acid, caustic, doctor, etc.), then assisting the action of separation of the chemical phase from the hydrocarbon phase by an electrostatic field.

Electrolytic mercaptan process: a process in which aqueous caustic solution is used to extract mercaptan derivatives (RSH) from refinery streams.

Electrostatic precipitators: devices used to trap fine dust particles (usually in the size range 30–60 microns) that operate on the principle of imparting an electric charge to particles in an incoming air stream and which are then collected on an oppositely charged plate across a high voltage field.

Eluate: the solutes, or analytes, moved through a chromatographic column (see *elution*).

Eluent: solvent used to elute sample.

Elution: a process whereby a solute is moved through a chromatographic column by a solvent (liquid or gas) or eluent.

Emission control: the use gas cleaning processes to reduce emissions.

Emissions: substances discharged into the air during combustion.

Emission standard: the maximum amount of a specific pollutant permitted to be discharged from a particular source in a given environment.

Emulsion: a two-phase system comprising of two liquids which are not homogenous when mixed. In the emulsion, one of the liquid is the constantly dispersed (the dispersed phase) as globules in the second phase (the continuous phase), such as oil in water. See: demulsifiers.

Emulsion breaking: the settling or aggregation of colloidal-sized emulsions from suspension in a liquid medium.

End-of-pipe emission control: the use of specific emission control processes to clean gases after production of the gases.

Energy: the capacity of a body or system to do work, measured in joules (SI units); also the output of fuel sources.

Energy balance: The difference between the energy produced by a fuel and the energy required to obtain it through agricultural processes, drilling, refining, and transportation.

Energy crops: crops grown specifically for their fuel value; include food crops such as corn and sugarcane, and nonfood crops such as poplar trees and switch grass.

Energy-efficiency ratio: a number representing the energy stored in a fuel as compared to the energy required to produce, process, transport, and distribute that fuel.

Energy from biomass: the production of energy from biomass.

Engler distillation: a standard test for determining the volatility characteristics of a gasoline by measuring the per cent distilled at various specified temperatures.

Enhanced oil recovery (EOR): crude oil recovery following recovery by conventional (i.e., primary and/or secondary) methods.Enhanced oil recovery (EOR) process: a method for recovering additional oil from a crude oil reservoir beyond that economically recoverable by conventional primary and secondary recovery methods. EOR methods are usually divided into three main categories: (1) *chemical flooding*: injection of water with added chemicals into a crude oil reservoir. The chemical processes include: surfactant flooding, polymer flooding, and alkaline flooding, (2) *miscible flooding*: injection into a crude oil reservoir of a material that is miscible, or can become miscible, with the oil in the reservoir. Carbon dioxide, hydrocarbons, and nitrogen are used, (3) thermal recovery: injection of steam into a crude oil reservoir, or propagation of a combustion zone through a reservoir by air or oxygen-enriched air injection. The thermal processes include: steam drive, cyclic steam injection, and in situ combustion.

Entrained bed: a bed of solid particles suspended in a fluid (liquid or gas) at such a rate that some of the solid is carried over (entrained) by the fluid.

Entrainer: a separating agent that is used to enhance the separation of closely boiling compounds in azeotropic distillation or in extractive distillation.

EPA: Environmental Protection Agency.

Ester: a compound formed by the reaction between an organic acid and an alcohol. ethoxylated alcohols (i.e., alcohols having ethylene oxide functional groups attached to the alcohol molecule). Ethanol (ethyl alcohol, alcohol, or grain-spirit): A clear, colorless, flammable oxygenated hydrocarbon; used as a vehicle fuel by itself (E100 is 100% ethanol by volume), blended with gasoline (E85 is 85% ethanol by volume), or as a gasoline octane enhancer and oxygenate (10% by volume); formed during fermentation of sugars; used as an intoxicant and as a fuel.

Evaporation: a process for concentrating nonvolatile solids in a solution by boiling off the liquid portion of the waste stream.

Expanding clays: clay minerals that expand or swell on contact with water, e.g., montmorillonite. Explosive limits: the limits of percentage composition of mixtures of gases and air within which an explosion takes place when the mixture is ignited.

Extract: the portion of a sample preferentially dissolved by the solvent and recovered by physically separating the solvent.

Extractive distillation: the separation of different components of mixtures which have similar vapor pressures by flowing a relatively high-boiling solvent, which is selective for one of the components in the feed, down a distillation column as the distillation proceeds; the selective solvent scrubs the soluble component from the vapor.

Extra heavy crude oil: an oil that has similar proeprties to tar sand bitumen (e.g., API gravity: <10°; sulfur: 2.0–5.0% w/w; metals: 400–1,200 ppm) that is mobile in the reservoir/deposit because the reservoir/deposit temperature higher than the pour point of oil.

Fabric filters: filters made from fabric materials and used for removing particulate matter from gas streams (see Baghouse).

Facies: one or more layers of rock that differs from other layers in composition, age or content.

FAST: fracture assisted steamflood technology.

Fast pyrolysis: a pyrolysis process that involves heating a feedstock, such as biomass rapidly (2 seconds) at temperatures ranging on the order of from 350 to 650°C (650 to 1,200°F).

Fat oil: the bottom or enriched oil drawn from the absorber as opposed to lean oil.

Faujasite: a natural lyoccurring silica-alumina (SiO_2-Al_2O_3) mineral.

FCC: fluid catalytic cracking.

FCCU: fluid catalytic cracking unit.

Feedstock: crude oil as it is fed to the refinery; a refinery product that is used as the raw material for another process; biomass used in the creation of a particular biofuel (e.g., corn or sugarcane for ethanol, soybeans or rapeseed for biodiesel); the term is also generally applied to raw materials used in other industrial processes.Fermentation: the conversion of carbon-containing compounds by micro-organisms for production of fuels and chemicals such as alcohols, acids or energy-rich gases.

Ferrocyanide process: a regenerative chemical treatment for mercaptan removal using caustic-sodium ferrocyanide reagent.

Ferrofining process: a mild hydrotreating process that was developed to treat distilled and solvent-refined lubricating oils; the process eliminates the need for acid and clay treatment.

Field-scale: the application of EOR processes to a significant portion of a field.

Filtration: the use of an impassable barrier to collect solids but which allows liquids to pass. Fiber products; the products derived from fibers of herbaceous and woody plant materials; examples include pulp, composition board products, and wood chips for export.

Fine materials: wood residues that are not suitable for chipping, such as planer shavings and sawdust.

Fingering: the formation of finger-shaped irregularities at the leading edge of a displacing fluid in a porous medium which move out ahead of the main body of fluid.

Fire point: the lowest temperature at which, under specified conditions in standardized apparatus, a crude oil product vaporizes sufficiently rapidly to form above its surface an air-vapor mixture which burns continuously when ignited by a small flame.

First contact miscibility: see miscibility.

Fischer-Tropsch process: a process for synthesizing hydrocarbons and oxygenated chemicals from a mixture of hydrogen and carbon monoxide.

Fixed-bed: a stationary bed (of catalyst) to accomplish a process (see fluid bed).

Five-spot: an arrangement or pattern of wells with four injection wells at the comers of a square and a producing well in the center of the square.

Flammability range: the range of temperature over which a chemical is flammable.

Flammable: a substance that will burn readily.

Flammable liquid: a liquid having a flash point below 37.8°C (100°F).Flammable solid: a solid that can ignite from friction or from heat remaining from its manufacture, or which may cause a serious hazard if ignited.

Flash point: the lowest temperature to which the product must be heated under specified conditions to give off sufficient vapor to form a mixture with air that can be ignited momentarily by a flame.

Flash vaporization: a relatively simple unit operation in which a heated liquid mixture is throttled through a valve in order to vaporize the liquid mixture. The vaporization is done in order to separate the constituents of the liquid mixture. Flash vaporization can be seen as a single stage.Flexible-fuel vehicle (flex-fuel vehicle): a vehicle that can run alternately on two or more sources of fuel; includes cars capable of running on gasoline and gasoline/ethanol mixtures, as well as cars that can run on both gasoline and natural gas.

Flexicoking: a modification of the fluid coking process insofar as the process also includes a gasifier adjoining the burner/regenerator to convert excess coke to a clean fuel gas.

Flocculants: chemicals that flocculate the water droplets and facilitate coalescence.

Flocculation threshold: the point at which constituents of a solution (e.g., asphaltene constituents or coke precursors) will separate from the solution as a separate (solid) phase.

Floc point: the temperature at which wax or solids separate as a definite floc.

Flood, flooding: the process of displacing crude oil from a reservoir by the injection of fluids.

Flue gases: the gaseous products of the combustion process mostly comprised of carbon dioxide, nitrogen, and water vapor.

Fluid bed: the use of an agitated bed of inert granular material to accomplish a process in which the agitated bed resembles the motion of a fluid.

Flue gas: gas from the combustion of fuel, the heating value of which has been substantially spent and which is, therefore, discarded to the flue or stack.

Fluid: a reservoir gas or liquid.

Fluid-bed: a bed (of catalyst) that is agitated by an upward passing gas in such a manner that the particles of the bed simulate the movement of a fluid and has the characteristics associated with a true liquid; c.f. fixed-bed.

Fluid catalytic cracking: cracking in the presence of a fluidized bed of catalyst.

Fluid coking: a continuous fluidized solids process that cracks feed thermally over heated coke particles in a reactor vessel to gas, liquid products, and coke.

Fluidized-bed combustion: a process used to burn low-quality solid fuels in a bed of small particles suspended by a gas stream (usually air that will lift the particles but not blow them out of the vessel. Rapid burning removes some of the offensive by-products of combustion from the gases and vapors that result from the combustion process).Fluidized-bed boiler: a large, refractory-lined vessel with an air distribution member or plate in the bottom, a hot gas outlet in or near the top, and some provisions for introducing fuel; the fluidized bed is formed by blowing air up through a layer of inert particles (such as sand or limestone) at a rate that causes the particles to go into suspension and continuous motion.

Fly ash: particulate matter produced from mineral matter in coal that is converted during combustion to finely divided inorganic material and which emerges from the combustor in the gases.

Foots oil: the oil sweated out of slack wax; named from the fact that the oil goes to the foot, or bottom, of the pan during the sweating operation.

Forest land: land in which at least 10% is stocked by forest trees of any size, including land that formerly had such tree cover and that will be naturally or artificially regenerated; includes transition zones, such as areas between heavily forested and non-forested lands that are at least 10% stocked with forest trees and forest areas adjacent to urban and built-up lands; also included are pinyon-juniper and chaparral areas; minimum area for classification of forest land is 1 acre.

Forest residues: material not harvested or removed from logging sites in commercial hardwood and softwood stands as well as material resulting from forest management operations such as pre-commercial thinnings and removal of dead and dying trees.

Forest health: a condition of ecosystem sustainability and attainment of management objectives for a given forest area; usually considered to include green trees, snags, resilient stands growing at a moderate rate, and endemic levels of insects and disease.

Formation: an interval of rock with distinguishable geologic characteristics.

Formation volume factor: the volume in barrel! That one stock tank barrel occupies in the formation at reservoir temperature and with the solution gas that is held in the oil at reservoir pressure.

Fossil fuel resources: a gaseous, liquid, or solid fuel material formed in the ground by chemical and physical changes (diagenesis) in plant and animal residues over geological time; natural gas, crude oil, coal, and oil shale.

Fouling: the name applied to the disposition of solids or phase separation which typically commences with the introduction into the refinery of a feedstock (crude oil or the blends of crude oils) that is incompatible with (in the current context) the other components of a blend of feedstocks. Fouling includes any of the following events that can occur during or after processing which include (1) deposit formation, (2) encrustation, (3) deposition, (4) scaling, (5) scale formation, (6) slagging, and (7) sludge formation, all of which can have an adverse effect on recovery operations, refinery operations, as well as transportation of crude oil, and crude oil products. See also macrofouling, microfouling.

Fractional composition: the composition of crude oil as determined by fractionation (separation) methods.

Fractional distillation: the separation of the components of a liquid mixture by vaporizing and collecting the fractions, or cuts, which condense in different temperature ranges.

Fractional flow: the ratio of the volumetric flow rate of one fluid phase to the total fluid volumetric flow rate within a volume of rock.

Fractional flow curve: the relationship between the fractional flow of one fluid and its saturator during simultaneous flow of fluids through rock.

Fracture: a natural or man-made crack in a reservoir rock.

Fracturing: the breaking apart of reservoir rock by applying very high fluid pressure at the rock face.

Fractionating column: a column arranged to separate various fractions of crude oil by a single distillation and which may be tapped at different points along its length to separate various fractions in the order of their boiling points.

Fractionation: the separation of crude oil into the constituent fractions using solvent or adsorbent methods; chemical agents such as sulfuric acid may also be used.

Frasch process: a process formerly used for removing sulfur by distilling oil in the presence of copper oxide.

Fuel cell: a device that converts the energy of a fuel directly to electricity and heat, without combustion.

Fuel cycle: the series of steps required to produce electricity. The fuel cycle includes mining or otherwise acquiring the raw fuel source, processing and cleaning the fuel, transport, electricity generation, waste management and plant decommissioning.

Fuel oil: also called heating oil is a distillate product that covers a wide range of properties; see also No. 1 to No. 4 fuel oils.

Fuel treatment evaluator (FTE): a strategic assessment tool capable of aiding the identification, evaluation, and prioritization of fuel treatment opportunities.

Fuel wood: wood used for conversion to some form of energy, primarily for residential use.

Fuller's earth: a clay which has high adsorptive capacity for removing color from oils; Attapulgus clay is a widely used fuller's earth.

Functional group: the portion of a molecule that is characteristic of a family of compounds and determines the properties of these compounds.

Furfural extraction: a single-solvent process in which furfural is used to remove aromatic, naphthene, olefin, and unstable hydrocarbons from a lubricating-oil charge stock.

Furnace: an enclosed chamber or container used to burn biomass in a controlled manner to produce heat for space or process heating.

Furnace oil: a distillate fuel primarily intended for use in domestic heating equipment.

Garbage: see waste.

Gas cap: a part of a hydrocarbon reservoir at the top that will produce only gas.

Gaseous pollutants: gases released into the atmosphere that act as primary or secondary pollutants.

Gasification: a chemical or thermal process used to convert carbonaceous material (such as coal, crude oil, and biomass) into gaseous components such as carbon monoxide and hydrogen; a process for converting a solid or liquid fuel into a gaseous fuel useful for power generation or chemical feedstock with an oxidant and steam.

Gasifier: a device for converting solid fuel into gaseous fuel; in biomass systems, the process is also referred to as 'pyrolitic distillation'.

Gasifier cold gas efficiency (CGE): the percentage of the coal heating value that appears as chemical heating value in the gasifier product gas.

Gasohol: a mixture of 10% v/v anhydrous ethanol and 90% v/v gasoline; 7.5% v/v anhydrous ethanol and 92.5% v/v gasoline; or 5.5% v/v anhydrous ethanol and 94.5% v/v gasoline; a term for motor vehicle fuel comprising between 80 to 90% v/v unleaded gasoline and 10 to 20% v/v ethanol (see also Ethyl alcohol).

Gas oil: a crude oil distillate with a viscosity and boiling range between those of kerosene and lubricating oil.

Gas-oil ratio: the ratio of the number of cubic feet of gas measured at atmospheric (standard) conditions to barrels of produced oil measured at stock tank conditions.

Gas-oil sulfonate: sulfonate made from a specific refinery stream, in this case the gas-oil stream.

Gasoline: fuel for the internal combustion engine that is commonly, but improperly, referred to simply as gas.

Gas processing: usually involves the use of several integrated unit processes to remove: (1) oil, (2) water, (3) elements such as sulfur, helium, and carbon dioxide, and (iv) natural gas liquids; also referred to as 'gas refining'.

Gas refining: see gas processing.

Gas reversion: a combination of thermal cracking or reforming of naphtha with thermal polymerization or alkylation of hydrocarbon gases carried out in the same reaction zone.

Gas to liquids (GTL): the process of refining natural gas and other hydrocarbons into longer-chain hydrocarbons, which can be used to convert gaseous waste products into fuels.

Gas turbine: a device in which fuel is combusted at pressure and the products of combustion expanded through a turbine to generate power (the Brayton Cycle); it is based on the same principle as the jet engine.

Gel point: the point at which a liquid fuel cools to the consistency of petroleum jelly.

Genetically modified organism (GMO): an organism whose genetic material has been modified through recombinant DNA technology, altering the phenotype of the organism to meet desired specifications.

Gilsonite: an asphaltite that is >90% bitumen.

Girbotol process: a continuous, regenerative process to separate hydrogen sulfide, carbon dioxide, and other acid impurities from natural gas, refinery gas, etc., using mono-, di-, or triethanolamine as the reagent.

Glance pitch: an asphaltite.

Glycol-amine gas treating: a continuous, regenerative process to simultaneously dehydrate and remove acid gases from natural gas or refinery gas.

Grahamite: an asphaltite.

Grassland pasture and range: all open land used primarily for pasture and grazing, including shrub and brush land types of pasture; grazing land with sagebrush and scattered mesquite; and all tame and native grasses, legumes, and other forage used for pasture or grazing; because of the diversity in vegetative composition, grassland pasture and range are not always clearly distinguishable from other types of pasture and range; at one extreme, permanent grassland may merge with cropland pasture, or grassland may often be found in transitional areas with forested grazing land.

Gravity: see API gravity.

Gravity drainage: the movement of oil in a reservoir that results from the force of gravity.

Gravity segregation: partial separation of fluids in a reservoir caused by the gravity force acting on differences in density.

Gravity-stable displacement: the displacement of oil from a reservoir by a fluid of a different density, where the density difference is utilized to prevent gravity segregation of the injected fluid.

Gray clay treating: a fixed-bed, usually fuller's earth, vapor-phase treating process to selectively polymerize unsaturated gum-forming constituents (diolefins) in thermally cracked gasoline.

Grain alcohol: see ethyl alcohol.

Gravimetric: gravimetric methods weigh a residue.

Gravity drainage: the movement of oil in a reservoir that results from the force of gravity.

Gravity segregation: partial separation of fluids in a reservoir caused by the gravity force acting on differences in density.

Grease car: a diesel-powered automobile rigged post-production to run on used vegetable oil.

Greenhouse effect: the effect of certain gases in the earth's atmosphere in trapping heat from the sun.

Greenhouse gases: gases that trap the heat of the sun in the Earth's atmosphere, producing the greenhouse effect. The two major greenhouse gases are water vapor and carbon dioxide. Other greenhouse gases include methane, ozone, chlorofluorocarbons, and nitrous oxide.

Green hydrogen: hydrogen gas produced from water electrolysis that is typically of higher purity than hydrogen gas produced from fossil fuels, also known as gray hydrogen; production of this 'type' of hydrogen has significantly lower greenhouse gas emissions than production of gray hydrogen, which is derived from fossil fuels without carbon capture. See also blue hydrogen, gray hydrogen.

Gray hydrogen: hydrogen gas produced by the steam reforming of fossil fuels.

See also Blue hydrogen, Green hydrogen. Grid: an electric utility company's system for distributing power.

Growing stock: a classification of timber inventory that includes live trees of commercial species meeting specified standards of quality or vigor; cull trees are excluded.

Guard bed: a bed of an adsorbent (such as, for example, bauxite) that protects a catalyst bed by adsorbing species detrimental to the catalyst.

Gulf HDS process: a regenerative fixed-bed process for the catalytic hydrocracking of viscous feedstocks to lower-boiling distillates with accompanying desulfurization by catalytic hydrogenation to low-sulfur heavy fuel oils or to high-quality feedstocks for catalytic cracking units.

Gulfining: a catalytic hydrogen treating process for cracked and straight-run distillates and fuel oils, to reduce sulfur content; improve carbon residue, color, and general stability; and effect a slight increase in gravity.

Gum: an insoluble tacky semi-solid material formed as a result of the storage instability and/or the thermal instability of crude oil and crude oil products.

Habitat: the area where a plant or animal lives and grows under natural conditions. Habitat includes living and non-living attributes and provides all requirements for food and shelter.

HAP(s): hazardous air pollutant(s).
Hardness: the concentration of calcium and magnesium in brine.
Hardwoods: usually broad-leaved and deciduous trees.
Hazardous waste: a waste with properties that make it potentially dangerous or harmful to human health or the environment; generally categorized as (1) ignitable waste which is waste that combusts or burns easily, (2) corrosive waste which is waste that can cause damage or corrode any substances with which it comes into contact, (3) reactive which is waste that may explode or violently react with air, water, or other substances, and (4) toxic waste, which is waste that contains or that releases dangerous toxins; can include things such as chemicals, heavy metals, or substances generated as by-products during commercial manufacturing processes, as well as discarded household products like paint thinners, cleaning fluids, and old batteries. See also: waste, non-hazardous waste.
HCPV: hydrocarbon pore volume.
Headspace: the vapor space above a sample into which volatile molecules evaporate. Certain methods sample this vapor.
Hearn method: a method used in reservoir simulation for calculating a pseudo relative permeability curve that reflects reservoir stratification.
Heating oil: see fuel oil.
Heating value: the maximum amount of energy that is available from burning a substance.
Heat recovery steam generator: a heat exchanger that generates steam from the hot exhaust gases from a combustion turbine.
Heavy crude oil (also referred to as 'heavy oil'): oil that is more viscous that conventional crude oil, has a lower mobility in the reservoir but can be recovered through a well from the reservoir by the application of a secondary or enhanced recovery methods; sometimes crude oil having an API gravity of less than 20°; often referred to as 'heavy oil'.
Heavy ends: the highest boiling portion of a crude oil fraction; see also Light ends.
Heavy fuel oil: fuel oil having a high density and viscosity; generally residual fuel oil such as No. 5 and No 6. fuel oil
Hectare: common metric unit of area, equal to 2.47 acres. 100 hectares = 1 square kilometer.
Herbaceous: non-woody type of vegetation, usually lacking permanent strong stems, such as grasses, cereals and canola (rape).
Heteroatom compounds: chemical compounds which contain nitrogen and/or oxygen and/or sulfur and /or metals bound within their molecular structure(s).
Heterogeneity: lack of uniformity in reservoir properties such as permeability.
HF alkylation: an alkylation process whereby olefins (C_3, C_4, C_5) are combined with *iso*-butane in the presence of hydrofluoric acid catalyst.
Higgins-Leighton model: stream tube computer model used to simulate waterflood.
Hortonsphere (Horton sphere): a spherical pressure-type tank used to store a volatile liquid which prevents the excessive evaporation loss that occurs when such products are placed in conventional storage tanks.
Hot filtration test: a test for the stability of a crude oil product.
Hot spot: an area of a vessel or line wall appreciably above normal operating temperature, usually as a result of the deterioration of an internal insulating liner which exposes the line or vessel shell to the temperature of its contents.
Houdresid catalytic cracking: a continuous moving-bed process for catalytically cracking reduced crude oil to produce high octane gasoline and light distillate fuels.
Houdriflow catalytic cracking: a continuous moving-bed catalytic cracking process employing an integrated single vessel for the reactor and regenerator kiln.
Houdriforming: a continuous catalytic reforming process for producing aromatic concentrates and high-octane gasoline from low-octane straight naphtha.
Houdry butane dehydrogenation: a catalytic process for dehydrogenating light hydrocarbons to their corresponding mono- or diolefins.

Houdry fixed-bed catalytic cracking: a cyclic regenerable process for cracking of distillates.

Houdry hydrocracking: a catalytic process combining cracking and desulfurization in the presence of hydrogen.

Huff-and-puff: a cyclic EOR method in which steam or gas is injected into a production well; after a short shut-in period, oil and the injected fluid are produced through the same well.

Hydration: the association of molecules of water with a substance.

Hydraulic fracturing: the opening of fractures in a reservoir by high-pressure, high-volume injection of liquids through an injection well.

Hydrocarbonaceous material: a material such as bitumen that is composed of carbon and hydrogen with other elements (heteroelements) such as nitrogen, oxygen, sulfur, and metals chemically combined within the structures of the constituents; even though carbon and hydrogen may be the predominant elements, there may be very few true hydrocarbons.

Hydrocarbon compounds: chemical compounds containing only carbon and hydrogen.

Hydrocarbon-producing resource: a resource such as coal and oil shale (kerogen) which produces derived hydrocarbons by the application of conversion processes; the hydrocarbons so-produced are not naturally occurring materials.

Hydrocarbon resource: resources such as crude oil and natural gas which can produce naturally occurring hydrocarbons without the application of conversion processes.

Hydrocarbons: organic compounds containing only hydrogen and carbon.

Hydrolysis: a chemical reaction in which water reacts with another substance to form one or more new substances.

Hydroconversion: a term often applied to hydrocracking

Hydrocracking: a catalytic high-pressure high-temperature process for the conversion of crude oil feedstocks in the presence of fresh and recycled hydrogen; carbon-carbon bonds are cleaved in addition to the removal of heteroatomic species.

Hydrocracking catalyst: a catalyst used for hydrocracking which typically contains separate hydrogenation and cracking functions.

Hydrodenitrogenation: the removal of nitrogen by hydrotreating.

Hydrodesulfurization: the removal of sulfur by hydrotreating.

Hydrofining process: a fixed-bed catalytic process to desulfurize and hydrogenate a wide range of feedstocks such as straight run products, cracked products, coker-derived naphtha, and distillate boiling-range streams thereby increasing the supply of solvent naphtha, and other naphtha-type products as well as feedstocks for catalytic reforming units.

Hydroforming process: a process in which naphtha is passed over a catalyst at elevated temperatures and moderate pressures, in the presence of added hydrogen or hydrogen-containing gases, to form high-octane motor fuel or aromatics.

Hydrogen blistering: blistering of steel caused by trapped molecular hydrogen formed as atomic hydrogen during corrosion of steel by hydrogen sulfide.

Hydrogen addition: an upgrading process in the presence of hydrogen, e.g., hydrogenation.

Hydrogenation: the chemical addition of hydrogen to a material. In nondestructive hydrogenation, hydrogen is added to a molecule only if, and where, unsaturation with respect to hydrogen exists.

Hydrogen embrittlement: the mechanical damage of a metal due to the penetration of hydrogen into the metal causing loss in ductility and tensile strength; can occur due to corrosion of steel by hydrogen sulfide when hydrogen atoms are generated; also known as hydrogen-assisted cracking or hydrogen-induced cracking.

Hydrogen network: system that produces, distributes, and utilizes hydrogen within the refinery complex.

Hydrogen network pinch analysis: a valuable analytical method to identify the theoretical minimum hydrogen requirements for a given refinery.

Hydrogen purification: the process of removing contaminants from hydrogen gas to produce a product that will satisfy the future needs of the refiner.

Hydrogen transfer: the transfer of inherent hydrogen within the feedstock constituents and products during processing.

Hydroprocesses: refinery processes designed to add hydrogen to various products of refining.

Hydroprocessing: a term often equally applied to hydrotreating and to hydrocracking; also often collectively applied to both.

Hydrotreating: the removal of heteroatomic (nitrogen, oxygen, and sulfur) species by treatment of a feedstock or product at relatively low temperatures in the presence of hydrogen.

Hydrovisbreaking: a non-catalytic process, conducted under similar conditions to visbreaking, which involves treatment with hydrogen to reduce the viscosity of the feedstock and produce more stable products than is possible with visbreaking.

Hydropyrolysis: a short residence time high-temperature process using hydrogen.

Hyperforming: a catalytic hydrogenation process for improving the octane number of naphtha through removal of sulfur and nitrogen compounds.

Hypochlorite sweetening: the oxidation of mercaptan derivatives (RSH) in a sour feedstock by agitation with aqueous, alkaline hypochlorite solution; used where avoidance of free-sulfur addition is desired, because of a stringent copper strip requirements and minimum expense is not the primary object.

Idle cropland: land in which no crops were planted; acreage diverted from crops to soil-conserving uses (if not eligible for and used as cropland pasture) under federal farm programs is included in this component.

Ignitability: characteristic of liquids whose vapors are likely to ignite in the presence of ignition source; also characteristic of non-liquids that may catch fire from friction or contact with water and that burn vigorously.

Illuminating oil: oil used for lighting purposes.

Immiscible: two or more fluids that do not have complete mutual solubility and co-exist as separate phases.

Immiscible carbon dioxide displacement: injection of carbon dioxide into an oil reservoir to effect oil displacement under conditions in which miscibility with reservoir oil is not obtained; see Carbon dioxide augmented waterflooding.

Immiscible displacement: a displacement of oil by a fluid (gas or water) that is conducted under conditions so that interfaces exist between the driving fluid and the oil.

Immunoassay: portable tests that take advantage of an interaction between an antibody and a specific analyte. Immunoassay tests are semi-quantitative and usually rely on color changes of varying intensities to indicate relative concentrations.

Incinerator: any device used to burn solid or liquid residues or wastes as a method of disposal.

Incompatibility: the immiscibility of crude oil products and also of different crude oils which is often reflected in the formation of a separate phase after mixing and/or storage; the phenomenon may involve any one of two of several possible events which are (1) phase separation, (2) precipitation of asphaltene constituents when a paraffinic crude oil is blended with a viscous crude oil, (3) precipitation of asphaltene constituents when a paraffinic crude oil product is blended with a viscous crude oil, (4) when the blend is heated in pipes leading to a reactor, and also through (5) the formation of degradation products and other undesirable changes in the original properties of crude oil products. When such phenomena occur, which may not be immediately at the time of the blend but can occur after an induction period, it is often referred to as instability of the blends. See also: instability.

Incremental ultimate recovery: the difference between the quantity of oil that can be recovered by EOR methods and the quantity of oil that can be recovered by conventional recovery methods.

Inclined grate: a type of furnace in which fuel enters at the top part of a grate in a continuous ribbon, passes over the upper drying section where moisture is removed, and descends into the lower burning section. Ash is removed at the lower part of the grate.

Indirect-injection engine: an older model of diesel engine in which fuel is injected into a pre-chamber, partly combusted, and then sent to the fuel-injection chamber.

Indirect liquefaction: conversion of biomass to a liquid fuel through a synthesis gas intermediate step.

Industrial wood: all commercial round wood products except fuel wood.

Industrial solid waste: waste generated by businesses from an industrial or manufacturing process or non-manufacturing activities managed as a separate waste stream. See also: Waste.

Infill drilling: drilling additional wells within an established pattern.

Infrared spectroscopy: an analytical technique that quantifies the vibration (stretching and bending) that occurs when a molecule absorbs (heat) energy in the infrared region of the electromagnetic spectrum.

Inhibitor: a substance, the presence of which, in small amounts, in a crude oil product prevents or retards undesirable chemical changes from taking place in the product, or in the condition of the equipment in which the product is used.

Inhibitor sweetening: a treating process to sweeten gasoline of low mercaptan content, using a phenylenediamine type of inhibitor, air, and caustic.

Initial boiling point: the recorded temperature when the first drop of liquid falls from the end of the condenser.

Initial vapor pressure: the vapor pressure of a liquid of a specified temperature and zero per cent evaporated.

Injection profile: the vertical flow rate distribution of fluid flowing from the wellbore into a reservoir.

Injection well: a well in an oil field used for injecting fluids into a reservoir.

Injectivity: the relative ease with which a fluid is injected into a porous rock.

In-line blending: the controlled proportioning of two or more component streams to produce a final blended product of closely defined quality from the beginning to the end of the batch which permits the blended product to be used immediately for the prescribed purpose.

In situ: in its original place; in the reservoir.

In situ combustion: an EOR process consisting of injecting air or oxygen-enriched air into a reservoir under conditions that favor burning part of the in situ crude oil, advancing this burning zone, and recovering oil heated from a nearby producing well.

Instability: the inability of a crude oil product to exist for periods of time without change to the product. See also Incompatibility.

Integrated gasification combine cycle (IGCC): a power plant in which a gasification process provides syngas to a combined cycle under an integrated control system.

Integrity: maintenance of a slug or bank at its preferred composition without too much dispersion or mixing.

Interface: the thin surface area separating two immiscible fluids that are in contact with each other.

Interfacial film: a thin layer of material at the interface between two fluids which differs in composition from the bulk fluids.

Interfacial tension: the strength of the film separating two immiscible fluids, e.g., oil and water or microemulsion and oil; measured in dynes (force) per centimeter or milli-dynes per centimeter.

Interfacial viscosity: the viscosity of the interfacial film between two immiscible liquids.

Interference testing: a type of pressure transient test in which pressure is measured over time in a closed-in well while nearby wells are produced; flow and communication between wells can sometimes be deduced from an interference test.

Interphase mass transfer: the net transfer of chemical compounds between two or more phases.

Iodine number: a measure of the iodine absorption by oil under standard conditions; used to indicate the quantity of unsaturated compounds present; also called iodine value.

Ion exchange: a means of removing cations or anions from solution onto a solid resin.

Ion exchange capacity: a measure of the capacity of a mineral to exchange ions in amount of material per unit weight of solid.

Ions: chemical substances possessing positive or negative charges in solution.

Isocracking: a hydrocracking process for conversion of hydrocarbons which operates at relatively low temperatures and pressures in the presence of hydrogen and a catalyst to produce more valuable, lower-boiling products.

Isoforming: a process in which olefinic naphtha is contacted with an alumina catalyst at high temperature and low pressure to produce isomers of higher octane number.

Iso-Kel process: a fixed-bed, vapor-phase isomerization process using a precious metal catalyst and external hydrogen.

Isomate process: a continuous, non-regenerative process for isomerizing C_5-C_8 normal paraffin hydrocarbons, using aluminum chloride-hydrocarbon catalyst with anhydrous hydrochloric acid as a promoter.

Isomax process: a two-stage, fixed-bed catalyst process (each stage has a separate hydrogen recycling system) that operates in a temperature range on the order of 205 to 370°C (400 to 700°F) and a hydrogen pressure on the order of 500 to 1,500 psi using, for example, with middle distillate feedstocks.

Isomerate process: a fixed-bed isomerization process to convert pentane, heptane, and heptane to high-octane blending stocks.

Isomerization: the conversion of a n-paraffin hydrocarbon derivatives (normal paraffins, straight-chain paraffins) into an iso-paraffin hydrocarbon derivatives (branched-chain paraffin hydrocarbons) having the same atomic composition.

Isopach: a line on a map designating points of equal formation thickness.

Iso-plus Houdriforming: a combination process using a conventional Houdriformer operated at moderate severity, in conjunction with one of three possible alternatives-including the use of an aromatic recovery unit or a thermal reformer; see Houdriforming.

Jet fuel: fuel meeting the required properties for use in jet engines and aircraft turbine engines.

Joule: Metric unit of energy, equivalent to the work done by a force of one Newton applied over distance of one meter (= 1 kg m2/s2). One joule (J) = 0.239 calories (1 calorie = 4.187 J).

Kaolinite: a clay mineral formed by hydrothermal activity at the time of rock formation or by chemical weathering of rock with high feldspar content; usually associated with intrusive granite rock with high feldspar content.

Kata-condensed aromatic compounds: compounds based on linear condensed aromatic hydrocarbon systems, e.g., anthracene and naphthacene (tetracene).

Kauri butanol number: a measurement of solvent strength for hydrocarbon solvents; the higher the kauri-butanol (KB) value, the stronger the solvency; the test method (ASTM D1133) is based on the principle that kauri resin is readily soluble in butyl alcohol but not in hydrocarbon solvents and the resin solution will tolerate only a certain amount of dilution and is reflected as a cloudiness when the resin starts to come out of solution; solvents such as toluene can be added in a greater amount (and thus have a higher KB value) than weaker solvents like hexane.

Kerogen: a complex carbonaceous (organic) material that occurs in sedimentary rock and shale; generally insoluble in common organic solvents.

Kerosene (kerosine): a fraction of crude oil that was initially sought as an illuminant in lamps; a precursor to diesel fuel.

K-factor: see characterization factor.

Kilowatt: (kW): a measure of electrical power equal to 1,000 watts. 1 kW = 3412 Btu/hr = 1.341 horsepower.

Kilowatt hour (kWh): a measure of energy equivalent to the expenditure of one kilowatt for one hour. For example, 1 kWh will light a 100-watt light bulb for 10 hours. 1 kWh = 3412 Btu.

Kinematic viscosity: the ratio of viscosity to density, both measured at the same temperature.

Knock: the noise associated with self-ignition of a portion of the fuel-air mixture ahead of the advancing flame front.

Kriging: a technique used in reservoir description for interpolation of reservoir parameters between wells based on random field theory.

LAER: lowest achievable emission rate; the required emission rate in non-attainment permits.

Lamp burning: a test of burning oils in which the oil is burned in a standard lamp under specified conditions in order to observe the steadiness of the flame, the degree of encrustation of the wick, and the rate of consumption of the kerosene.

Lamp oil: see kerosene.

Landfill gas: a type of biogas that is generated by decomposition of organic material at landfill disposal sites. Landfill gas is approximately 50% methane. See also biogas.

Laterite: the name applied to both a soil and a rock type rich in iron and aluminum and is commonly considered to have formed in hot and wet tropical areas. Nearly all laterites are of rusty-red coloration, because of high iron oxide content.

Leaded gasoline: gasoline containing tetraethyl lead or other organometallic lead antiknock compounds.

Lean gas: the residual gas from the absorber after the condensable gasoline has been removed from the wet gas.

Lean oil: absorption oil from which gasoline fractions have been removed; oil leaving the stripper in a natural-gasoline plant.

Lewis acid: a chemical species which can accept an electron pair from a base.

Lewis base: a chemical species which can donate an electron pair.

Light ends: the lower-boiling components of a mixture of hydrocarbons; see also heavy ends, light hydrocarbons.

Light hydrocarbons: hydrocarbons with molecular weights less than that of heptane (C_7H_{16}).

Light oil: the products distilled or processed from crude oil up to, but not including, the first lubricating-oil distillate.

Light petroleum: crude oil having an API gravity greater than 20°.

Lignin: structural constituent of wood and (to a lesser extent) other plant tissues, which encrusts the walls and cements the cells together.

Ligroine (Ligroin): a saturated crude oil naphtha boiling in the range of 20 to 135°C (68 to 275°F) and suitable for general use as a solvent; also called benzine or crude oil ether.

Linde copper sweetening: a process for treating gasoline and distillates with a slurry of clay and cupric chloride.

Liquid petrolatum: see white oil.

Liquefied petroleum gas: propane, butane, or mixtures thereof, gaseous at atmospheric temperature and pressure, held in the liquid state by pressure to facilitate storage, transport, and handling.

Liquid chromatography: a chromatographic technique that employs a liquid mobile phase.

Liquid/liquid extraction: an extraction technique in which one liquid is shaken with or contacted by an extraction solvent to transfer molecules of interest into the solvent phase.

Liquid sulfur dioxide-benzene process: a mixed-solvent process for treating lubricating-oil stocks to improve viscosity index; also used for dewaxing.

Lithology: the geological characteristics of the reservoir rock.

Live cull: a classification that includes live cull trees; when associated with volume, it is the net volume in live cull trees that are 5.0 inches in diameter and larger.

Live steam: steam coming directly from a boiler before being utilized for power or heat.

Liver: the intermediate layer of dark-colored, oily material, insoluble in weak acid and in oil, which is formed when acid sludge is hydrolyzed.

Logging residues: the unused portions of growing-stock and non-growing-stock trees cut or killed logging and left in the woods.

Lorenz coefficient: a permeability heterogeneity factor.

Lower-phase micro emulsion: a microemulsion phase containing a high concentration of water that, when viewed in a test tube, resides near the bottom with oil phase on top.

Lube: see lubricating oil.

Lube cut: a fraction of crude oil of suitable boiling range and viscosity to yield lubricating oil when completely refined; also referred to as lube oil distillates or lube stock.

Lubricating oil: a fluid lubricant used to reduce friction between bearing surfaces.

M85: n alcohol fuel mixture containing 85% methanol and 15% gasoline by volume. Methanol is typically made from natural gas, but can also be derived from the fermentation of biomass.

Macrofouling: fouling of refinery equipment and pipes by the deposition of coarse matter from either organic or biological or inorganic origin; theses deposits foul the surfaces of heat exchangers and may cause deterioration of the relevant heat transfer coefficient as well as flow blockages. See also: fouling, microfouling.

MACT: maximum achievable control technology. Applies to major sources of hazardous air pollutants.

Mahogany acids: oil-soluble sulfonic acids formed by the action of sulfuric acid on crude oil distillates. They may be converted to their sodium soaps (mahogany soaps) and extracted from the oil with alcohol for use in the manufacture of soluble oils, rust preventives, and special greases. The calcium and barium soaps of these acids are used as detergent additives in motor oils; see also brown acids and sulfonic acids.

Major source: a source that has a potential to emit for a regulated pollutant that is at or greater than an emission threshold set by regulations.

Maltene fraction (maltenes fraction or maltenes): that fraction of crude oil that is soluble in, for example, pentane or heptane; deasphaltened oil; also the term arbitrarily assigned to the pentane-soluble portion of crude oil that is relatively high boiling (>300°C, 760 mm) (see also Petrolenes).

Marine engine oil: oil used as a crankcase oil in marine engines.

Marine gasoline: fuel for motors in marine service.

Marine sediment: the organic biomass from which crude oil is derived.

Marsh: an area of spongy waterlogged ground with large numbers of surface water pools.

Marshes usually result from: (1) an impermeable underlying bedrock; (2) surface deposits of glacial boulder clay; (3) a basin-like topography from which natural drainage is poor; (4) very heavy rainfall in conjunction with a correspondingly low evaporation rate; (5) low-lying land, particularly at estuarine sites at or below sea level.

Marx-Langenheim model: mathematical equations for calculating heat transfer in a hot water or steam flood.

Mass spectrometer: an analytical technique that *fractures* organic compounds into characteristic 'fragments' based on functional groups that have a specific mass-to-charge ratio.

Mayonnaise: low-temperature sludge; a black, brown, or gray deposit having a soft, mayonnaise-like consistency; not recommended as a food additive!

MCL: maximum contaminant level as dictated by regulations.

MDL: See Method detection limit.

Medicinal oil: highly refined, colorless, tasteless, and odorless crude oil used as a medicine in the nature of an internal lubricant; sometimes called liquid paraffin.

Megawatt: (MW): a measure of electrical power equal to one million watts (1,000 kW).

Membrane technology: gas separation processes utilizing membranes that permit different components of a gas to diffuse through the membrane at significantly different rates.

MEK-(methyl ethyl ketone): a colorless liquid ($CH_3COCH_2CH_3$) used as a solvent; as a chemical intermediate; and in the manufacture of lacquers, celluloid, and varnish removers.

MEK deoiling: a wax-deoiling process in which the solvent is generally a mixture of methyl ethyl ketone and toluene.

MEK dewaxing: a continuous solvent dewaxing process in which the solvent is generally a mixture of methyl ethyl ketone and toluene.

MEOR: microbial enhanced oil recovery.

Methanol: see methyl alcohol.

Method detection limit: the smallest quantity or concentration of a substance that the instrument can measure.

Methyl t-butyl ether: an ether added to gasoline to improve its octane rating and to decrease gaseous emissions; see oxygenate.

Mercapsol process: a regenerative process for extracting mercaptan derivatives (RSH), utilizing aqueous sodium (or potassium) hydroxide containing mixed cresols as solubility promoters.

Mercaptans: organic compounds having the general formula R-SH.

Metagenesis: the alteration of organic matter during the formation of crude oil that may involve temperatures above 200°C (390°F); see also Catagenesis and Diagenesis.

Metal hydride (often represented simply as MH): a chemical that is synthesized by the reaction of a metal with hydrogen.

Methyl alcohol (methanol; wood alcohol): a colorless, volatile, inflammable, and poisonous alcohol (CH_3OH) traditionally formed by destructive distillation of wood or, more recently, as a result of synthetic distillation in chemical plants; a fuel typically derived from natural gas, but which can be produced from the fermentation of sugars in biomass.

Methyl ethyl ketone: see MEK.

Mica: a complex aluminum silicate mineral that is transparent, tough, flexible, and elastic.

Micellar fluid (surfactant slug): an aqueous mixture of surfactants, co-surfactants, salts, and hydrocarbons. The term micellar is derived from the word micelle, which is a submicroscopic aggregate of surfactant molecules and associated fluid.

Micelle: the structural entity by which asphaltene constituents are dispersed in crude oil.

Microcarbon residue: the carbon residue determined using a thermogravimetric method. See also Carbon residue.

Microcrystalline wax: wax extracted from certain crude oil residua and having a finer and less apparent crystalline structure than paraffin wax.

Microemulsion: a stable, finely dispersed mixture of oil, water, and chemicals (surfactants and alcohols).

Microemulsion or micellar/emulsion flooding: an augmented waterflooding technique in which a surfactant system is injected in order to enhance oil displacement toward producing wells.

Microfouling: deposition of solids in refinery equipment or pipes in which the solid is (1) particulate fouling, which is the accumulation of particles on a surface, (2) chemical reaction fouling, such as decomposition of organic matter on heating surfaces, (3) solidification fouling, which occurs when components of a flowing fluid with a high-melting point freeze onto a subcooled surface, (4) corrosion fouling, which is caused by corrosion, (5) biofouling, which can often ensure after biocorrosion which is due to the action of bacteria or algae, and (6) composite fouling, whereby fouling involves more than one foulant or fouling mechanism. See also: fouling, macrofouling.

Microorganisms: animals or plants of microscopic size, such as bacteria.

Microscopic displacement efficiency: the efficiency with which an oil displacement process removes the oil from individual pores in the rock.

Mid-boiling point: the temperature at which approximately 50% of a material has distilled under specific conditions.

Middle distillate: distillate boiling between the kerosene and lubricating oil fractions.

Middle-phase micro emulsion: a micro emulsion phase containing a high concentration of both oil and water that, when viewed in a test tube, resides in the middle with the oil phase above it and the water phase below it. Migration (primary) the movement of hydrocarbons (oil and natural gas) from mature, organic-rich source rocks to a point where the oil and gas can collect as droplets or as a continuous phase of liquid hydrocarbon.

Migration (secondary): the movement of the hydrocarbons as a single, continuous fluid phase through water-saturated rocks, fractures, or faults followed by accumulation of the oil and gas in sediments (traps, *q.v.) from which further migration is prevented.*

Mill residue: wood and bark residues produced in processing logs into lumber, plywood, and paper.

Mineral hydrocarbons: crude oil hydrocarbons, considered *mineral* because they come from the earth rather than from plants or animals.

Mineral oil: the older term for crude oil; the term was introduced in the nineteenth century as a means of differentiating crude oil (rock oil) from whale oil which, at the time, was the predominant illuminant for oil lamps.

Minerals: naturally occurring inorganic solids with well-defined crystalline structures.

Mineral seal oil: a distillate fraction boiling between kerosene and gas oil.

Mineral wax: yellow to dark brown, solid substances that occur naturally and are composed largely of paraffins; usually found associated with considerable mineral matter, as a filling in veins and fissures or as an interstitial material in porous rocks.

Minimum miscibility pressure (MMP): see miscibility.

Miscibility: an equilibrium condition, achieved after mixing two or more fluids, which is characterized by the absence of interfaces between the fluids: (1) *first-contact miscibility:* miscibility in the usual sense, whereby two fluids can be mixed in all proportions without any interfaces forming. Example: At room temperature and pressure, ethyl alcohol and water are first -contact miscible. (2) multiple-contact miscibility (dynamic miscibility): miscibility that is developed by repeated enrichment of one fluid phase with components from a second fluid phase with which it comes into contact. (3) *minimum miscibility* pressure: the minimum pressure above which two fluids become miscible at a given temperature, or can become miscible, by dynamic processes.

Miscible flooding: see EOR process.

Miscible fluid displacement (miscible displacement): an oil displacement process in which is an oil displacement process in which an alcohol, a refined hydrocarbon, a condensed crude oil gas, carbon dioxide, liquefied natural gas, or even exhaust gas is injected into an oil reservoir, at pressure levels such that the injected gas or fluid and reservoir oil are miscible; the process may include the concurrent, alternating, or subsequent injection of water.

Mitigation: identification, evaluation, and cessation of potential impacts of a process product or by-product.

Mixed-phase cracking: the thermal decomposition of higher-boiling hydrocarbons to gasoline components.

Mobility: a measure of the ease with which a fluid moves through reservoir rock; the ratio of rock permeability to apparent fluid viscosity.

Mobility buffer: the bank that protects a chemical slug from water invasion and dilution and assures mobility control.

Mobility control: ensuring that the mobility of the displacing fluid or bank is equal to or less than that of the displaced fluid or bank.

Mobility ratio: ratio of mobility of an injection fluid to mobility of fluid being displaced.

Modified alkaline flooding: the addition of a co-surfactant and/or polymer to the alkaline flooding process.

Modified/unmodified diesel engine: traditional diesel engines must be *modified* to heat the oil before it reaches the fuel injectors in order to handle straight vegetable oil. Modified, any diesel engine can run on veggie oil; without modification, the oil must first be converted to biodiesel.

Modified naphtha insolubles (MNI): an insoluble fraction obtained by adding naphtha to crude oil; usually the naphtha is modified by adding paraffin constituents; the fraction might be equated to asphaltenes *if* the naphtha is equivalent to n-heptane, but usually it is not

Moisture content: (MC): The weight of the water contained in wood, usually expressed as a percentage of weight, either oven-dry or as received.

Moisture content, dry basis: moisture content expressed as a percentage of the weight of oven-wood, i.e.: [(weight of wet sample – weight of dry sample) / weight of dry sample] × 100.

Moisture content, wet basis: moisture content expressed as a percentage of the weight of wood as-received, i.e.: [(weight of wet sample – weight of dry sample) / weight of wet sample] × 100.

Molecular sieve: a synthetic zeolite mineral having pores of uniform size; it is capable of separating molecules, on the basis of their size, structure, or both, by absorption or sieving.

Motor octane method: a test for determining the knock rating of fuels for use in spark-ignition engines; see also research octane method.

Moving-bed catalytic cracking: a cracking process in which the catalyst is continuously cycled between the reactor and the regenerator.

MSDS: material safety data sheet.

MTBE: methyl tertiary butyl ether is highly refined high octane light distillate used in the blending of gasoline.

Municipal solid waste: A waste type consisting of everyday items that are discarded by the public; the term 'garbage' can also refer specifically to food waste, as in a garbage disposal; the two are sometimes collected separately. See also: waste

NAAQS (National Ambient Air Quality Standards) standards that exist for the pollutants known as the criteria air pollutants: nitrogen oxides (NO_x), sulfur oxides (SO_x), lead, ozone, particulate matter, less than 10 microns in diameter, and carbon monoxide (CO).

Naft: A pre-Christian era (Greek) term for that evolved into the name 'naphtha'.

Napalm: a thickened gasoline used as an incendiary medium that adheres to the surface it strikes.

Naphtha: a generic term applied to refined, partly refined, or unrefined crude oil products and liquid products of natural gas, the majority of which distills below 240°C (464°F); the volatile fraction of crude oil which is used as a solvent or as a precursor to gasoline.

Naphthenes: cycloparaffins.

Native asphalt: see bitumen.

Natural asphalt: see bitumen.

Natural gas: the naturally occurring gaseous constituents that are found in many crude oil reservoirs; also there are also those reservoirs in which natural gas may be the sole occupant.

Natural gas liquids (NGL): the hydrocarbon liquids that condense during the processing of hydrocarbon gases that are produced from oil or gas reservoir; see also Natural gasoline.

Natural gasoline: a mixture of liquid hydrocarbons extracted from natural gas suitable for blending with refinery gasoline.

Natural gasoline plant: a plant for the extraction of fluid hydrocarbon, such as gasoline and liquefied petroleum gas, from natural gas.

NESHAP (National Emissions Standards for Hazardous Air Pollutants): emission standards for specific source categories that emit or have the potential to emit one or more hazardous air pollutants; the standards are modeled on the best practices and most effective emission reduction methodologies in use at the affected facilities.

Neutralization: a process for reducing the acidity or alkalinity of a waste stream by mixing acids and bases to produce a neutral solution; also known as pH adjustment.

Neutral oil: a distillate lubricating oil with viscosity usually not above 200 sec at 100°F.

Neutralization number: the weight, in milligrams, of potassium hydroxide needed to neutralize the acid in 1 g of oil; an indication of the acidity of an oil.

Nitrogen fixation: the transformation of atmospheric nitrogen into nitrogen compounds that can be used by growing plants.

Nitrogen oxides (NOx): the products of combustion that contribute to the formation of smog and ozone.

Non-forest land: land that has never supported forests and lands formerly forested where use of timber management is precluded by development for other uses; if intermingled in forest areas, unimproved roads and non-forest strips must be more than 120 feet wide, and clearings, etc., must be more than 1 acre in area to qualify as non-forest land.

Non-asphaltic road oil: any of the non-hardening crude oil distillates or residual oils used as dust layers; oil that has a sufficiently low viscosity to be applied without heating and, together with asphaltic road oils, are sometimes referred to as dust palliatives.

Non-attainment area: any area that does not meet the national primary or secondary ambient air quality standard established (by the Environmental Protection Agency) for designated pollutants, such as carbon monoxide and ozone.

Non-hazardous waste: a waste that is not dangerous but harms the environment; must be disposed of appropriately to comply with the rules; includes household waste, such as food and bathroom waste, and corporate waste, such as waste from factories and farms. Waste might be obnoxious, but it is not harmful. See waste, hazardous waste.

Non-industrial private: an ownership class of private lands where the owner does not operate wood processing plants.

Non-ionic surfactant: a surfactant molecule containing no ionic charge.

Non-Newtonian: a fluid that exhibits a change of viscosity with flow rate.

NOx: the oxides of nitrogen.

Nuclear magnetic resonance spectroscopy: an analytical procedure that permits the identification of complex molecules based on the magnetic properties of the atoms they contain.

No. 1 fuel oil: very similar to kerosene and is used in burners where vaporization before burning is usually required and a clean flame is specified.

No. 2 fuel oil: also called domestic heating oil; has properties similar to diesel fuel and heavy jet fuel; used in burners where complete vaporization is not required before burning.

No. 4 fuel oil: a light industrial heating oil and is used where preheating is not required for handling or burning; there are two grades of No. 4 fuel oil, differing in safety (flash point) and flow (viscosity) properties.

No. 5 fuel oil: a heavy industrial fuel oil which requires preheating before burning.

No. 6 fuel oil: a heavy fuel oil and is more commonly known as Bunker C oil when it is used to fuel ocean-going vessels; preheating is always required for burning this oil.

Observation wells: wells that are completed and equipped to measure reservoir conditions and/or sample reservoir fluids, rather than to inject produced reservoir fluids.

Octane barrel yield: a measure used to evaluate fluid catalytic cracking processes; defined as (RON + MON)/2 times the gasoline yield, where RON is the research octane number and MON is the motor octane number.

Octane number: a number indicating the anti-knock characteristics of gasoline.

Off-gas (offgas, off gas): a gas that is produced as a by-product of crude oil refining (as well as other industrial processes) or that is emitted by a manufactured object or material; refinery off-gases from fluid catalytic cracking units could be a source of ethylene ($CH_2 = CH_2$). See also: Refinery gas, Refinery off-gas.

Oil bank: see bank.

Oil breakthrough (time): the time at which the oil-water bank arrives at the producing well.

Oil from tar sand: synthetic crude oil.

Oil mining: application of a mining method to the recovery of bitumen.

Oil originally in place (OOIP): the quantity of crude oil existing in a reservoir before oil recovery operations begin.

Oils: that portion of the maltenes that is not adsorbed by a surface-active material such as clay or alumina.

Oil sand: see tar sand.

Oil sand bitumen: See Tar sand bitumen,Oil shale: a fine-grained impervious sedimentary rock which contains an organic material called kerogen.

Olefin: synonymous with *alkene.*

OOIP: see oil originally in place.

Open-loop biomass: biomass that can be used to produce energy and bio-products even though it was not grown specifically for this purpose; include agricultural livestock waste, residues from forest harvesting operations and crop harvesting.

Optimum salinity: the salinity at which a middle-phase microemulsion containing equal concentrations of oil and water results from the mixture of a micellar fluid (surfactant slug) with oil.

Organic sedimentary rocks: rocks containing organic material such as residues of plant and animal remains/decay.

Orifice-plate mixer (orifice mixer): a mixer in which two or more liquids are pumped through an orifice constriction to cause turbulence and consequent mixing action.

Overhead: the portion of the feedstock which is vaporized and removed during distillation.

Override: the gravity-induced flow of a lighter fluid in a reservoir above another heavier fluid.

Oxidation: a process which can be used for the treatment of a variety of inorganic and organic substances.

Oxidized asphalt: see air-blown asphalt.

Ozokerite (Ozocerite): a naturally occurring wax; when refined also known as ceresin.

Oxygenate: an oxygen-containing compound that is blended into gasoline to improve its octane number and to decrease gaseous emissions; a substance which, when added to gasoline, increases the amount of oxygen in that gasoline blend; includes fuel ethanol, methanol, and methyl tertiary butyl ether (MTBE).

Oxygenated gasoline: gasoline with added ethers or alcohols, formulated according to the Federal Clean Air Act to reduce carbon monoxide emissions during winter months.

Oxygen scavenger: a chemical which reacts with oxygen in injection water, used to prevent degradation of polymer.

Pale oil: a lubricating oil or a process oil refined until its color, by transmitted light, is straw to pale yellow.

Paraffinum liquidum: see liquid petrolatum.

Paraffin wax: the colorless, translucent, highly crystalline material obtained from the light lubricating fractions of paraffin crude oils (wax distillates).

Particle density: the density of solid particles.

Particulate: a small, discrete mass of solid or liquid matter that remains individually dispersed in gas or liquid emissions.

Particulate emissions: particles of a solid or liquid suspended in a gas, or the fine particles of carbonaceous soot and other organic molecules discharged into the air during combustion.

Particulate matter (particulates): particles in the atmosphere or on a gas stream that may be organic or inorganic and originate from a wide variety of sources and processes.

Particle size distribution: the particle size distribution (of a catalyst sample) expressed as a percent of the whole.

Partitioning: in chromatography, the physical act of a solute having different affinities for the stationary and mobile phases.

Partition ratios, K: the ratio of total analytical concentration of a solute in the stationary phase, CS, to its concentration in the mobile phase, CM.

Pattern: the areal pattern of injection and producing wells selected for a secondary or enhanced recovery project.

Pattern life: the length of time a flood pattern participates in oil recovery.

Pay zone thickness: the depth of a tar sand deposit from which bitumen (or a product) can be recovered.

Penex process: a continuous, non-regenerative process for isomerization of C_5 and/or C_6 fractions in the presence of hydrogen (from reforming) and a platinum catalyst.

Pentafining: a pentane isomerization process using a regenerable platinum catalyst on a silica-alumina support and requiring outside hydrogen.

Pepper sludge: the fine particles of sludge produced in acid treating which may remain in suspension.

Peri-condensed aromatic compounds: Compounds based on angular condensed aromatic hydrocarbon systems, such as phenanthrene, chrysene, picene, and other similar condenses aromatic hydrocarbon derivatives.

Permeability: the ease of flow of the water through the rock.

Petrol: a term commonly used in some countries for gasoline.

Petrolatum: a semisolid product, ranging from white to yellow in color, produced during refining of residual stocks; see petroleum jelly.

Petrolenes: the term applied to that part of the pentane-soluble or heptane-soluble material that is low boiling (<300°C, <570°F, 760 mm) and can be distilled without thermal decomposition (see also Maltenes).

Petroleum: a naturally occurring mixture of gaseous, liquid, and solid hydrocarbon constituents. See: conventional crude oil, crude oil.

Petroleum asphalt: see asphalt.

Petroleum ether: see ligroine.

Petroleum jelly: a translucent, yellowish to amber or white, hydrocarbon substance (melting point: 38 to 54°C, 100 to 130°F) having almost no odor or taste, derived from crude oil and used principally in medicine and pharmacy as a protective dressing and as a substitute for fats in ointments and cosmetics; also used in many types of polishes and in lubricating greases, rust preventives, and modeling clay; obtained by dewaxing heavy lubricating-oil stocks.

Petroleum refinery (crude oil refinery): see refinery.

Petroleum refining (crude oil refining): a complex sequence of events that result in the production of a variety of products.

Petroleum sulfonate (crude oil sulfonate): a surfactant used in chemical flooding prepared by sulfonating selected crude oil fractions.

Petroporphyrins: see porphyrins.

Phase: a separate fluid that co-exists with other fluids; gas, oil, water, and other stable fluids such as micro emulsions are all called phases in EOR research.

Phase behavior: the tendency of a fluid system to form phases as a result of changing temperature, pressure, or the bulk composition of the fluids or of individual fluid phases. Phase diagram – a graph of phase behavior. In chemical flooding a graph showing the relative volume of oil, brine, and sometimes one or more micro emulsion phases. In carbon dioxide flooding, conditions for formation of various liquid, vapor, and solid phases.

Phase properties: types of fluids, compositions, densities, viscosities, and relative amounts of oil, microemulsion, or solvent, and water formed when a micellar fluid (surfactant slug) or miscible solvent (e.g., carbon dioxide) is mixed with oil.

Phase separation: the formation of a separate phase that is usually the prelude to coke formation during a thermal process; the formation of a separate phase as a result of the instability/incompatibility of crude oil and crude oil products; the precipitation of asphaltene constituents when a paraffinic crude oil is blended with a viscous crude oil or when the blend is heated in pipes leading to a reactor leading to the formation of degradation products and other undesirable changes in the original properties of crude oil products. When such phenomena occur, which may not be immediately at the time of the blend but can occur after an induction period, it is often referred to as instability of the blends. See also: Incompatibility.

pH adjustment: neutralization.

Phosphoric acid polymerization: a process using a phosphoric acid catalyst to convert propene, butene, or both, to gasoline or petrochemical polymers.

Photoionization: a gas chromatographic detection system that utilizes an *detector (PID)* ultraviolet lamp as an ionization source for analyte detection. It is usually used as a selective detector by changing the photon energy of the ionization source.

Photosynthesis: the process by which chlorophyll-containing cells in green plants concert incident light to chemical energy, capturing carbon dioxide in the form of carbohydrates.

PINA analysis: a method of analysis for paraffins, *iso*-paraffins, naphthenes, and aromatics.

PIONA analysis: a method of analysis for paraffins, *iso*-paraffins, olefins, naphthenes, and aromatics.

Pipe still: a still in which heat is applied to the oil while being pumped through a coil or pipe arranged in a suitable firebox.

Pipestill gas: the most volatile fraction that contains most of the gases that are generally dissolved in the crude. Also known as pipestill light ends.

Pipestill light ends: see *pipestill gas.*

Pitch: the nonvolatile, brown to black, semi-solid to solid viscous product from the destructive distillation of many bituminous or other organic materials, especially coal.

Platforming: a reforming process using a platinum-containing catalyst on an alumina base.

PNA: a polynuclear aromatic compound.

NA: a polynuclear aromatic compound.

PNA analysis: a method of analysis for paraffins, naphthenes, and aromatics.

Polar aromatics: resins; the constituents of crude oil that are predominantly aromatic in character and contain polar (nitrogen, oxygen, and sulfur) functions in their molecular structure(s).

Pollutant: a chemical (or chemicals) introduced into the land water and air systems of that is (are) not indigenous to these systems; also an indigenous chemical (or chemicals) introduced into the land water and air systems in amounts greater than the natural abundance.

Pollution: the introduction into the land water and air systems of a chemical or chemicals that are not indigenous to these systems or the introduction into the land water and air systems of indigenous chemicals in greater-than-natural amounts.

Polyacrylamide: very high molecular weight material used in polymer flooding.

Polycyclic aromatic hydrocarbons (PAHs): polycyclic aromatic hydrocarbons are a suite of compounds comprised of two or more condensed aromatic rings. They are found in many crude oil mixtures, and they are predominantly introduced to the environment through natural and anthropogenic combustion processes.

Polyforming: a process charging both C_3 and C_4 gases with naphtha or gas oil under thermal conditions to produce gasoline.

Polymer: in EOR, any very high molecular weight material that is added to water to increase viscosity for polymer flooding.

Polymer augmented waterflooding: waterflooding in which organic polymers are injected with the water to improve areal and vertical sweep efficiency.

Polymer gasoline: the product of polymerization of gaseous hydrocarbons to hydrocarbons boiling in the gasoline range.

Polymerization: the combination of two olefin molecules to form a higher molecular weight paraffin.

Polymer stability: the ability of a polymer to resist degradation and maintain its original properties.

Polynuclear aromatic compound: an aromatic compound having two or more fused benzene rings, e.g., naphthalene, phenanthrene.

Polysulfide treating: a chemical treatment used to remove elemental sulfur from refinery liquids by contacting them with a non-regenerable solution of sodium polysulfide.

PONA analysis: a method of analysis for paraffins (P), olefins (O), naphthenes (N), and aromatics (A).

Pore diameter: the average pore size of a solid material, e.g., catalyst.

Pore space: a small hole in reservoir rock that contains fluid or fluids; a four inch cube of reservoir rock may contain millions of inter-connected pore spaces.

Pore volume: total volume of all pores and fractures in a reservoir or part of a reservoir; also applied to catalyst samples.

Porosity: the percentage of rock volume available to contain water or other fluid.

Porphyrins: organometallic constituents of crude oil that contain vanadium or nickel; the degradation products of chlorophyll that became included in the protopetroleum.

Positive bias: a result that is incorrect and too high.

Possible reserves: reserves where there is an even greater degree of uncertainty but about which there is some information.

Potential reserves: reserves based upon geological information about the types of sediments where such resources are likely to occur and they are considered to represent an educated guess.

Pour point: the lowest temperature at which oil will pour or flow when it is chilled without disturbance under definite conditions.

Powerforming: a fixed-bed naphtha-reforming process using a regenerable platinum catalyst.

Power-law exponent: an exponent used to model the degree of viscosity change of some non-Newtonian liquids.

Precipitation number: the number of milliliters of precipitate formed when 10 ml of lubricating oil is mixed with 90 ml of crude oil naphtha of a definite quality and centrifuged under definitely prescribed conditions.

Preflush: a conditioning slug injected into a reservoir as the first step of an EOR process.

Pressure cores: cores cut into a special coring barrel that maintains reservoir pressure when brought to the surface; this prevents the loss of reservoir fluids that usually accompanies a drop in pressure from reservoir to atmospheric conditions.

Pressure gradient: rate of change of pressure with distance.

Pressure maintenance: augmenting the pressure (and energy) in a reservoir by injecting gas and/or water through one or more wells.

Pressure pulse test: a technique for determining reservoir characteristics by injecting a sharp pulse of pressure in one well and detecting it surrounding wells.

Pressure transient testing: measuring the effect of changes in pressure at one well on other we in a field.

Primary oil recovery: oil recovery utilizing only naturally occurring forces.

Primary wood-using mill: A mill that converts round wood products into other wood products; common examples are sawmills that convert saw logs into lumber and pulp mills that convert pulpwood round wood into wood pulp.

Primary structure: the chemical sequence of atoms in a molecule.

Primary tracer: a chemical that, when inject into a test well, reacts with reservoir fluids form a detectable chemical compound.

Probable reserves: mineral reserves mineral that are nearly certain but about which a slight doubt exists.

Process heat: heat used in an industrial process rather than for space heating or other housekeeping purposes.

Producer gas: fuel gas that has a high proportion of carbon monoxide (CO) and hydrogen (H_2) that is produced by burning a solid fuel with insufficient air or by passing a mixture of air and steam through a burning bed of solid fuel.

Producibility: the rate at which oil or gas can produced from a reservoir through a wellbore.

Producing well: a well in an oil field used for removing fluids from a reservoir.

Propane asphalt: see solvent asphalt.

Propane deasphalting: solvent deasphalting using propane as the solvent.

Propane decarbonizing: a solvent extraction process used to recover catalytic cracking feed from heavy fuel residues

Propane dewaxing: a process for dewaxing lubricating oils in which propane serves as solvent.

Propane fractionation: a continuous extraction process employing liquid propane as the solvent; a variant of propane deasphalting.

Protopetroleum: a generic term used to indicate the initial product formed changes have occurred to the precursors of crude oil.

Proved reserves: mineral reserves that have been positively identified as recoverable with current technology.

PSD: prevention of significant deterioration.

PTE: potential to emit; the maximum capacity of a source to emit a pollutant, given its physical or operation design, and considering certain controls and limitations.

Pulpwood: Round wood, whole-tree chips, or wood residues that are used for the production of wood pulp.

Pulse-echo ultrasonic borehole televiewer: well-logging system wherein a pulsed, narrow acoustic beam scans the well as the tool is pulled up the borehole; the amplitude of the reflecting beam is displayed on a cathode-ray tube resulting in a pictorial representation of wellbore.

Purge and trap: a chromatographic sample introduction technique in volatile components that are purged from a liquid medium by bubbling gas through it. The components are then concentrated by 'trapping' them on a short intermediate column, which is subsequently heated to drive the components on to the analytical column for separation.

Purge gas: typically helium or nitrogen, used to remove analytes from the sample matrix in purge/trap extractions.

Pyrobitumen: see asphaltoid.

Pyrolysis: the thermal decomposition of biomass at high temperatures (greater than 400°F, or 200°C) in the absence of air; the end product of pyrolysis is a mixture of solids (char), liquids (oxygenated oils), and gases (methane, carbon monoxide, and carbon dioxide) with proportions determined by operating temperature, pressure, oxygen content, and other conditions; exposure of a feedstock to high temperatures in an oxygen-poor environment.

Pyrophoric: substances that catch fire spontaneously in air without an ignition source.

Quad: One quadrillion Btu (10^15 Btu) = 1.055 exajoules (EJ), or approximately 172 million barrels of oil equivalent.

Quadrillion: 1×10^{15}

Quench: the sudden cooling of hot material discharging from a thermal reactor.

RACT: reasonably available control technology standards that are implemented in areas of non-attainment to reduce emissions of volatile organic compounds and nitrogen oxides.

Raffinate: that portion of the oil which remains undissolved in a solvent refining process.

Ramsbottom carbon residue: see carbon residue.

Raw crude oil: crude oil direct from the wellbore, before it is treated in a gas separation plant; typically contains nonhydrocarbon contaminants.

Raw materials: minerals extracted from the earth prior to any refining or treating.

Recovery boiler: a pulp mill boiler in which lignin and spent cooking liquor (black liquor) is burned to generate steam.

Rectification zone: the zone in a distillation tower in which the more volatile component(s) is (are) removed through contacting the rising vapor with the down-flowing liquid; at the stripping section, the down-flowing liquid is stripped of the more volatile component by the rising vapor; the distillation tower can be equipped with trays or packings or both; also known as the rectification section or the rectifying section.

Recycle ratio: the volume of recycle stock per volume of fresh feed; often expressed as the volume of recycle divided by the total charge.

Recycle stock: the portion of a feedstock which has passed through a refining process and is recirculated through the process.

Recycling: the use or reuse of chemical waste as an effective substitute for a commercial products or as an ingredient or feedstock in an industrial process.

Red liquor: black liquor.

Reduced crude: a residual product remaining after the removal, by distillation or other means, of an appreciable quantity of the more volatile components of crude oil.

Refinery: a series of integrated unit processes by which crude oil can be converted to a slate of useful (salable) products.

Refinery gas: a gas (or a gaseous mixture) produced as a result of crude oil refining operations; the refinery gas streams are composed primarily of paraffin hydrocarbons (alkanes) and olefin derivatives, both of which have carbon derivatives in the C1 to C4 range; some streams also contain varying amounts of other chemicals including ammonia (NH_3), hydrogen (H_2), nitrogen (N_2), hydrogen sulfide (H_2S), mercaptan derivatives (e.g. CH3SH), carbon monoxide (CO), carbon dioxide (CO2), ethylene (CH2 = CH2), propylene (CH3CH = CH2), 1,3-butadiene (CH2 = CHCH = CH2), and/or benzene (C6H6). See also: off-gas.

Refinery off-gas: a gas that is produced as a by-product of crude oil refining (as well as other industrial processes) or that is emitted by a manufactured object or material; refinery off-gases from fluid catalytic cracking units could be a source of ethylene ($CH_2 = CH_2$). See also: off-gas, refinery gas.

Refining: the processes by which crude oil is distilled and/or converted by application of physical and chemical processes to form a variety of products are generated.

Reformate: the liquid product of a reforming process.

Reformed gasoline: gasoline made by a reforming process.

Reforming: The conversion of hydrocarbons with low octane numbers into hydrocarbons having higher octane numbers, such as the conversion of a n-paraffin into an iso-paraffin. See: catalytic reforming.

Reformulated gasoline (RFG): gasoline designed to mitigate smog production and to improve air quality by limiting the emission levels of certain chemical compounds such as benzene and other aromatic derivatives; often contains oxygenates.

Refractory lining: a lining, usually ceramic, capable of resisting and maintaining high temperatures.

Refuse-derived fuel (RDF): fuel prepared from municipal solid waste; non-combustible materials such as rocks, glass, and metals are removed, and the remaining combustible portion of the solid waste is chopped or shredded; the combustible portion of municipal solid waste after removal of glass and metals.

Reid vapor pressure: a measure of the volatility of liquid fuels, especially gasoline.

Regeneration: the or reactivation of a catalyst by burning off the coke deposits.

Regenerator: a reactor for catalyst reactivation.

Relative permeability: the permeability of rock to gas, oil, or water, when any two or more are present, expressed as a fraction of the sir phase permeability of the rock.

Renewable energy sources: solar, wind, and other non-fossil fuel energy sources.

Rerunning: the distillation of an oil which has already been distilled.

Research octane method (ROM): a test for determining the knock rating, in terms octane numbers, of fuels for use in spark-ignition engines; see also Motor Octane Method.

Reserves: well-identified resources that can be profitably extracted and utilized with existing technology.

Reservoir: a rock formation below the earth's surface containing crude oil or natural gas; a domain where a pollutant may reside for an indeterminate time.

Reservoir simulation: analysis and prediction of reservoir performance with a computer model.

Residual asphalt: see straight-run asphalt.

Residual fuel oil: obtained by blending the residual product(s) from various refining processes with suitable diluent(s) (usually middle distillates) to obtain the required fuel oil grades.

Residual oil: the crude oil that remains after in situ after oil recovery operations. See Residuum;

Residual resistance factor: the reduction in permeability of rock to water caused by the adsorption of polymer.

Residues: bark and woody materials that are generated in primary wood-using mills when round wood products are converted to other products.

Residuum (resid; pl: residua): the residue obtained from crude oil after nondestructive distillation has removed all the volatile materials from crude oil, e.g., an atmospheric (345°C, 650°F+) residuum.

Resins: that portion of the maltenes that is adsorbed by a surface-active material such as clay or alumina; the fraction of deasphaltened oil that is insoluble in liquid propane but soluble in n-heptane.

Resistance factor: a measure of resistance to flow of a polymer solution relative to the resistance to flow of water.

Resource: the total amount of a commodity (usually a mineral but can include non-minerals such as water and crude oil) that has been estimated to be ultimately available.

Retention: the loss of chemical components due to adsorption onto the rock's surface, precipitation, or to trapping within the reservoir.

Retention time: the time it takes for an eluate to move through a chromatographic system and reach the detector. Retention times are reproducible and can therefore be compared to a standard for analyte identification.

Rexforming: a process combining platforming with aromatics extraction, wherein low octane raffinate is recycled to the platformer.

Rich oil: absorption oil containing dissolved natural gasoline fractions.

Riser: the part of the bubble-plate assembly which channels the vapor and causes it to flow downward to escape through the liquid; also the vertical pipe where fluid catalytic cracking reactions occur. Resid (also referred to as resids, residua): the non-volatile residue remaining after atmospheric distillation (hence: atmospheric resid) or vacuum distillation (hence: vacuum resid); the properties are on the order of: API gravity: 5–15°; sulfur: 2.0–5.0% w/w; metals: 100–1,000 ppm; resids do not occur naturally; manufactured products produced by refinery distillation and are typically immobile under ambient conditions.

Rock asphalt: bitumen which occurs in formations that have a limiting ratio of bitumen-to-rock matrix.

Rock matrix: the granular structure of a rock or porous medium.

Rotation: period of years between establishment of a stand of timber and the time when it is considered ready for final harvest and regeneration.

Round wood products: logs and other round timber generated from harvesting trees for industrial or consumer use.

Rubbish: see waste.

Run-of-the-river reservoirs: reservoirs with a large rate of flow-through compared to their volume.

Salinity: the concentration of salt in water.

Sand: a course granular mineral mainly comprising quartz grains that is derived from the chemical and physical weathering of rocks rich in quartz, notably sandstone and granite.

Sand face: the cylindrical wall of the wellbore through which the fluids must flow to or from the reservoir.

Sandstone: a sedimentary rock formed by compaction and cementation of sand grains; can be classified according to the mineral composition of the sand and cement.

SARA analysis: a method of fractionation by which crude oil is separated into saturates, aromatics, resins, and asphaltene fractions.

SARA separation: see SARA analysis.

Saturated steam: steam at boiling temperature for a given pressure.

Saturates: paraffins and cycloparaffins (naphthenes).

Saturation: the ratio of the volume of a single fluid in the pores to pore volume, expressed as a percent and applied to water, oil, or gas separately; the sum of the saturations of each fluid in a pore volume is 100%.

Saybolt Furol viscosity: the time, in seconds (Saybolt Furol Seconds, SFS), for 60 ml of fluid to flow through a capillary tube in a Saybolt Furol viscometer at specified temperatures between 70 and 210°F; the method is appropriate for high-viscosity oils such as transmission, gear, and heavy fuel oils.

Saybolt Universal viscosity: the time, in seconds (Saybolt Universal Seconds, SUS), for 60 ml of fluid to flow through a capillary tube in a Saybolt Universal viscometer at a given temperature.

Scale wax: the paraffin derived by removing the greater part of the oil from slack wax by sweating or solvent deoiling.

Screen factor: a simple measure of the viscoelastic properties of polymer solutions.

Screening guide: a list of reservoir rock and fluid properties critical to an EOR process.

Scrubber: a device that uses water and chemicals to clean air pollutants from combustion exhaust.

Scrubbing: purifying a gas by washing with water or chemical; less frequently, the removal of entrained materials.

Secondary oil recovery: application of energy (e.g., water flooding) to recovery of crude oil from a reservoir after the yield of crude oil from primary recovery diminishes.

Secondary pollutants: a pollutant (chemical species) produced by interaction of a primary pollutant with another chemical or by dissociation of a primary pollutant or by other effects within a particular ecosystem.

Secondary recovery: oil recovery resulting from injection of water, or an immiscible gas at moderate pressure, into a crude oil reservoir after primary depletion.

Secondary structure: the ordering of the atoms of a molecule in space relative to each other.

Secondary tracer: the product of the chemical reaction between reservoir fluids and an injected primary tracer.

Secondary wood processing mills: a mill that uses primary wood products in the manufacture of finished wood products, such as cabinets, moldings, and furniture.

Sediment: an insoluble solid formed as a result of the storage instability and/or the thermal instability of crude oil and crude oil products.

Sedimentary: formed by or from deposits of sediments, especially from sand grains or silts transported from their source and deposited in water, as sandstone and shale; or from calcareous remains of organisms, as limestone.

Sedimentary strata: typically consist of mixtures of clay, silt, sand, organic matter, and various minerals; formed by or from deposits of sediments, especially from sand grains or silts transported from their source and deposited in water, such as sandstone and shale; or from calcareous remains of organisms, such as limestone.

Selective solvent: a solvent which, at certain temperatures and ratios, will preferentially dissolve more of one component of a mixture than of another and thereby permit partial separation.

Separation process: an upgrading process in which the constituents of crude oil are separated, usually without thermal decomposition, e.g., distillation and deasphalting.

Separator-Nobel dewaxing: a solvent (trichloroethylene) dewaxing process.

Separatory funnel: glassware shaped like a funnel with a stoppered rounded top and a valve at the tapered bottom, used for liquid/liquid separations.

Sepiolite (also known by the German name meerschaum): a soft white clay mineral, often used to make tobacco pipes; chemically the mineral is a complex magnesium silicate, a typical chemical formula for which is $Mg_4Si_6O_{15}(OH)_2.6H2O$, it can be present in fibrous, fine-particulate, and solid forms.

Shear: mechanical deformation or distortion, or partial destruction of a polymer molecule as it flows at a high rate.

Shear rate: a measure of the rate of deformation of a liquid under mechanical stress.

Shear-thinning: the characteristic of a fluid whose viscosity decreases as the shear rate Increases.

Shell fluid catalytic cracking: a two-stage fluid catalytic cracking process in which the catalyst is regenerated.

Shell still: a still formerly used in which the oil was charged into a closed, cylindrical shell and the heat required for distillation was applied to the outside of the bottom from a firebox.

Sidestream (side-stream): a liquid stream taken from any one of the intermediate plates of a bubble tower.

Sidestream stripper: a device used to perform further distillation on a liquid stream from any one of the plates of a bubble tower, usually by the use of steam.

Single well tracer: a technique for determining residual oil saturation by injecting an ester, allowing it to hydrolyze; following dissolution of some of the reaction product in residual oil the injected solutions produced back and analyzed.

Slack wax: the soft, oily crude wax obtained from the pressing of paraffin distillate or wax distillate.

Slime: a name used for crude oil in ancient texts.

Slim tube testing: laboratory procedure for the determination of minimum miscibility pressure using long, small-diameter, sand-packed, oil- saturated, stainless steel tube.

Sludge: a semi-solid to solid product which results from the storage instability and/or the thermal instability of crude oil and crude oil products.

Slug: a quantity of fluid injected into a reservoir during enhanced oil recovery.

Slurry hydroconversion process: a process in which the feedstock is contacted with hydrogen under pressure in the presence of a catalytic coke-inhibiting additive; in the process, the feedstock is premixed with the nano-sized catalyst particles, and the mixture is then introduced into the reactor and the hydrogen gas is injected from the bottom of the reactor. As for all catalysts, the activity of slurry catalyst is influenced by (1) the type and number of the active sites, (2) the dispersion of the catalyst in the feedstock, (3) the type of metal used in the catalyst, and (4) the concentration of the catalyst.

Slurry phase reactors: tanks into which wastes, nutrients, and microorganisms are placed.

Smoke point: a measure of the burning cleanliness of jet fuel and kerosine.

Sodium hydroxide treatment: see Caustic wash.

Sodium plumbite: a solution prepared front a mixture of sodium hydroxide, lead oxide, and distilled water; used in making the doctor test for light oils such as gasoline and kerosine.

Solubility parameter: a measure of the solvent power and polarity of a solvent.

Solutizer-steam regenerative process: a chemical treating process for extracting mercaptan derivatives (RSH) from gasoline or naphtha, using solutizers (potassium iso-butyrate, potassium alkyl phenolate) in strong potassium hydroxide solution.

Solvent: a liquid in which certain kinds of molecules dissolve. While they typically are liquids with low boiling points, they may include high-boiling liquids, supercritical fluids, or gases.

Solvent asphalt: the asphalt produced by solvent extraction of residua or by light hydrocarbon (propane) treatment of a residuum or an asphaltic crude oil.

Solvent deasphalting: a process for removing asphaltic and resinous materials from reduced crude oils, lubricating-oil stocks, gas oils, or middle distillates through the extraction or precipitant action of low-molecular-weight hydrocarbon solvents; see also Propane deasphalting.

Solvent decarbonizing: see propane decarbonizing.

Solvent deresining: see solvent deasphalting.

Solvent dewaxing: a process for removing wax from oils by means of solvents usually by chilling a mixture of solvent and waxy oil, filtration or by centrifuging the wax which precipitates, and solvent recovery.

Solvent extraction: a process for separating liquids by mixing the stream with a solvent that is immiscible with part of the waste but that will extract certain components of the waste stream.

Solvent gas: an injected gaseous fluid that becomes miscible with oil under. reservoir conditions and improves oil displacement.

Sonic log: a well log based on the time required for sound to travel through rock, useful in determining porosity.

Solvent naphtha: a refined naphtha of restricted boiling range used as a solvent; also called crude oil naphtha; crude oil spirits.

Solvent refining: see solvent extraction.
Sonication: a physical technique employing ultrasound to intensely vibrate a sample media in extracting solvent and to maximize solvent/analyte interactions.
Sonic log: a well log based on the time required for sound to travel through rock, useful in determining porosity.
Sour crude oil: crude oil containing an abnormally large amount of sulfur compounds; see also sweet crude oil.
SOx: the oxides of sulfur.
Soxhlet extraction: an extraction technique for solids in which the sample is repeatedly contacted with solvent over several hours, increasing extraction efficiency.
Spalling: the term used to describe the breaking of a solid (such as coke) into smaller pieces.
Spontaneous ignition: ignition of a fuel, such as coal, under normal atmospheric conditions; usually induced by climatic conditions.
Specific gravity: the mass (or weight) of a unit volume of any substance at a specified temperature compared to the mass of an equal volume of pure water at a standard temperature; see also density.
Spent catalyst: catalyst that has lost much of its activity due to the deposition of coke and metals.
Stabilization: the removal of volatile constituents from a higher boiling fraction or product (q.v. stripping); the production of a product which, to all intents and purposes, does not undergo any further reaction when exposed to the air.
Stabilizer: a fractionating tower for removing light hydrocarbons from an oil to reduce vapor pressure particularly applied to gasoline.
Stand (of trees): a tree community that possesses sufficient uniformity in composition, constitution, age, spatial arrangement, or condition to be distinguishable from adjacent communities.
Standpipe: the pipe by which catalyst is conveyed between the reactor and the regenerator.
Stationary phase: in chromatography, the porous solid or liquid phase through which an introduced sample passes. The different affinities the stationary phase has for a sample allow the components in the sample to be separated, or resolved.
Steam cracking: a conversion process in which the feedstock is treated with superheated steam.
Steam distillation: distillation in which vaporization of the volatile constituents is effected at a lower temperature by introduction of steam (open steam) directly into the charge.
Steam drive injection (steam injection): EOR process in which steam is continuously injected into one set of wells (injection wells) or other injection source to effect oil displacement toward and production from a second set of wells (production wells); steam stimulation of production wells is *direct steam stimulation* whereas steam drive by steam injection to increase production from other wells is *indirect steam stimulation*.
Steam stimulation: injection of steam into a well and the subsequent production of oil from the same well.
Steam methane reforming (steam-methane reforming, SMR): a process in which methane from natural gas is heated, with steam, usually with a catalyst, to produce a mixture of carbon monoxide and hydrogen; natural gas is the most common feedstock in steam reformers; naphtha and refinery off-gas may also be used. The furnace is where the process of liberating hydrogen from natural gas and steam begins. Thus:

CH4 + H2O CO + 3H2

See also: steam naphtha, Steam reforming.
Steam-naphtha reforming: a catalytic process for the production of hydrogen; the reforming reactions include (1) dehydrogenation, (2) dehydrocyclization, (3) isomerization and (4) hydrocracking which are performed on both the metallic and acidic parts of the catalyst. See also: Steam methane reforming, steam reforming.
Steam reforming: a process for the generation of hydrogen-rich synthesis gas from light carbohydrates. The feed materials natural gas, liquid gas or naphtha are endothermically converted

with water steam into synthesis gas in catalytic tube reactors. See also: See also: Steam methane reforming, Steam naphtha reforming.

Steam turbine: a device for converting energy of high-pressure steam (produced in a boiler) into mechanical power which can then be used to generate electricity.

Stiles method: a simple approximate method for calculating oil recovery by waterflood that assumes separate layers (stratified reservoirs) for the permeability distribution.

Storage stability (or storage instability): the ability (inability) of a liquid to remain in storage over extended periods of time without appreciable deterioration as measured by gum formation and the depositions of insoluble material (sediment).

Straight-run asphalt: the asphalt produced by the distillation of asphaltic crude oil.

Straight-run products: obtained from a distillation unit and used without further treatment.

Straight vegetable oil (SVO): any vegetable oil that has not been optimized through the process of transesterification.

Strata: layers including the solid iron-rich inner core, molten outer core, mantle, and crust of the earth.

Straw oil: pale paraffin oil of straw color used for many process applications.

Stripper well: a well that produces (strips from the reservoir) oil or gas.

Stripping: a means of separating volatile components from less volatile ones in a liquid mixture by the partitioning of the more volatile materials to a gas phase of air or steam (q.v. stabilization).

Sulfite liquor: the spent cooking liquor from sulfite pulping; usually called 'brown liquor', but the terms 'red liquor', and 'thick liquor' are also used (compared to black liquor in the kraft process); pulp washers, using countercurrent flow, remove the spent cooking chemicals and degraded lignin and hemicellulose.

Sulfonic acids: acids obtained from crude oil or a crude oil product with strong sulfuric acid.

Sulfuric acid alkylation: an alkylation process in which olefins (C_3, C_4, and C_5) combine with *iso*-butane in the presence of a catalyst (sulfuric acid) to form branched chain hydrocarbons used especially in gasoline blending stock.

Supercritical fluid: an extraction method where the extraction fluid is present at a pressure and temperature above its critical point.

Superheated steam: steam which is hotter than boiling temperature for a given pressure.

Surface active material: a chemical compound, molecule, or aggregate of molecules with physical properties that cause it to adsorb at the interface between *two* immiscible liquids, resulting in a reduction of interfacial tension or the formation of a microemulsion.

Surfactant: a type of chemical, characterized as one that reduces interfacial resistance to mixing between oil and water or changes the degree to which water wets reservoir rock.

Suspensoid catalytic cracking: a non-regenerative cracking process in which cracking stock is mixed with slurry of catalyst (usually clay) and cycle oil and passed through the coils of a heater.

Sustainable: an ecosystem condition in which biodiversity, renewability, and resource productivity are maintained over time.

SW-846: an EPA multi-volume publication entitled *Test Methods for Evaluating Solid Waste, Physical/Chemical Methods*; the official compendium of analytical and sampling methods that have been evaluated and approved for use in complying with the RCRA regulations and that functions primarily as a guidance document setting forth acceptable, although not required, methods for the regulated and regulatory communities to use in responding to RCRA-related sampling and analysis requirements. SW-846 changes over time as new information and data are developed.

Sweated wax: a crude oil-based wax that has been freed from oil by having been passed through a sweater.

Sweating: the separation of paraffin oil and low-melting wax from paraffin wax.

Sweep efficiency: the ratio of the pore volume of reservoir rock contacted by injected fluids to the total pore volume of reservoir rock in the project area. (*See also* areal sweep efficiency *and* vertical sweep efficiency.)

Sweet crude oil: crude oil containing little sulfur; see also sour crude oil.

Sweetening: the process by which crude oil products are improved in odor and color by oxidizing or removing the sulfur-containing and unsaturated compounds.

Swelling: increase in the volume of crude oil caused by absorption of EOR fluids, especially carbon dioxide. Also increase in volume of clays when exposed to brine.

Swept zone: the volume of rock that is effectively swept by injected fluids.

Synthesis gas (syngas): a gas produced by the gasification of a solid or liquid fuel that consists primarily of carbon monoxide and hydrogen.

Synthetic crude oil (syncrude): a hydrocarbon product produced by the conversion of coal, oil shale, or tar sand bitumen that resembles conventional crude oil; can be refined in a crude oil refinery.

Synthetic ethanol: ethanol produced from ethylene, a crude oil by-product.

Tar: the volatile, brown to black, oily, viscous product from the destructive distillation of many bituminous or other organic materials, especially coal; a name used arbitrarily for crude oil in ancient texts.

Target analyte: target analytes are compounds that are required analytes in U. S. EPA analytical methods. BTEX and PAHs are examples of crude oil-related compounds that are target analytes in U. S. EPA Methods.

Tar sand (also known as bituminous sand): a formation in which the organic material (bitumen) exists as a filling in veins and fissures in fractured rocks or impregnating relatively shallow sand, sandstone, and limestone strata; a sandstone reservoir. The extremely viscous, black hydrocarbonaceous, material cannot be retrieved through a well by conventional or enhanced oil recovery techniques. The term 'tar sand', also refers to the rock types that contain an extremely viscous hydrocarbon which is not recoverable in its natural state by conventional oil well production methods including currently used enhanced recovery techniques; see Bituminous sand.

Tar sand bitumen: the viscous material that occurs in a tar sand deposit – in Canada the tar sand deposits are referred to as oil sand deposits'; Bitumen properties are typically API gravity: <10°; sulfur: 2.0–5.0% w/w; metals: 400–1,200 ppm; the bitumen is immobile in the deposit; deposit lower than the pour point of the bitumen and extreme heating methods or innovative methods are required for recovery of the bitumen.

Tertiary structure: the three-dimensional structure of a molecule.

Tetraethyl lead (TEL): an organic compound of lead, $Pb(CH_3)_4$, which, when added in small amounts, increases the antiknock quality of gasoline.

Thermal coke: the carbonaceous residue formed as a result of a non-catalytic thermal process; the Conradson carbon residue; the Ramsbottom carbon residue.

Thermal cracking: a process which decomposes, rearranges, or combines hydrocarbon molecules by the application of heat, without the aid of catalysts.

Thermal polymerization: a thermal process to convert light hydrocarbon gases into liquid fuels.

Thermal process: any refining process which utilizes heat, without the aid of a catalyst.

Thermal recovery: see EOR process.

Thermal reforming: a process using heat (but no catalyst) to effect molecular rearrangement of low-octane naphtha into gasoline of higher antiknock quality.

Thermal stability (thermal instability): the ability (inability) of a liquid to withstand relatively high temperatures for short periods of time without the formation of carbonaceous deposits (sediment or coke).

Thermochemical conversion: Use of heat to chemically change substances from one state to another, e.g., to make useful energy products.

Thermofor catalytic cracking: a continuous, moving-bed catalytic cracking process.

Thermofor catalytic reforming: a reforming process in which the synthetic, bead-type catalyst of coprecipitated chromia (Cr_2O_3) and alumina (Al_2O_3) flows down through the reactor concurrent with the feedstock.

Thermofor continuous percolation: a continuous clay treating process to stabilize and decolorize lubricants or waxes.

Thick liquor: See black liquor, brown liquor, red liquor, sulfite liquor.

Thief zone: any geologic stratum not intended to receive injected fluids in which significant amounts of injected fluids are lost; fluids may reach the thief zone due to an improper completion or a faulty cement job.

Thin layer chromatography (TLC): a chromatographic technique employing a porous medium of glass coated with a stationary phase. An extract is spotted near the bottom of the medium and placed in a chamber with solvent (mobile phase). The solvent moves up the medium and separates the components of the extract, based on affinities for the medium and solvent.

Throttling device: the generic name of any device or process that dissipates pressure energy by irreversibly converting it into thermal energy. See also: throttling valve.

Throttling valve: a type of valve that can be used to start, stop and regulate the flow of fluid through a rotodynamic pump. When the flow of a pump is regulated using a throttling valve, the system curve is changed. The operating point moves to the left on the pump curve when the flow is decreased. See also: throttling device valve.

Timberland: forest land that is producing or is capable of producing crops of industrial wood, and that is not withdrawn from timber utilization by statute or administrative regulation.

Time-lapse logging: the repeated use of calibrated well logs to quantitatively observe changes in measurable reservoir properties over time.

Tipping fee: a fee for disposal of waste.

Ton (short ton): 2,000 pounds.

Tonne (Imperial ton, long ton, shipping ton): 2,240 pounds; equivalent to 1,000 kilograms or in crude oil terms approximately 7.5 barrels of oil.

Topped crude: crude oil that has had volatile constituents removed up to a certain temperature, e.g., 250oC+ (480oF+) topped crude; not always the same as a residuum.

Topping: the distillation of crude oil to remove light fractions only

Topping and back pressure turbines: Turbines which operate at exhaust pressure considerably higher than atmospheric (non-condensing turbines); often multistage with relatively high efficiency.

Topping cycle: a cogeneration system in which electric power is produced first. The reject heat from power production is then used to produce useful process heat.

Torrefaction: a thermochemical process that is used to decrease the amount of water and volatile content of biomass which gives the biomass a higher energy density, hydrophobic behavior, elimination of biological activity, easier grindability, and a more homogeneous composition.

Total petroleum hydrocarbons (TPH): the family of several hundred chemical compounds that originally come from crude oil.

Tower: equipment for increasing the degree of separation obtained during the distillation of oil in a still.

TPH E: gas chromatographic test for TPH extractable organic compounds.

TPH V: gas chromatographic test for TPH volatile organic compounds.

TPH-D(DRO): gas chromatographic test for TPH diesel-range organics.

TPH-G(GRO): gas chromatographic test for TPH gasoline-range organics.

Trace element: those elements that occur at very low levels in a given system.

Tracer test: a technique for determining fluid flow paths in a reservoir by adding small quantities of easily detected material (often radioactive) to the flowing fluid, and monitoring their appearance at production wells. Also used in cyclic injection to appraise oil saturation.

Transesterification: the chemical process in which an alcohol reacts with the triglycerides in vegetable oil or animal fats, separating the glycerin and producing biodiesel.

Transmissibility (transmissivity): an index of producibility of a reservoir or zone, the product of permeability and layer thickness.

Traps: sediments in which oil and gas accumulate from which further migration is prevented.

Traveling grate: a type of furnace in which assembled links of grates are joined together in a perpetual belt arrangement. Fuel is fed in at one end and ash is discharged at the other.

Treatment: any method, technique, or process that changes the physical and/or chemical character of crude oil.

Triaxial borehole seismic survey: a technique for detecting the orientation of hydraulically induced fractures, wherein a tool holding three mutually seismic detectors is clamped in the borehole during fracturing; fracture orientation is deduced through analysis of the detected microseismic perpendicular events that are generated by the fracturing process.

Trickle hydrodesulfurization: a fixed-bed process for desulfurizing middle distillates.

Trillion: 1×10^{12}

True boiling point (true boiling range): the boiling point (boiling range) of a crude oil fraction or a crude oil product under standard conditions of temperature and pressure.

True boiling point curve (TBP curve): the composition of any crude oil sample is approximated by a true boiling point curve. The method used is a batch distillation operation, using a large number of stages – typically in excess of 60, and high reflux to distillate ratio (in excess of 5). The temperature at any point on the temperature-volumetric yield curve represents the true boiling point of the hydrocarbon material present at the given volume percent point distilled. True boiling point distillation curves are generally determined for the whole crude oil and not for any of the crude oil products.

Tube-and-tank cracking: an older liquid-phase thermal cracking process.

Turbine: a machine for converting the heat energy in steam or high-temperature gas into mechanical energy. In a turbine, a high velocity flow of steam or gas passes through successive rows of radial blades fastened to a central shaft.

Turn down ratio: the lowest load at which a boiler will operate efficiently as compared to the boiler's maximum design load.

Ultimate analysis: elemental composition.

Ultimate recovery: the cumulative quantity of oil that will be recovered when revenues from further production no longer justify the costs of the additional production.

Ultrafining process: a fixed-bed catalytic hydrogenation process to desulfurize naphtha and upgrade distillates by essentially removing sulfur, nitrogen, and other materials.

Ultraforming process: a low-pressure naphtha-reforming process employing on-stream regeneration of a platinum-on-alumina catalyst and producing high yields of hydrogen and high-octane-number reformate.

Unassociated molecular weight: the molecular weight of asphaltenes in an non-associating (polar) solvent, such as dichlorobenzene, pyridine, or nitrobenzene.

Unconformity: a surface of erosion that separates younger strata from older rocks.

Unifining process: a fixed-bed catalytic process to desulfurize and hydrogenate refinery distillates.

Unionfining process: a hydrotreating technology that is employed to produce low-sulfur, color stable diesel fuel as well as to reduce the amount of aromatic constituents in the product and (in the case of a diesel product) to improve the cetane number.

Unisol process: a chemical process for extracting mercaptan sulfur and certain nitrogen compounds from sour gasoline or distillates using regenerable aqueous solutions of sodium or potassium hydroxide containing methanol.

Unit process: one of a grouped operation in a refinery system that can be defined and separated from others.

Universal viscosity: see Saybolt Universal viscosity.

Unresolved complex: the thousands of compounds that a gas chromatograph *mixture (UCM)* is unable to fully separate.

Unstable: usually refers to a crude oil product that has more volatile constituents present or refers to the presence of olefin and other unsaturated constituents.

UOP alkylation: a process using hydrofluoric acid (which can be regenerated) as a catalyst to unite olefins with *iso*-butane.

UOP copper sweetening: a fixed-bed process for sweetening gasoline by converting mercaptan derivatives (RSH) to disulfide derivatives (RSSR) by contact with ammonium chloride and copper sulfate in a bed.

UOP fluid catalytic cracking: a fluid process of using a reactor-over-regenerator design.

Upgrading: the conversion of crude oil to value-added salable products.

Upper-phase microemulsion: a microemulsion phase containing a high concentration of oil that, when viewed in a test tube, resides on top of a water phase.

Urea dewaxing: a continuous dewaxing process for producing low-pour-point oils, and using urea which forms a solid complex (adduct) with the straight-chain wax paraffins in the stock; the complex is readily separated by filtration.

Vacuum distillation: a secondary distillation process which uses a partial vacuum to lower the boiling point of residues from primary distillation and extract further blending components; distillation under reduced pressure.

Vacuum residuum: a residuum obtained by distillation of a crude oil under vacuum (reduced pressure); that portion of crude oil which boils above a selected temperature such as 510°C (950°F) or 565°C (1050°F).

Vapor-phase cracking: a high-temperature, low-pressure conversion process.

Vapor-phase hydrodesulfurization: a fixed-bed process for the desulfurization and hydrogenation of naphtha.

Vertical sweep efficiency: the fraction of the layers or vertically distributed zones of a reservoir that are effectively contacted by displacing fluids.

Visbreaking: a process for reducing the viscosity of heavy feedstocks by controlled thermal decomposition.

Viscosity: a measure of the ability of a liquid to flow or a measure of its resistance to flow; the force required to move a plane surface of area one square meter over another parallel plane surface one meter away at a rate of one meter per second when both surfaces are immersed in the fluid; the higher the viscosity, the slower the liquid flows.

VGC (viscosity-gravity constant): an index of the chemical composition of crude oil defined by the general relation between specific gravity, sg, at 60°F and Saybolt Universal viscosity, SUV, at 100°F:a = 10sg - 1.

0752 log (SUV – 38)/10sg – log (SUV – 38)

The constant, a, is low for the paraffin crude oils and high for the naphthenic crude oils.

VI (viscosity index): an arbitrary scale used to show the magnitude of viscosity changes in lubricating oils with changes in temperature.

Viscosity-gravity constant: see VGC.

Viscosity index: see VI.

Vicious feedstock: a term which includes (1) distillation residua, (2) the viscous non-volatile portion from other (non-distillation) processes, (3) heavy crude oil, (4) extra heavy crude oil, and (5) tar sand bitumen; in the distillation process, the temperature of the distillation is usually maintained below 350°C (660°F) since the rate of thermal decomposition of crude oil constituents is minimal below this temperature but the rate of thermal decomposition of crude oil constituents is substantial above 350°C (660°F). See also: residua, heavy crude oil, extra heavy crude oil, and tar sand bitumen.

VOC (VOCs): volatile organic compound(s); volatile organic compounds are regulated because they are precursors to ozone; carbon-containing gases and vapors from incomplete gasoline combustion and from the evaporation of solvents.

Volatile compounds: a relative term that may mean (1) any compound that will purge, (2) any compound that will elute before the solvent peak (usually those < C6), or (3) any compound that will not evaporate during a solvent removal step.

Volatile organic compounds (VOCs): the name given to light organic hydrocarbons which escape as vapor from fuel tanks or other sources, and during the filling of tanks. VOCs contribute to smog.

Volumetric sweep: the fraction of the total reservoir volume within a flood pattern that is effectively contacted by injected fluids.

VSP: vertical seismic profiling, a method of conducting seismic surveys in the borehole for detailed subsurface information.

Waste: any unwanted and unusable materials that are commonly regarded as a substance which is of no use; also known as garbage, rubbish, debris, dross. The four major categories of waste are (1) municipal solid waste, (2) industrial waste, (3) agricultural waste, and (4) hazardous waste.

Waste streams: the unused solid or liquid by-products of a process.

Waste vegetable oil (WVO): grease from the nearest fryer which is filtered and used in modified diesel engines, or converted to biodiesel through the process of transesterification and used in any diesel-fueled vehicle.

Water-cooled vibrating grate: a boiler grate made up of a tuyere grate surface mounted on a grid of water tubes interconnected with the boiler circulation system for positive *cooling; the structure is supported by flexing plates allowing the grid and grate to move in a vibrating action; ash is automatically discharged.

Waterflood: injection of water to displace oil from a reservoir (usually a secondary recovery process).

Waterflood mobility ratio: mobility ratio of water displacing oil during waterflooding. (See *also* mobility ratio.)

Waterflood residual: the waterflood residual oil saturation; the saturation of oil remaining after waterflooding in those regions of the reservoir that have been thoroughly contacted by water.

Watershed: the drainage basin contributing water, organic matter, dissolved nutrients, and sediments to a stream or lake.

Watson characterization factor: see characterization factor.

Watt: the common base unit of power in the metric system; one watt equals one joule per second, or the power developed in a circuit by a current of one ampere flowing through a potential difference of one volt. One watt = 3.412 Btu/hr.

Wax: see mineral wax and paraffin wax.

Wax distillate: a neutral distillate containing a high percentage of crystallizable paraffin wax, obtained on the distillation of paraffin or mixed-base crude, and on reducing neutral lubricating stocks.

Wax fractionation: a continuous process for producing waxes of low oil content from wax concentrates; see also MEK deoiling.

Wax manufacturing: a process for producing oil-free waxes.

Weathered crude oil: crude oil which, due to natural causes during storage and handling, has lost an appreciable quantity of its more volatile components; also indicates uptake of oxygen.

Wellbore: the hole in the earth comprising a well.

Well completion: the complete outfitting of an oil well for either oil production or fluid injection; also the technique used to control fluid communication with the reservoir.

Wellhead: the portion of an oil well above the surface of the ground.

Wet gas: gas containing a relatively high proportion of hydrocarbons which are recoverable as liquids; see also lean gas.

Wet scrubbers: devices in which a counter-current spray liquid is used to remove impurities and particulate matter from a gas stream.

Wettability: the relative degree to which a fluid will spread on (or coat) a solid surface in the presence of other immiscible fluids.

Wettability number: a measure of the degree to which a reservoir rock is water-wet or oil-wet, based on capillary pressure curves.

Wettability reversal: the reversal of the preferred fluid wettability of a rock, e.g., from water-wet to oil-wet, or vice versa.

Wheeling: the process of transferring electrical energy between buyer and seller by way of an intermediate utility or utilities.

White oil: a generic tame applied to highly refined, colorless hydrocarbon oils of low volatility, and covering a wide range of viscosity.

Whole-tree harvesting: a harvesting method in which the whole tree (above the stump) is removed.

Wobbe index (or Wobbe number): the calorific value of a gas divided by the specific gravity.

Wood alcohol: see methyl alcohol.

Yarding: the initial movement of logs from the point of felling to a central loading area or landing.

Zeolite: a crystalline aluminosilicate used as a catalyst and having a particular chemical and physical structure.

Conversion Factors

GENERAL

1 acre = 43,560 sq ft
1 acre foot = 7758.0 bbl
1 atmosphere = 760 mm Hg = 14.696 psia = 29.91 in. Hg
1 atmosphere = 1.0133 bars = 33.899 ft. H_2O
1 barrel (oil) = 42 gal = 5.6146 *cu* ft
1 barrel (water) = 350 lb at 60°F
1 barrel per day = 1.84 cu cm per second
1 Btu = 778.26 ft-lb
1 centipoise × 2.42 = lb mass/(ft) (hr), viscosity
1 centipoise × 0.000672 = lb mass/(ft) (sec), viscosity
1 cubic foot = 28,317 cu cm = 7.4805 gal
Density of water at 60°F = 0.999 gram/cu cm = 62.367 lb/cu ft = 8.337 lb/gal
1 gallon = 231 cu in. = 3,785.4 cu cm = 0.13368 cu ft
1 horsepower-hour - 0.7457 kwhr = 2544.5 Btu
1 horsepower = 550 ft-lb/sec = 745.7 watts
1 inch = 2.54 cm
1 meter = 100 cm = 1,000 mm = 10 microns = 10 angstroms (A)
1 ounce = 28.35 grams
1 pound = 453.59 grams = 7,000 grains
1 square mile = 640 acres

CONCENTRATION CONVERSIONS

1 part per million (1 ppm) = 1 microgram per liter (1 μg/L)
1 microgram per liter (1 μg/L) = 1 milligram per kilogram (1 mg/kg)
1 microgram per liter (μg/L) × 6.243×10^8 = 1 lb per cubic foot (1 lb/ft^3)
1 microgram per liter (1 μg/L) × 10^{-3} = 1 milligram per liter (1 mg/L)
1 milligram per liter (1 mg/L) × 6.243×10^5 = 1 pound per cubic foot (1 lb/ft^3)
I gram mole per cubic meter (1 g mol/m^3) × 6.243×10^5 = 1 pound per cubic foot (1 lb/ft^3)
10,000 ppm = 1% w/w
1 ppm hydrocarbon in soil × 0.002 = 1 lb of hydrocarbons per ton of contaminated soil

WEIGHT CONVERSION

1 ounce (1 oz) = 28.3495 grams (18.2495 g)
1 pound (1 lb) = 0.454 kilogram
1 pound (1 lb) = 454 grams (454 g)
1 kilogram (1 kg) = 2.20462 pounds (2.20462 lb)
1 stone (English, 1 st) = 14 pounds (14 lb)
1 ton (US; 1 short ton) = 2,000 lbs
1 ton (English; 1 long ton) = 2,240 lbs
1 metric ton = 2204.62262 pounds
1 tonne = 2204.62262 pounds

TEMPERATURE CONVERSIONS

°F = (°C × 1.8) + 32
°C = (°F – 32)/1.8
(°F – 32) × 0.555 = °C
Absolute zero = –273.15°C
Absolute zero = –459.67°F

AREA

1 square centimeter (1 cm^2) = 0.1550 square inches
1 square meter 1 (m^2) = 1.1960 square yards
1 hectare = 2.4711 acres
1 square kilometer (1 km^2) = 0.3861 square miles
1 square inch (1 $inch^2$) = 6.4516 square centimeters
1 square foot (1 ft^2) = 0.0929 square meters
1 square yard (1 yd^2) = 0.8361 square meters
1 acre = 4046.9 square meters
1 square mile (1 mi^2) = 2.59 square kilometers

OTHER APPROXIMATIONS

14.7 pounds per square inch (14.7 psi) = 1 atmosphere (1 atm)
1 kiloPascal (kPa) × 9.8692 × 10^{-3} = 14.7 pounds per square inch (14.7 psi)
1 yd^3 = 27 ft^3
1 US gallon of water = 8.34 lbs
1 imperial gallon of water – 10 lbs
1 yd^3 = 0.765 m^3
1 acre-inch of liquid = 27,150 gallons = 3.630 ft^3
1 ft depth in 1 acre (in-situ) = 1,613 × (20 to 25 % excavation factor) = ~2,000 yd^3
1 yd^3 (clayey soils-excavated) = 1.1 to 1.2 tons (US)
1 yd^3 (sandy soils-excavated) = 1.2 to 1.3 tons (US)

SI METRIC CONVERSION FACTORS

(E = exponent; i.e., E + 03 = 10^3 and E – 03 = 10^{-3}

acre-foot × 1.233482	E + 03 = meters cubed
barrels × 1.589873	E – 01 = meters cubed
centipoise × 1.000000	E – 03 = pascal seconds
Darcy × 9.869233	E – 01 = micro meters squared
feet × 3.048000	E – 01 = meters
pounds/acre-foot × 3.677332	E – 04 = kilograms/meters cubed
pounds/square inch × 6.894757	E + 00 = kilo pascals
dyne/cm × 1.000000	E + 00 = mN/m
parts per million × 1.000000	E + 00 = milligrams/kilograms

CONVERTING VOLUMES OF GAS

Multiply Flow Of	By	To Obtain Flow Of
Natural Gas	0.625	Propane
	0.547	Butane
	0.775	Air
Propane	1.598	Natural Gas
	0.874	Butane
	1.237	Air
Butane	1.826	Natural Gas
	1.140	Propane
	1.414	Air
Air	1.290	Natural gas
	0.808	Propane
	0.707	Butane

NATURAL GAS CONVERSION TABLE

ft^3	Ccf	Mcf	MMcf	therm	dekatherm	Btu	MMBtu	kJ	kWh
0.0009	0.00001	0.000001	0	0.00001	0.0001	0.9482	0.000001	1	0
0.001	0.00001	0.000001	0.000000001	0.00001	0.0001	1	0.000001	1.055	0
1	0.01	0.001	0.000001	0.01	0.001	1000	0.001	1054	0.293
100	1	0.1	0.0001	1	0.1	100000	0.1	105461.5	29
1000	10	1	0.001	10	1	1000000	1	1054615	293
1000000	10000	1000	1	10000	1000	1.00E+09	1000	1054615000	293071

* Assumed 1 cf = 1,000 Btu, for approximate reference only

Index